7 Vog

D0438571

ETHOLOGY
THE BIOLOGY OF BEHAVIOR

ETHOLOGY
THE BIOLOGY OF BEHAVIOR

IRENÄUS EIBL-EIBESFELDT

Max Planck Institute for Behavioral Physiology

translated by ERICH KLINGHAMMER

Purdue University

HOLT, RINEHART AND WINSTON

New York
Chicago
San Francisco
Atlanta
Dallas
Montreal
Toronto
London
Sydney

To my esteemed teacher and friend
Dr. Konrad Lorenz with gratitude

PREFACE

Ethology—the comparative study of behavior—has in recent years achieved a significance far beyond the realm of biology. The realization that phylogenetic adaptations determine the behavior of animals in a definable manner has increasingly led even those sciences which deal exclusively with man to search for the biological bases of human behavior. All behavioral sciences are based on the assumption that predictions about behavior can be made if a sufficient number of relevant variables are known. It is their goal to investigate such lawful relationships. Animals possess predictable behavioral programs. How these programs are acquired is a matter of dispute. There exist schools of thought which hold tenaciously to the view that animals and man learn all their behavior during the course of ontogenetic development. Ethology has refuted the generality of this assertion. It was first shown in animals that a basic repertoire of behavior patterns matures during the course of development. Emergence of these movement patterns, like that of cells and organs, is guided by phylogenetically acquired developmental blueprints. The impulse to study behavior anew came from biology, and the applicability—in principle—of working hypotheses derived from the study of animal behavior to human behavior is today an accepted fact. We know that even human behavior is determined to a certain extent by phylogenetic adaptations, and that these adaptations are of utmost practical as well as theoretical importance for the sciences dealing with man; consider, for example, the implications for education and sociology.

Phylogenetic adaptations may determine the course of behavior in various ways. Some behavior patterns when performed show what may be called an inborn skill. Such "instinctive activities" are as much a part of an animal as all his anatomical structures. They need not be fully functional at the time of an animal's birth, but instead may mature gradually, and this occurs even if opportunities to learn by individual trial and error or by the imitation of others are withheld. Comparisons of closely related species frequently reveal behavior patterns that are identical in form, pointing to a common phylogenetic origin. Sometimes it is possible to reconstruct their evolutionary path by comparative methods (p. 408).

In addition to inborn skills, many animals have the phylogenetically acquired ability to react to stimulation with specific actions that are appropriate and of advantage to the species. Since the behavior occurs in response to certain specific stimuli or stimulus configurations, a special *innate releasing mechanism* (p. 431) is required.

Inborn *drive mechanisms* cause the spontaneous activity of organisms. Furthermore, we know of *inborn dispositions to learn:* Not every animal is able to learn everything, or at all times, equally well. It appears as if learning takes place in sensitive periods, during which certain aspects are preferred, or as if specific innate learning capacities were largely preprogrammed (p. 217).

This ethological knowledge, based on animal studies, can contribute to a better understanding of human behavior, and K. Lorenz recognizes this as "essentially the most important task" of the branch of science which he founded. This is so because in species which are mentally more advanced the behavior to conspecifics is determined to a greater extent by innate components and less by acquired achievements than is their behavior toward the environment. "That this is unhappily so even in man," Lorenz (1951) writes, "is expressed drastically by the discrepancy between the enormous success in controlling the external environment and the crushing inability to solve intra-specific problems."

With disorders of our own social behavior so acute today, it would be of prime importance for education and the study of the psychology of peoples if we could discover which disorders of social behavior can be influenced by education and which cannot. In this way ethology extends into the province of social sciences and philosophy. D. Ploog (1964) has pointed out the numerous points of contact between ethology and psychiatry.

The subject matter of ethology was presented in textbook form for the first time by N. Tinbergen (1951). In 1952 the German translation of the original English version became available. To date there exists no revised edition of this basic work. However, especially during the last fifteen years, decisive advances have been made in the area of behavioral research, thus calling for a new review.

Of those books which discuss the ethological point of view substantially, W. H. Thorpe (1963), G. Tembrock (1961), P. R. Marler and W. H. Hamilton (1966), R. A. Hinde (1966), and J. Altmann (1966) must be mentioned. Because ethology is a young area of research, no overall framework as yet exists. Thus it can be debated how extensively hormone physiology or sensory physiology should be covered. Depending on whether a student of behavior is oriented toward a comparative-morphological or comparative-physiological point of view, his study could be expected to reflect his particular preference. In basing this book on specifically *Lorenzian* views, I have nevertheless attempted to present all other current viewpoints as well.

As a colleague of Konrad Lorenz it has been my privilege to be a part of the development of ethology in one of its decisive phases. As early as 1946 I was introduced by Otto Koenig, at the Biological Station Wilhelminenberg near Vienna, to the problems and methodology of this research area.

Since 1949 I have worked with Konrad Lorenz, whom I followed in 1951 to the Max Planck Institute for Behavioral Physiology in Seewiesen, near Munich, Germany. My special interests were directed toward the phylogenetic adaptations in the behavior of mammals and to the question of how, during the course of development in young animals, innate and acquired behavioral elements became linked into new functional units. In addition, I investigated by comparative methods, the function and phylogeny of expressive movements of vertebrates and became acquainted with many different groups of animals. I received many stimulating new ideas from research expeditions into the tropics, to which I was repeatedly invited by my friend Hans Hass. The abundant animal life in the coral reefs led me to a deeper appreciation of the ecology of behavior. Ethological studies on various vertebrates then led to additional problems concerning man, problems I am presently studying with Hans Hass.

While making plans for this book I derived great value from the stimulating discussions with my colleagues at the Max Planck Institute for Behavioral Physiology, where practically all the important lines of behavior research are represented. I wish to thank all the members of this institution. This book is based on a previous book of mine which first appeared in the Handbuch der Biologie (Akademische Verlagsgesellschaft Athenaion), and has here been expanded and adapted to meet the needs of students.

I wish to thank my esteemed teacher and friend Konrad Lorenz for all his encouragement and help. I want to thank especially my friend Dr. Hans Hass for the many hours of stimulating discussion as well as for inviting me to participate in the expeditions to the tropical seas which were filled with so many new impressions for me. I also want to thank Otto Koenig for his support while I was on the staff of the Biological Station Wilhelminenberg from 1946 to 1949; Otto Koehler, whose frank and encouraging criticism was of immense benefit to me; as well as my friends B. Hassenstein, E. H. Hess, E. Klinghammer, P. Leyhausen, P. Marler, J. Nicolai, E. S. Reese, H. Sielmann, H. Schöne, N. Tinbergen, and W. Wickler. I remember with great pleasure my collaboration with H. Sielmann while making films.

Special thanks are due all colleagues who contributed pictures and observations to this book, especially to the illustrator Hermann Kacher, the Piper Publishing Company for their care in the preparation of the original German edition, and Holt, Rinehart and Winston, Inc., for their help in preparing the English edition.

As an Austrian citizen it is my duty as well as pleasure to thank Germany, the country where I am a guest, and the Max Planck Gesellschaft, the Deutsche Forschungsgemeinschaft, the A. v. Gwinner Foundation, and the Fritz-Thyssen Foundation for the generous manner in which they have supported my work.

I. E.-E.

Seewiesen, West Germany
January 1970

CONTENTS

1 A SHORT HISTORY OF ETHOLOGY

Behavior consists of patterns in time. Investigations of behavior deal with sequences that, in contrast to bodily characteristics, are not always visible. It is true that even the development of an organism can be thought of as a sequence and it is possible to investigate its growth as a kind of "behavior." However, in respect to our own time perception, bodily structures appear to be static, and they can always be fixed as an anatomical preparation. Behavior patterns, however, must be artificially transformed into spatial structures by means of motion picture film and sound tape if they are to become a preparation and a remaining document.

Behavior is usually expressed as the movement of muscles, sometimes also in the activity of glands or migration of pigments. Growth, swelling, and turgor movements, which are elicited and directed by means of specific stimuli, comprise the behavior repertoire of plants, of which C. Darwin (1872), to my knowledge, was the first investigator (see F. Gessner [1942]). In this book we shall be concerned exclusively with the behavior of animals and man. Just as one can speak of the study of behavior of single-celled organisms, one may also study the behavior of the cells of an organism to determine the releasing stimuli for the movement of leucocytes, how scleroblasts store calcium, or how myxomycetes migrate to form a sporophore. Each cell possesses a behavior program and reacts to certain specific and directing stimuli.

When an ethologist observes an animal performing a certain action, he may ask why the animal behaves in this and not some other manner. If we hear a bird sing, we may ask just why it does so. In what way does his song contribute to the preservation of that particular species; that is, what selective advantage does the song confer upon the animal? With questions about the specific adaptation or function of behavior, the ethologist closely approaches the interests of the ecologist.

Once the function of a behavior is known, one can ask how it developed. During the course of development in the young animal the process of differentiation and integration of behavior patterns can frequently be observed directly. Ethologists employ the methods of comparative morphology (W. Wickler [1961a, 1967b]) to investigate behavioral phylogeny.

Behavior always has a cause. External sensory stimuli are just as responsible for its manifestation as internal drive mechanisms of a central nervous system nature, hormones, and internal sensory stimuli. These physiological causes of behavior are also the subject of ethological investigation, making in this way a bridge with physiology, whose methods are often employed. However, where physiologists generally try to get an understanding of the simplest behavior patterns (for example, heartbeat, respiration, muscular reflexes, and function of isolated muscle fibers) ethologists—who come mainly from zoology—investigate primarily the behavior of the total organism and its relationships with the organic and inorganic environment. They usually investigate the complex and well-integrated functioning of various muscle groups and less frequently the function of isolated parts that were removed from the whole. They work, in other words, at a higher level of integration than the physiologists, although their areas of research are by no means sharply defined.

Students of the biology of behavior share the particular concern with the behavior of the whole organism with some schools of psychology. There is nothing mystical about the term "wholeness," a point that is made clear especially by the cybernetic study of behavior. Feedback systems operate within the whole organism, but are nevertheless subject to causal analysis (B. Hassenstein, 1966).

Psychologists and ethologists have developed their formulations about behavior from different points of view. Psychology is derived from philosophy. Relatively early it became involved in the mechanism–vitalism controversy, the result of which K. Lorenz (1950a, 1950b, 1957) had discussed at some length. Vitalists were excellent observers of animals as a rule, but they did not look for a causal analysis of behavior. They considered the behavior of the whole organism to be unanalyzable by mechanistic methods. Instead, they postulated as final causes entelechial soul-like factors and unfailing inexplicable instincts. "We consider an instinct but we do not explain it," said J. A. Bierens de Haan (1940). The American school of purposive psychology also contains a strong vitalistic bias. The investigators of this group emphasize that behavior is purposefully directed toward a specific goal. In their view, this goal directs the activity as a specific motivation. The animal is motivated by expectancies that require no further explanation (W. McDougall 1936; E. C. Tolman 1932; E. S. Russell 1938). Finally, a number of Gestalt psychologists employ concepts of wholeness (*Ganzheit*) in the manner in which Driesch uses "entelechy" (F. Krueger 1948).

The mechanistic schools, on the other hand, have been convinced since Descartes that all behavior can be derived in the final analysis from physical laws. The concept of wholeness, in their view misused by the vitalists, is rejected by them. Furthermore, they ignore subjective phenomena and describe only that which is objectively observable, asserting that one cannot make reliable statements about the experiences of other organisms. They adhere to a "psychology without the mind." They all search for elements with which they can build up even complex behavior. A. Bethe had already turned against a subjectivistic psychology in 1898. J. Loeb (1913) in his theory of tropisms endeavored to establish a purely mechanistic or machinelike explanation of the behavior of animals. Reflexology, founded and developed by W. Bechterew (1913) and I. P. Pavlov (1927), explains all behavior on the basis of conditioned and unconditioned reflexes and asserts that complex sequences of behavior are nothing but chain reflexes (J. Loeb 1913; H. E. Ziegler 1920). The concept of instinct as misapplied by the vitalists was rejected by the reflexologists. Similar is the position of some American behaviorists who repudiate terms such as feeling, attention, and will, asserting instead that one can only determine stimuli and reactions and the laws governing their interactions. On the whole this assertion has validity when we deal with an animal whose subjective "inner experience" is forever closed to us because of reasons implied in the theory of cognition. We may assume that subjective phenomena play a role in the behavior of animals, but we cannot say anything specific about them. The temptation to reason by analogy from our own experiences to similar ones in higher animals is constantly present, particularly in those who know animals well, but such reasoning can be no substitute for proof and becomes less valid the more dissimilar a species is from our own. About the subjective experiences of our fellowman, however, we obtain objective data not only through self-observation but also from the reports of others. In spite of the fact that definite statements cannot be made about an animal's subjective experiences, many concepts of a subjective nature are still used today. Physiologists speak of hunger and thirst. This kind of shorthand description is useful if for no other reason than that it is generally understood.

Of the representative American behaviorists only J. B. Watson (1930), E. L. Thorndike (1911), K. S. Lashley (1938), and B. F. Skinner (1953) will be mentioned. Because the founder of behaviorism, J. B. Watson, stressed the influence of the environment, subsequent investigations tended to focus on learning phenomena and frequently overlooked the inherited and innate bases of behavior.

These mechanistic schools have been accused of a certain one-sidedness—justifiably so where an attempt was made to elevate one explanatory principle to the level of an all-inclusive one. Thus J. Loeb wrongly generalized his discoveries into his theory of tropisms. His turbellaria and insects, which normally move toward or away from light, moved in circles when blinded in one eye. From this he concluded that normally the equally strong stimulation of sensory organs on the left and right would neutralize opposing muscle contractions on both sides of the body. On the basis of such simple reflexes he attempted to explain all oriented movements as tropisms, which in fact they are not.

For the longest time reflexologists and behaviorists overlooked spontaneity of behavior, which was not readily observable in their particular experimental situations. For them all behavior consisted of reactions to stimulation. They clung overtenaciously to an experimental method once it was found successful (for example, the maze experiment), and this resulted in a certain one-sidedness. Some extreme proponents of behaviorism consider all behavior the result of learning processes. They state that the environment alone determines the behavior of the animal during the course of his ontogeny, a view that will be discussed in more detail later. Behaviorists and reflexologists have made contributions to the study of behavior through the rigorous application of the scientific method; and this is especially true in respect to the phenomena of learning.

Ethology emerged as another discipline exploring these research areas in the natural sciences. It developed out of zoology, especially through the work of K. Lorenz and N. Tinbergen, and is based on the discovery of phylogenetic adaptations in behavior. However, the knowledge that some behavior is relatively uninfluenced by individual experience is much older. As early as 1716 F. A. v. Pernau knew that animals possessed innate skills in addition to those they acquired, behavior patterns that they did not have to learn by imitation or other forms of training. He described the behavior patterns of various birds and showed which species had to learn their songs from their parents and which were able to sing their species-typical songs without prior exposure to it upon becoming sexually mature. H. S. Reimarus (1773) wrote in a similar vein:

How do the spider and the ant lion go about finding means of supporting themselves? Both can do no other than to live by catching flying and creeping insects; although they are slower in their own movements than is the prey which they seek out. But the former already felt within the ability and the drive to artfully weave a net, before she as much as had seen or tasted a gnat, fly, or bee; and now that one has been caught in her net she knows how to secure and devour it. . . . The ant lion, on the other hand, who can hardly move in the dry sand, mines a hollow funnel by burrowing backward, in expectation of ants and other worms which tumble down, or buries them with a rain of sand which it throws up in order to cover them and bring them into his reach. . . . Since these animals possess by nature such skills in their voluntary actions which serve the preservation of themselves and their kind, and where many variations are possible; so they possess by nature certain innate skills. . . . A great number of their artistic drives are performed without error at birth without external experience, education, or example and are thus inborn naturally and inherited. . . . One part of these artistic drives is not expressed until a certain age and condition has been reached, or is even performed only once in a lifetime, but even then it is done by all in a similar manner and with complete regularity. For these reasons these skills are not acquired by practice. . . . But not everything is determined completely in the drives of the animals, and frequently they adjust, of their own volition, their actions to meet various circumstances in various and extraordinary ways. . . . For if everything and all of their natural powers were to be determined completely, that is, would possess the highest degree of determination, they would be lifeless and mechanical rather than endowed with the powers of living animals.

D. A. Spalding (1873) demonstrated the maturation of innate behavior patterns when he raised swallows in cages so small that they could not flap their wings. In spite of this the birds flew excellently when first given the opportunity to do so. Innate behavior patterns have also been reported by R. A. F. Réaumur (1734–1742), A. J. Rösel v. Rosenhof (1746–1761), B. Altum (1868), G. Peckham and E. Peckham (1904), and J. H. Fabré (1879–1910).

W. James (1890) presented a thoroughly mechanistic definition of instinct. He called instincts the correlates of the organs. Just as an animal has certain organs, it also possesses an inborn ability to use them, and this is founded upon a given neural organization. C. Lloyd Morgan (1894, 1900) expressed himself in a similar manner when he said that the structure of the central nervous system, which underlies instincts, is the result of phylogenetic evolution.

As the direct forerunners of ethology, however, we consider C. Darwin (1872), C. O. Whitman (1899, 1919), O. Heinroth (1910), and W. Craig (1918). In his work on the expressive movements of man and animals Darwin was the first to introduce the comparative phylogenetic method into the study of behavior. Heinroth investigated the systematics of closely related species of ducks and geese, and Whitman did the same for pigeons and doves. In their search for systematically useful characteristics, they encountered predictable inborn behavior patterns, which were characteristic for certain systematic categories in the same way as morphological characteristics.

By their graded similarities they could reveal the degree of close or distant relationship of the taxa. Heinroth called these behavior patterns "arteigene Triebhandlungen."

W. Craig (1918) was the first to distinguish the stereotyped *consummatory action,* from the more variable initial *appetitive behavior,* which is a search for the appropriate releasing stimulus situation.

Great influence was exerted upon the development of ethology by J. v. Uexküll (1921), who conducted experiments to investigate the interrelations between organisms and their environment. He showed that an animal can perceive only a limited part of its potential environment with its sense organs. Some of these perceived characteristics of the environment serve as specific cues. According to Uexküll only those objects serve as cues which are of significance in the life of an animal, thereby becoming the bearers of meaning for the subject (see p. 64).

The appearance of an object (cue bearer) in the perceptual field of a subject always has an effect which imparts functional significance (effector cues) to the object. The effector cue or meaning extinguishes the receptor cue or meaning and brings the action to an end. The effector cue may be objectively extinguished (eaten if it is food), or it may be subjectively extinguished by satiation, when the stimulus filter of the sense organ is turned off. As soon as the effector cue of the object extinguishes its perceptual cue, the functional cycle passing from the object through the subject and again to the object is completed (J. v. Uexküll 1937:34).

Figuratively speaking, each animal subject grasps its object with the two arms of a forceps (receptor and effector) and impresses upon it receptor and effector cues (meaning) (Fig. 1). Uexküll illustrates this schema of a functional cycle with the example of the tick. Mated females climb bushes and wait until a mammal passes by. When they perceive the odor of butyric acid, which is secreted by the skin glands of all mammals, they release their grip

Figure 1. Functional cycle. (After J. v. Uexkull [1921].)

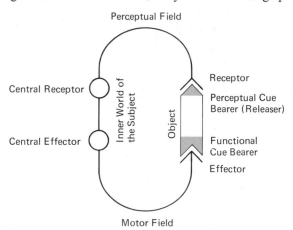

and drop. If they should hit something warm, they begin to search for an area of skin free of hair, bore through with their head, and suck in blood until full. However, other warm liquids are accepted as well, as was shown during experiments with artificial membranes. If we now fit into the schema of the functional cycle the tick as the subject and the mammal as the object, we can see that three functional cycles run off according to plan:

The skin glands of the mammal constitute the cue bearer for the first cycle, because the stimulus of butyric acid releases in the receptor organ specific receptor cues which are projected outward as an olfactory cue. These processes in the receptor organ induce (how is not known) in the effector organ appropriate impulses which cause the release of the legs and the dropping down. The falling tick imparts to the hair of the mammal the effector cue of the collision, which in turn releases a tactual cue by means of which the olfactory cue of butyric acid becomes extinguished. The new cue releases the search for a hair-free spot where the cue—warmth—takes over, which results in the boring through (J. v. Uexküll and G. Kriszat 1934, new ed. 1963:28).

K. Lorenz (Fig. 2) was the first to appreciate the full significance of these discoveries (1935, 1937). Aided by a wealth of his own observations as a basis for induction he proposed a synthesis which forms the basis of ethology. The core of his investigations was at first the innate motor pattern. He recognized the spontaneity that underlies instinctive movements (p. 32), a physiological event of great importance which had been overlooked by the classical reflex-

Figure 2. Konrad Lorenz.
(Photograph: H. Kacher.)

ologists. He investigated the key stimuli that release a specific behavior prior to all experience (p. 64) and studied the phylogenesis and ontogenesis of innate behavior patterns. In the "instinct—learning intercalation" he found a new mode of integration of innate and acquired components (p. 208), and in the phenomenon of imprinting (p. 226) he discovered an inborn disposition to learn, of special significance. Over and over he emphasized the importance of these discoveries for the sciences dealing with man.

In 1937 the first volume of *Zeitschrift für Tierpsychologie,* consisting of 289 pages, appeared. Now the yearly volume is three times this size. In 1948 the journal *Behaviour* (Brill, Holland) began publication, and since 1953 the *British Journal for Animal Behavior* (*Animal Behavior* since 1958) and since 1966 *Revue du Comportement Animal* has joined those journals, which deal primarily with the new science of behavior. Today one generally talks about the comparative study of behavior (*vergleichende Verhaltenforschung*) or ethology (ethos habit, manner). The term "ethology" had previously been used by biologists and covered what today falls under the heading "ecology" (L. Dollo 1895, 1909). J. S. Mill (1843) understood by this term "the exact science of human nature." From 1907 to 1940 the *Zoological Record* carried a section on ethology—meaning the study of behavior—for each class of animals. Since 1951 (see N. Tinbergen [1951] and Fig. 3) the term has become generally accepted as referring to this specific branch of the natural

Figure 3. Niko Tinbergen. (Photograph B. Tschanz.)

sciences. The main emphasis of ethological research initially dealt with the "study of instinct" but without being limited to this. Ethology is a natural science, a branch of biology, from which it took the comparative method for the study of behavioral morphology and the analytic method for the causal analysis of behavioral physiology. Its philosophical base is a critical realism. Its orientation is neo-Darwinistic and it enjoys a fruitful exchange of ideas with other schools of behavior, especially with behaviorism, and increasingly with Russian schools (L. V. Krushinskii 1962).

2 THE ETHOGRAM:
A BEHAVIORAL INVENTORY

Science begins with the description and categorization of the events it studies. The basis of each ethological investigation is the *ethogram,* the precise catalogue of all the behavior patterns of an animal. For this catalogue one selects functional units of behavior which are neither too small nor too large. In practice it is not too difficult to find such easily recognizable functional units which are constant in form, such as scratching, chewing, and head-up tail-up. The exhortation that each behavioral study of an animal species begin with the establishment of a behavior inventory was made as early as 1906 by H. S. Jennings (1906), who called them *action systems.*

It seems axiomatic that the species studied should be accurately classified. However, with some animals this is not so easily done. In such cases W. Wickler (1960a) suggests that a specimen of the species studied be sent to a museum, and that the museum's name and address as well as the catalogue number of the specimen be included in the publication. Should the need arise, other investigators can then reexamine and if necessary reclassify the specimen. A. Seitz (1940) and L. R. Aronson (1949) originally were thought to be studying the same cichlid fish, *Tilapia macrocephala* Bleeker. Their results did not agree. Whether or not they actually worked with the same species can no longer be determined, because neither of the investigators preserved sample specimens.

The description of a behavior pattern should include each detail of the event. Such a *physical* description is never complete in actuality, because the observer usually omits what is not important to him.

For this reason motion picture film has become the ethologist's most important means of documentation.

On film the behavior patterns become fixed and can be preserved for later comparison. In addition, fast and slow motion allow for the analysis of data that would not normally be accessible to direct observation. The slow-motion technique has frequently been used to make visible for the human eye events that run off too quickly to be perceived. The value of the speeding-up (Zeitraffer) technique has hardly been recognized in ethology. This is especially true for human ethology (p. 413). The Institute for Scientific Film (Institut für den Wissenschaftlichen Film) in Göttingen, Germany, has been assembling an archive of technically perfect motion picture films (16 mm) of behavior sequences which is called the Encyclopaedia Cinematographica (G. Wolf 1957a, 1957b).[1] (In the United States these films and many others are now available from Audio-Visual Services, 6 Willard Building, The Pennsylvania State University, University Park, Pennsylvania 16802. Copies of the English translation of film titles can be obtained and films may be rented for a small charge.)

Frequently behavior is described according to its *function*. In this case one focuses on the goal rather than on the movement co-ordinations leading to it. "Carrying in" or "nest building" are functional terms. However, they already include an interpretation of the observer, and therefore this procedure involves a certain risk (R. A. Hinde 1959).

In order to record a sequence of movement patterns with respect to duration, frequency, and relative position in time to one another, without interrupting the continuous observations, one may use multichannel-event records. Each of a number of previously selected behavior patterns is represented by a particular key which activates a pen. Pressing down the key records the event on a paper roll, which moves at one of several constant rates of speed.

Most observations are made on captive animals, and this has certain disadvantages. Lack of an opportunity to hunt, explore, and so on, may lead to distortions in behavior—especially in mammals that normally are quite active. Pacing back and forth retracing the same paths, swinging to and fro, and other movement stereotypes can often be observed (M. Holzapfel 1938, 1939). Such behavior may have various causes. An armadillo in the Amsterdam Zoo stopped its stereotyped movements at once when the bottom if its hitherto bare cage was covered with a layer of earth, 20 cm deep, so that the animal could bury itself at night in order to sleep.

---

[1] Since 1964, commentaries on these films, of great use to the student of behavior, have been published in Publications to Scientific Films (Publikationen zu wissenschaftlichen Filmen) by the Institut für den Wissenschaftlichen Film, Göttingen, Germany.

The stereotyped movements reappeared when the floor was cleaned of the earth. H. Hediger (1942) discussed a large number of additional behavior aberrations found in captivity. Higher mammals frequently suffer from a lack of opportunity for various activities, which led many zoos to initiate what might be termed "work therapy." Breeding frequently fails because the animals do not mate or do not raise their young. Observations in the natural environment often lead to a correction of the problems. O. Koenig (1951) was able to induce his bearded titmice to breed, but the parents threw their young out of the nest shortly after hatching. The cause was found in the overabundance of food he had provided: The parents stuffed their young full in a very short time, and when they returned to feed them more the young no longer gaped; this can never happen in the wild, because the parents spend a certain length of time in search of food. In the wild, young titmice that do not gape are either sick or dead and are removed from the nest by the parents. The captive bearded titmice behaved toward their nongaping young as if they were dead. It was sufficient to offer food to the captive birds in small, infrequent portions to correct this problem.

Many other disturbances of behavior in captive animals have similar causes. If one knows the animal well they can be corrected. It is by no means true that captive animals always behave abnormally and that observations in captivity are therefore of little value, as is sometimes asserted. A number of excellent studies attest to the opposite (H. Kummer 1957, 1968; H. Kummer and F. Kurt 1965). Details of behavior can be observed only during close and continuing contact with a particular species. Distortions can be minimized if one maintains the animals relatively free in their natural

Figure 4. Konrad Lorenz in the midst of his free-ranging geese. (Photograph: I. Eibl-Eibesfeldt.)

Figure 5. Baroness Jane van Lawick-Goodall feeds one of the chimpanzee males that she tamed in the wild. (Photograph: Baron van La-wick.)

environment. K. Lorenz chose this means when he settled his jack-daws near his home in Altenberg and permitted them to fly free. His graylag geese and ducks have been settled by raising and feeding them year-round near a small lake, where they move about freely (Fig. 4). C. R. Carpenter introduced a group of rhesus monkeys to the island of Cayo Santiago (Puerto Rico) in 1938. This colony has since been almost constantly observed (p. 359). P. Krott and K. Krott (1963) raised European brown bears in the wild and observed them there. It is rare to be able to observe a larger number of animals in the wild for longer periods of time. J. Goodall (1963, 1965) has been camping for years in a valley that is inhabited by chimpanzees near Kigoma, Tanzania (Tanganyika). Owing to her great patience, the chimpanzees gradually became used to her presence and today they move without fear around her campsite. She can feed or groom them and they even solicit her to play with them (Fig. 5). Additional examples that illustrate the value of observations in the wild are given by the works of M. Altmann (1952), J. A. King (1955), J. Adamson (1960), L. Crisler (1962), N. Tinbergen (1963), S. L. Washburn and I. DeVore (1961), I. DeVore (1965), G. B. Schaller (1963), W. Kühme (1965), and F. R. Walther (1965).

Observations in the wild and in captivity complement each other, and a discussion of the relative merits of each method would be superfluous.

The unbiased observation and recording of behavior patterns is also a prerequisite for the scientific study of man from an ethological point of view. That the ethology of human behavior is only now beginning may be seen from the fact that today there exists hardly one motion picture document of natural human behavior which is published and on file (pp. 412 ff).

3 THE FIXED ACTION PATTERN (INBORN SKILLS)

Fixed action pattern and its taxis component

In the behavioral repertoire of an animal one encounters recognizable, and therefore "form-constant" movements that do not have to be learned by the animal and provide, like morphological characteristics, distinguishing features of a species. In a manner of speaking we are confronted with "innate skills." Such innate move-ment coordinations have been called *fixed action patterns* (fixed patterns) or *instinctive movements* (K. Lorenz 1953); K. Lorenz and N. Tinbergen 1938)—and the German term—Erbkoordination (inherited coordination)—denotes that the innateness of these sequences is the deciding criterion. (In this translation the term "fixed action pattern" will be used, because it is now a well-established term in the English-language ethological literature.) This innateness is not, however, recognized in the stereotyped sequence of the pattern, as M. Konishi (1966) erroneously attributes to ethologists, but by means of specific experiments that will be discussed later (p. 19). The form constancy may be a strong indication that the movement is inherited, especially when closely related species show similar movement patterns. We know, however, especially through the study of bird songs (J. Nicolai 1959a, 1964) that learned behavior patterns also possess a high degree of stereotypy, so that this criterion alone cannot serve as a definition of the fixed action pattern (W. Wickler 1961c).

Although this has been emphasized repeatedly by ethologists, they are accused, as recently as 1967 (see Z. Y. Kuo [1967]), of uncritically concluding, on the basis of the stereotyped characteristic and species specificity only, that it is probably genetically determined.

Innate behavior patterns may already be fully functional at the time of hatching or birth. A newly hatched chick is soon able to walk, peck at seeds, scratch on the ground, and drink. It flees to the mother hen when raptors appear, it calls loudly when it has lost contact with its mother, and it shakes itself when it has become wet. These and many other behavior patterns are present from the time of hatching. The same holds true for a recently hatched duckling, but its behavior deviates in important details from that of the chicken. The duckling runs to the water, swims, dives, feeds below the surface, and oils its feathers. These differences in the behavior of chicken and duck must be rooted in inheritance, because even a chick hatched and raised under a duck retains the chick characteristics, while a duckling hatched under a chicken runs to the water and feeds on the bottom with its bill, which is adapted for straining, in spite of the foster mother's efforts to entice it away.

Not all inborn behavior patterns are fully developed at the time of hatching. Some develop gradually as the animal grows older, such as the complex courtship movements (head-up tail-up, grunt whistle, and so on) of ducks (p. 107). Since these behavior patterns develop in each male even if raised in isolation from other ducks, so that he has no opportunity to imitate these complex behavior sequences, we assume that they, too, were inherited as a phylogenetic adaptation and merely require a longer period of maturation.

We will discuss later in more detail what is meant by the form constancy of fixed action patterns (p. 41). At this time it should be pointed out that a fixed action pattern can occur at several levels of intensity, ranging from mere intention movements, which indicate what an animal is about to do, to completely executed actions. However, even then the typical pattern is recognizable, in the same manner in which a rhythm, whether repeated slowly or rapidly, can be recognized as the same provided the relative spacing of the sound impulses remains the same.

Fixed action patterns usually proceed without any indication of insight into the species-preserving function of the activity on the part of the animal, as is clearly shown by inappropriate actions (p. 23). When the inner readiness to act (p. 44) coincides with the appropriate releasing stimulus situation (p. 64) in an animal, then a particular fixed action pattern will run its course almost automatically. Thus a dog hiding a bone in the living room shows the movements of covering it as if earth were available; this is the way his behavior is genetically preprogrammed to be adaptive in nature. He will turn several times before lying down, although there is no grass to be trampled.

Fixed action patterns normally occur with orienting movements or taxis superimposed on them. In contrast to the fixed action patterns, taxes require continuous directing stimuli to be noticeable. The unity of taxis and fixed action pattern is the basis of *instinctive activity*. Instinctive activities can be quite complex. By the incorporation of various taxis components they acquire a high degree of adaptability and variability suitable to a particular situation. However, it can be shown upon closer examination that even here we are dealing with genetically programmed behavior sequences. While constructing a cocoon the spider *Cupiennius salei* first produces a base plate, then a raised rim which provides the opening into which the eggs are deposited. Having laid the eggs, the female closes this opening. If she is disrupted while spinning her cocoon, after the base plate has been completed, she will not produce a new base plate half an hour later when she builds a new cocoon, but instead spins only a few threads and then continues with construction of the rim, so that the bottom of the cocoon remains open. If one counts the number of spinning movements she performed for the previous base plate and for the new substitute cocoon, the number roughly equals the number normally used to build a complete cocoon. She has available, so to speak, a limited number of spinning movements— approximately 6400 dabbing movements. This number of movements is performed, even if, under abnormal circumstances, she is no longer able to secrete any threads. This has happened when the glands dry up as a result of hot lights used during filming. In such instances the spider still produces her behavior program. After the appropriate number of ineffectual dabbing movements she will lay her eggs, which will then drop to the ground. Then she continues as if she were closing the rudimentary cocoon. Finally, the spider tries to remove this partial structure from the substratum (M. Melchers 1964).

The spider is therefore not affected by the success of her efforts. This can also be seen when such a spider is placed upon a half-completed structure. The existing structure is not taken into account. Instead, she continues as if she were sitting on her own cocoon. In this manner structures may be produced which are unsuitable for receiving the eggs (M. Melchers 1960, 1963).

The complex cocoon structure built by the caterpillar of the moth *Platysamia cecropia* is precisely programmed. The larvae spins three layers of a cocoon. Displacement experiments showed that it is not incapable of continuing a cocoon that was already begun or of producing a new outer layer in a new place when it had already begun its own elsewhere. It is unable to repeat any of the layers. In this species the behavior is determined by the amount of substance in the glands. If the spinning glands are filled to 60 percent of capacity, the animal will begin with the construction of the inner cocoon layer, with an accompanying change in the frequency of rotation about the body axis (W. G. van der Kloot and C. M. Williams 1953).

The manner in which taxis and fixed action pattern are combined into one instinctive activity has been shown by K. Lorenz and N. Tinbergen (1938) in the egg-rolling movement of the graylag goose. If a graylag goose is presented with an egg outside her nest, she will reach out with her bill over and beyond the egg and pull it in with the underside of the bill, balancing it carefully back into the nest (Fig. 6). This behavior may be broken down into two components. If one removes the egg after the rolling movement has been started, then the movement continues in vacuo. The bird behaves as if the egg were still there. However, the lateral balancing movements cease and the neck is pulled back in a straight line to the nest. This movement, which once released will continue in the absence of additional external stimuli, is the fixed action pattern. The lateral balancing movements are the orienting movements or taxis components which are also inborn but are discontinued in the absence of the releasing stimuli. Taxis and fixed action pattern are related to each other as the steering mechanism and the engine of a car. Each change of direction requires an external impulse, but the engine, once started, will continue without an external impulse.

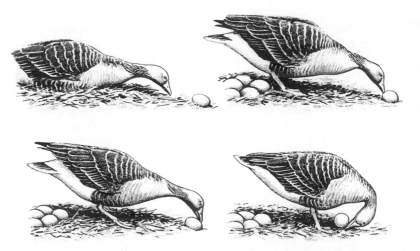

Figure 6. Egg-rolling movement of the graylag goose. (After K. Lorenz and N. Tinbergen [1938].)

Figure 7. Prey-catching frog: (1) aiming (taxis); (2) tongue flick (fixed action pattern). (After N. Tinbergen [1951].)

While a male stickleback fans his eggs he is positioned head down in front and above them. When fanning spontaneously and without eggs he is horizontal to the ground. Fixed action pattern and taxis may mature at different times during ontogenesis. A newborn mouse will at first show scratching movements in the air with its hind legs, without touching its body.

In the original concept of a taxis (orienting movement), the movements of the animal were combined with orienting movements. N. Tinbergen (1951) suggested that only the simple turning movement be called a taxis. This separation of taxis and fixed action pattern becomes especially obvious when they are not coupled but occur one after the other. This occurs in a prey-catching frog who will turn his body prior to the attack, with its snout pointing directly toward the prey (Fig. 7). This turning movement or alignment with the prey is the taxis, while the action strike comprises the fixed action pattern (N. Tinbergen 1951).

The original concept of taxis is different from Tinbergen's formulation in this important respect in that Tinbergen uses the term to merely describe the event, while A. Kühn tried to characterize the physiological mechanisms involved.

In line with the distinction between the fixed action pattern and taxis one must also distinguish between stimuli that release and those that steer or orient fixed action patterns (N. Tinbergen and D. J. Kuenen 1939).

Deprivation experiment— raising an animal while withholding specific information

We have stated that behavior patterns may be inborn in an animal. How can the innateness of behavior be experimentally demonstrated? In recent years this particular question has fanned a controversy between ethologists and some behaviorists who subscribe to an environmentalistic position.

Ethologists maintain that the question about innate and acquired components of animals' behavior can be answered if one raises an animal in isolation from all other conspecifics, thus preventing any possible imitation, and by making it impossible to learn the behavior in question by trial and error. The value of such *Kaspar–Hauser experiments* has been questioned by some investigators (D. S. Lehrman 1953; T. C. Schneirla 1956, 1966; R. A. Hinde 1966). They argued that it is impossible to raise an animal without any experience, because it is always a part of an environment, even within the egg or the uterus, and that it is always experiencing something while interacting with its environment. A paper by Z. Y. Kuo (1932) is frequently cited as an example of how a chick within the egg "learns" the movement coordination of pecking at food: The head of the 3-day-old embryo rests on the heart and is at first passively raised and lowered. At the same time the yolk sac is said to provide tactile stimulation of the head, because it is moved by amnion contractions that are synchronous with the heart beat. One day later the embryo bends its head actively when touched and opens and closes its beak. During these movements liquid is pressed into the mouth, which is swallowed from the tenth day on. In this way the initially isolated movements of swallowing, nodding, and beak opening become integrated into a stereotyped behavior pattern—as Kuo sees it, by "experience." D. S. Lehrman (1953) equated this with learning and thus gave a clearer formulation to this position. The chick is able to peck at food immediately after hatching.

In response to these observations of Kuo, K. Lorenz (1961) raised the question why other species of birds, which have similar experiences with their heartbeat within the egg, do not peck but gape while other species, ducks, for example, strain the mud for food and still others, such as doves, put their bill into the mouth of their parents.

R. A. Hinde (1966) accepts Kuo's interpretation up to a certain degree and accuses us of making fun of his hypotheses. He writes: "Such suggestions have been greeted with ridicule by a number of writers (for example, K. Lorenz, 1961; I. Eibl-Eibesfeldt, 1961; W. H. Thorpe 1963), but it is difficult to see why" (p. 327).

We must reject this reproach, because we have admitted the possibility of such learning (for example, I. Eibl-Eibesfeldt [1961:706]). Kuo's example was rejected as a speculation that was not experimentally supported, especially because it has long been known (W. Preyer 1885) that at the time the thorax is able to move the chick's head passively, the connection between sensory and motor neurons in the spinal cord is not yet made, which escaped Kuo as well as all others who uncritically quoted him in support of their hypotheses. It is further known as a result of the careful investigations of V. Hamburger (1963, 1966) and R. Oppenheim (1966) that this tactile self-stimulation is irrelevant for the development of this behavior. Thus, as a result of removal of the amniotic membrane there is no change in activity, and the hypotheses of Kuo are disproved.[1] Theoretically the possibility remains that in other animals embryos stimulate themselves and that they learn from this experience. However, even if this could be shown in a specific instance, it would by no means do away with the concept of innate behavior. The important fact that requires an explanation is the adaptation of behavior patterns to certain environmental situations. Behavior is molded toward them as to fit or even duplicate models. As K. Lorenz (1961b, 1965) has explained, such an adaptation or copy can come about only when the organism at some time obtained information about such environmental contingencies to which its behavior is adapted. A specific adaptation requires the interaction with a specific code of information. The acquisition of information can take place in the course of the development of the young animal, for example, through an active dialogue with its environment, and we know today that some adaptations in animals are the result of traditions passed on by some model (p. 22). In all these instances the experience that has been gained is stored in the central nervous system of the individual.

A specific adaptation, however, is quite often the result of phylogenetic evolution. In that case the species has come to terms with its environment. Natural selection is the "teacher," so to speak, and the acquired "experience" has become preserved in the genome of the species and become decoded during ontogenesis.

The expression "acquisition of information" as applied to phylogenetic adaptation may seem strange to those who think of the chance events that are involved in increasing the chance of survival for an animal. It is at first difficult to speak of acquisition of information when a change in phenotype has come about through the loss of a part of a chromosome, which then results in adaptiveness. The individual in question did, in fact, not acquire information about its environment. If one looks at the level of the species, however, we gain a new perspective. When the more advantageous genome begins to increase in frequency within a population, the increased fitness at the species level is comparable to an acquisition of information,

---

[1] In a recent publication Kuo denies that he ever believed in learning within the egg. He emphasizes, instead, that self-stimulation processes undoubtedly play an important role in the ontogeny of behavior of the bird embryo, in line with the expressed sentiment stated above, and that pecking has its behavioral precursors before actual hatching (Z. Y. Kuo, 1967:50, 108).

because from that moment on an interaction with the environment begins by means of natural selection.

That adaptation always presupposes that such an interaction has not been understood by those critics who have been concerned with these questions (for example, A. D. Blest [1966]).

The path by which a particular adaptation came about can be discovered with the aid of a deprivation experiment. All that needs to be done is to withhold from a growing animal a specific kind of information to which the behavior pattern being investigated is adapted. If the subject still shows appropriate, adaptive behavior, then we know that this specific adaptativeness is the result of phylogenetic evolution.

Such behavior is adaptive as a result of inheritance, as distinguished from adaptive modifications of behavior which are acquired. If the phylogenetic adaptation is on the motor level, we are dealing with fixed action patterns. It may, however, also lie on the receptor side as a selective stimulus filter by means of which an animal reacts with specific behavior patterns to specific stimuli or stimulus configurations prior to all necessary experience (p. 64). In addition, there are innate dispositions to learn (p. 217) and motivating mechanisms that result from phylogenetic adaptation, causing the animal to act as a result of inner drives. Instead of using the term phylogenetically adapted, we often say innate or "instinctive."

L. Carmichael (1926, 1927, 1928) raised tadpoles under permanent narcosis (acetone chloroform) until such time as control animals could swim well. When he withdrew the narcotic, the experimental animals swam as well, although they had been unable to practice. When A. Fromme (1941) repeated these experiments, the experimental animals did not swim as well as the controls, but were nevertheless able to do so.

J. Grohmann (1939) raised pigeons in cages so small that they were unable to flap their wings. He released them when their control siblings were able to fly. In spite of being prevented from practicing they flew well.

In these instances the experiments showed that the behavior in question is present as a phylogenetic adaptation, probably in the form of coordinating central mechanism. This can also be verified for more complex behavior sequences.

The squirrel *Sciurus vulgaris* L. buries nuts in the ground each fall, employing a quite stereotyped sequence of movement. It picks a nut, climbs down to the ground, and searches for a place at the bottom of a tree trunk or a large boulder. At the base of such a conspicuous landmark it will scratch a hole by means of alternating movements of the forelimbs and place the nut in it. Then the nut is rammed into place with rapid thrusts of its snout, covered with dirt by sweeping motions and tamped down with the forepaws.

One cannot decide, by observing it, to what degree this behavior is innate and to what degree it is acquired. However, one can easily withhold the relevant information which a squirrel would require in order to learn how to collect its winter stores. The animal is hand raised in isolation with liquid food and placed into a cage with a bare floor. It can neither observe another squirrel burying nuts nor can it practice the burying of food. In addition, it never experiences times of starvation; thus it is unable to learn that food hidden by chance can be useful in times of need.

If an animal so raised is tested when fully grown, one finds that it masters the entire hiding actions on the first attempt. If one presents it with nuts, it will eat some first. Upon satiation, additional nuts are not dropped but are carried about in the mouth as if in search of something. Vertical structures seem to attract the squirrel. At a corner in a room, perhaps the squirrel will deposit the nut, push it into the corner with its snout, and finally make the covering and tamping-down movements with the front legs, although it did not dig a hole (I. Eibl-Eibesfeldt 1963). The entire behavior sequence therefore is preprogrammed as phylogenetic adaptation. Shrikes (*Lanius collurio*) removed the stingers from hymenoptera before eating them, and this knowledge is just as innate as the recognition of which bees and wasps sting (E. Gwinner 1961).

The ability of sexually mature ring doves (*Streptopelia roseogrisea*) to feed newly hatched squabs in species-typical manner need not be acquired individually by chance learning but is innate (E. Kling-hammer and E. H. Hess 1964), in the same way honeybees will perform the complex waggle dance (p. 136) without the need to learn the code. Young bees that were put together in groups shortly after hatching develop the dance in 7 days (M. Lindauer 1952). The spider *Aranea diademata* will spin threads without a recognizable pattern shortly after hatching; after the first molt, however, they produce their artistic net. It can be shown that this behavior is not learned by placing them into small glass tubes in which they can just turn around but cannot string threads. When released after their first molt they will spin their nets, just as perfectly as previously unconfined animals (G. Mayer 1952).

Good examples of phylogenetically adapted behavior patterns are the songs of some birds. In these behavior patterns, serving the purpose of communication, it is clear that a complex store of infor-mation must be acquired to produce the appropriate song. This information can be acquired by listening to a sample of the song. Theoretically it would also be possible too that the correct song becomes rewarded by the appropriate behavior of a respondant. To date no such instance is known. If these two possibilities are excluded for the moment and a bird still produces its species-typical song, then the conclusion is unavoidable that a phylogenetic adaptation is present.

F. Sauer (1954) raised whitethroats (*Sylvia communis*) singly in soundproof chambers. They still developed all 25 species-typical songs. M. Konishi (1963) deafened chickens by operative procedures. They could still produce the species-specific calls. Doves also produce their calls innately. Oregon juncos (*Junco oregonus* and *Junco phaeonotus*) and the blackheaded grosbeak (*Phcucticus melanocephalus*) sing in species-typical fashion provided they can hear themselves, even if raised in isolation (M. Konishi 1964, 1965a). Sound-isolated chaffinches (*Fringilla coelebs*) develop a song that is similar to the species song in the number of syllables and total length but lacking the characteristic patterning into three stanzas. This must be learned, but the animals possess an innate knowledge about which song to imitate. If offered tape recordings of several bird songs, they imitate only those which resemble their species song in tone quality and form of strophes. The sequence of the strophes is not preprogrammed, because species-typical songs with experimentally reversed strophes are imitated (W. H. Thorpe 1954, 1958, 1961). These last two examples are especially interesting because they show that the phylogenetic adaptation is not always present as a fixed action pattern. It may also be present in a specific "learning schedule"— here the innate knowledge of the song of the particular species (see also p. 225).

In the bullfinch this learning schedule consists of the young birds' habit of imitating only their father. J. Nicolai (1959) once had a bullfinch male raised by canaries. This bird sang like a canary and passed on this song to its young, and these in turn passed it on to their young. The wydah birds (*Viduinae*), who are breeding parasites of the grass finches (*Estrildidae*), in whose nests they lay their eggs, possess an innate territory song and an acquired courtship song which they learn from their foster parent species. They imitate the latter so perfectly that the imitation cannot be distinguished from the original (J. Nicolai 1964; see also Fig. 8).

A behavior sequence usually consists of parts which in turn can be broken down into simpler functional units (p. 42), so one must always state clearly the level of integration at which one is operating. Thus an animal could possess a learning mechanism as a phylogenetic adaptation by means of which the coordination of two antagonistic muscles is learned. These learned units, however, could be integrated into new functional units on the basis of an innate program. In this case behavior at the higher level of integration would be considered as phylogenetic adaptation. In this way, learning processes could theoretically serve as a means of decoding phylogenetically

Figure 8. Top: Greeting song of the female Estrildid finch *(Granatina granatina)*, bottom: imitation of this song by a male widow bird *(Tetraenura regia)*. (From J. Nicolai [1964].)

acquired information that has been stored in the genome.

The strengths as well as the limitations of the deprivation experiment have been discussed by K. Lorenz (1961) and stated in some specific rules which are presented here in abbreviated and augmented form:

1. The question asked deals with the origins of the adaptiveness of a behavior, so it is a prerequisite for each deprivation experiment that the species-preserving function of a behavior be known. This requires a thorough knowledge of the biology of the experimental animal.

2. The experiments must be set up in such a manner that only the adaptation to be investigated is disrupted (I. Eibl-Eibesfeldt and

W. Wickler 1962). If one wants to know if a particular strophe of a bird's song is inborn in its entirety, then one should prevent the bird only from hearing this particular strophe. If one wants to assess the ability of an organism to respond to optical stimuli or stimulus configurations, then it is well not to raise the animal in complete darkness, because this may lead to atrophy of the retina (A. H. Riesen 1960) and to the disruption of all visual reactions. In other words, one must be aware that there are not only various levels of behavioral integration, but also levels of adaptation, and that during ontogenesis a diversity of factors may influence later behavior. Rats that were exposed during the first 10 days of life to mechanical, electrical, and cold stimuli later showed not only increased resistance to the effects of food deprivation and cold stress, but they also learned faster (J. P. Scott 1962). In this connection the experimental hypotheses are frequently stated too broadly and generally. Thus B. F. Riess (1954) asked whether the nest-building behavior in its entirety is inborn or learned. With such complex behavior sequences it is preferable to work at a lower level of integration. In reply to R. A. Hinde's (1966) recently repeated argument that one can never raise an animal devoid of all experience, it can be said that such an attempt would miss the point. The experimental design must be at the appropriate level.

3. The deprivation experiment informs us only about what does not have to be learned. Although we should always strive to disrupt only a particular adaptation, as stated in rule 2, it is not always possible to prevent a disruption of the total condition of an animal. We know that spontaneously occurring fixed action patterns are reduced in intensity, perhaps because the endogenous energy production is disrupted (p. 56). This results in an increased threshold of responsiveness to the releasing stimuli. Furthermore, innate releasing mechanisms (p. 60) frequently lose their selectivity, and social inhibitions are often lost. As an example of such disruptions K. Lorenz reported that his hand-reared red-backed shrikes showed the movement of impaling prey but had to learn to direct this movement toward thorns. He cited this as an example of an instinct-learning intercalation until G. Kramer demonstrated that shrikes that had been raised on a more adequate diet aimed for the thorns innately. This has since been verified by additional experiments by K. Lorenz and U. v. Saint-Paul (1968).

4. In the testing situation the experimental animal must be presented with all the relevant releasing stimuli for the behavior. This dictum was overlooked by B. F. Riess (1954) in his tests for nest-building behavior when he placed inexperienced rats in a strange experimental cage from whose walls paper strips were hanging. The rats, which had until then lived isolated in wire

cages without nesting material and powdered food, did pull down the paper strips, but instead of building nests they scattered them. Riess concluded from their behavior that nest building must be learned as a result of experience with solid objects. They would discover that nest material gathered by chance protects against cold, and based on this experience they would then build nests. He overlooks here that even experienced rats will not build at first in a strange environment but will explore instead. Rats which I (1963) raised according to Riess's method, but which were tested in their home cage, built nests in their sleeping corner, and showed all the species-typical nest-building movements (Figs. 9 and 10). The rat must therefore first have a nesting place before it can appropriately deposit material. A recent experiment by F. Wehmer (1965) has confirmed this.

That even skilled and experienced experimenters are prone to commit such errors is shown by the work of D. S. Lehrman (1955), who concluded on the basis of his experiments that ring doves do not recognize or feed their young innately but must learn to do so. He explained that they at first look down upon their newly hatched young, which stir under them and which then touch the

Figure 9. Nest building by a naive female rat: (a) laying-back move-ment; (b) pulling in; (c) and (d) splitting. (From Scientific Film B757, I. Eibl-Eibesfeldt [1958b].)

Figure 10. Nest building by naive female rats with crepe paper strips: (a) pulling paper from the holder; (b) pulling the paper in during nest building; (c) and (d) one sleeping nest each built by naive female rats: (c) made from split straw; (d) made from crepe paper strips. (Parts (a) and (b) from Scientific Film B757; photograph: I. Eibl-Eibesfeldt.)

crop region of the parent. Under the influence of the hormone prolactin, the crop is swollen with cropmilk and becomes sensitive. If the young succeeded by trial and error in releasing the vomiting of cropmilk, this would then quickly reinforce the behavior of the parent, who would recognize the young as a stimulus leading to a reduction of tension in the crop. This subsequently would lead parents to actively approach the young. Experiments seem to support this view: Of 12 experienced ring doves injected with prolactin, 10 fed the 7-day-old squabs. On the other hand, 12 similarly treated inexperienced ring doves did not feed young of this age with which they were presented. E. Klinghammer and E. H. Hess (1964) obtained conflicting results when they placed newborn mourning doves under incubating experienced and inexperienced ring doves. All fed the young; inexperienced doves, who had as yet no cropmilk in the crop, fed instead a clear liquid. Lehrman used 7-day-old squabs that were already well feathered, and Klinghammer and Hess suspect that the newly hatched downy young presented the more appropriate releaser for parental feeding behavior. This suggests that doves recognize and feed newly hatched young innately. Lehrman's conclusions are then limited to the statement that only experienced ring doves recognized 7-day-old squabs as young, and he has since accepted this interpretation.

5. In view of the fact that results obtained while working with one species are often checked with another species, K. Lorenz emphasizes the almost axiomatic rule that agreements of experimental results in respect to inherited behavior patterns can only be expected when genetically similar animals are used. If these precautions are taken, the deprivation experiment is the appropriate way to demonstrate phylogenetic adaptations in behavior. We often use the terms inborn, innate, or instinctive instead of phylogenetically adapted as a conveniant shorthand description, even in instances in which the behavior matures during the course of development of the young animal (p. 16). To be exact, it is not a certain behavior that is inborn but the developmental "blueprint," the norm of a reaction. Characteristics in themselves are not inherited but are developed within the bounds of inherited variations. The term "inborn" has frequently been understood only in a negative sense as "unlearned" (D. O. Hebb 1953; R. A. Hinde 1966). In line with our discussion it should be clear that we define the term on the basis of the origin of the adaptation. It is without question the merit of our critics (especially D. S. Lehrman) to have compelled ethologists to clarify these concepts. Furthermore, fruitful discussions and a closing of the gap between points of view took place. Even T. C. Schneirla (1965, 1966), who then, as now, argues against the possibility that inborn and acquired components of behavior can be distinguished, makes certain concessions. "A concept of interactionism, therefore," he writes, "that implies engagements of separate and disjunctive developmental entities seems invalid, *at least for the early stages*" (p. 283, italics mine). On the other hand, even T. C. Schneirla (1965) feels compelled to distinguish "maturation" and "experience" during the embryonic development of behavior. But he does not look for the source of the developmental "blueprint." More recently he emphasizes, as does R. A. Hinde (1966), that embryonic development ought to be studied, and that one can then discover how environmental influences (always described as "experience") affect behavior during each stage of development. A similar position is taken by G. Gottlieb and Z. Y. Kuo (1965), who move away from an extreme environmentalist position when they propose the term "self-stimulation" instead of "experience." In that case all that remains of the old milieu-theoretical (environmentalistic) position is the statement that self-stimulation is a factor in behavior development. How this position is incompatible with the

concept of the inborn, as Gottlieb and Kuo still believe, is difficult to understand. Perhaps they did not understand us. No
ethologist has ever asserted that diverse environmental influences,
especially those of the internal environment or self-stimulation
processes during embryonal development, play no role during
development.

After all, we know from the pioneer work of H. Spemann in
experimental embryology that organic substances secreted from
tissues of newt embryos stimulate neighboring tissues to develop
into specific organs. Thus the eye cup induces lens formation in
the vertebrate epidermis above it. If one transplants the eye cup
of a newt embryo into the ventral region, then lens formation is
stimulated in the new location. In this manner genetically encoded
information becomes decoded during the process of self-differentiation. This is also valid for the embryology of phylogenetically
adapted behavior. In both instances one can activate by specific
factors such as temperature or chemical stimulation the potentials
given by the specific genetic code. Thus monkey and rat females
can show permanently manifested male behavior characteristics
(for example, increased aggressive behavior) if they are given male
hormones during a sensitive phase of embryonal or early development (W. C. Young 1965; G. W. Harris 1964). In rats this
sensitive period, in which hormone treatment can exert this
sex-reversing influence, ends one day after birth. Females that
were treated with testosterone before this time showed no normal
sexual behavior even after they were castrated and treated with
estrogen and progesterone, which releases heat in normal females.
The hormonal influences then fixate a male (or in the reverse case
the female) role during a sensitive phase and later in life sexual
hormones primarily activate the sexual behavior patterns preformed
in the sensitive period. Those who desire a better understanding
of these relationships will find A. Kühn (1955) very helpful. One
and the same phenotypic expression can come about in several
ways and no one can deduce by appearance alone the course that
development has taken. It has been possible, however, to determine when and during which stage inherited factors, and when
environmental factors, act upon developmental events. The same
is true for the development of behavior. Specific neural structures
that underlie a behavior develop like other organs on the basis
of a developmental code contained in the genome. Z. Y. Kuo
(1967) took issue with this view. First he emphasized that strictly
speaking there is no species-specific behavior, because not even
one step is like the other even in the same animal (see the discussion of form constancy, p. 41). Morphological structures limit
the potential of an animal's behavior:

Morphological structures and their functional capabilities act as determining factors of behavior only in a negative way, that is, they merely set a limit to certain body movements (for example, a dog can only snarl at or bite its enemy but cannot throw a stone at him) (Z. Y. Kuo 1967:13).

In another place he writes:

There are to be found some common factors in behavior such as those due to some common morphological characteristics of the species. For example, morphological structures of the limbs determine the modes of locomotion; the oral structure determines the modes of eating and drinking; the vocal apparatus determines the characteristics of voice and singing (p. 23).

He is even led to make such statements as:

The fact that the human hand has a far greater flexibility in movement, dexterity and range of potential capacities than those of any other primate is sufficient, in our view, to explain why human beings became the most creative and the most resourceful creatures on earth even long before human language was developed. Some primates are almost human. But not quite. The hands tell the difference. I often speculate that if we could succeed in exchanging brains between a human neonate and a gorilla neonate and raise them in an identical environment with complete absence of human language and culture, the human child would grow up to behave with human characteristics and the gorilla with the characteristics of its own species because the skeletal framework of the body and the fine structures of the hands of the two different species are different (p. 188).

On p. 195 Kuo writes:

If the species known as *Homo sapiens* is so far superior to all the other species throughout the animal kingdom, it is not because it has a human brain per se, but because it possesses a pair of human hands and because the human vocal mechanisms have developed a most complex spoken and written language.

Phylogenic adaptations that preprogram behavior apparently cannot exist in the central nervous system, according to Kuo. We would not deal with this unscientific speculation in such detail had it not appeared in publications that are widely read by students.

Sometimes the objection is raised that while it may be theoretically possible to distinguish innate from acquired components of behavior, in practice this would be of little value, because at best only extreme cases could be assigned to one or the other category; intermediate cases would constitute the majority. This statement can be shown to be wrong if the behavioral repertoire of various species is investigated. If we consider the courtship behavior of the mallard duck, for example, we find an array of highly specific, innate courtship movements, but not a single learned one or one that is substantially modified through learning. In many instances there seems to have been a strong selection for resistance to modifiability. We find this as a rule when a species must depend on the proper functioning of the behavior. When learning is involved, additional control mechanisms are needed to ensure that what is learned will serve the preservation of the species.

Physiological characteristics of the fixed action pattern

For a long time the concept of the classical reflex influenced our conception of the nature of a movement sequence: According to this concept each act is a response to external or internal stimuli. Afferent nerve endings are stimulated and pass the excitation on to the central nervous system. From there the excitation is carried, often via intermediate neurons, to an effector neuron that in turn excites efferent pathways leading to an effector, a muscle or a gland, which is then activated. This course of excitation is called a *reflex arc,* and the process is called a *reflex.* In monosynaptic reflexes of mammals the wave of excitation is said to pass directly from the sensory to the motor neuron. The excitations activate the same muscle from whose proprioceptor they have been released. All other reflexes transverse additional intermediate neurons, and the excitation of many neurons can activate a specific organ, just as the excitation of a few receptors can activate many organs.

More complex movements are called *chain reflexes* in the reflex theory, whereby one reflex provides the stimulation for the release of the next. The release of one reflex may inhibit or facilitate others. For each reflex arc there are inherited "unconditioned" releasing stimuli. By means of learning processes new stimuli can become conditioned stimuli or releasers (p. 250), or a given stimulus can become hooked up to new reaction sequences. These processes of stimulus and response selection frequently occur in combinations, and these newly acquired reaction sequences are called *conditioned reflexes.* The participation of such reflex processes in the structure of behavior cannot be denied. However, it is not true that each movement is the result of an afferent impulse. T. Graham Brown (1911, 1912) established the theory that the quadrupedal walk is a central automatism after he had discovered that two completely deafferented antagonistic leg muscles of a cat showed rhythmic movement. E. v. Holst (1935, 1936) then demonstrated in a number of experiments that an inborn movement sequence can be centrally coordinated without participation of afferent stimulation.

According to the classical reflex theory the regular undulating movement of an eel occurs by the participation of internal sensory organs, the proprioceptors. The contraction of one muscle segment is said to release the contraction of the adjacent segments via these proprioceptors. If this were true, then an eel whose central nervous system does not receive impulses from the periphery should no longer show undulating movements. E. v. Holst showed that this is not the case. If one separates the spinal cord of an eel from the brain, by a cut behind the head, one obtains a spinal preparation that can be kept alive for awhile by artificial respiration. If one now cuts all the dorsal roots of the spinal cord, which alone are capable of transmitting impulses from the sensory organs to the spinal cord, the eel will show undulations once the operative shock has subsided. A purely mechanical transmission of the undulating movement is also excluded. If the central third of the eel's body is mechanically restrained so that it can no longer move, an undulating movement from the first segment will appear in the posterior third after the same amount of time that it would take if the central segment had taken part in the movement.

The experiments prove, first, that an endogenous production of excitatory potential in the central nervous system exists and, second, that these central impulses are also centrally coordinated. Such central automatisms also seem to be the basis of the respiratory movements of the gill covers in goldfish. E. D. Adrian and F. J. J. Buytendijk (1931) recorded rhythmic impulse patterns from the isolated respiratory center in the medulla, which corresponded to the frequency of movement of the gill covers. Rhythmic electrical impulses of the normal crawling rhythm of the earthworm were also found by E. v. Holst in the isolated ventral cord. The movement rhythm is apparently centrally produced, but proprioceptors also participate. It is generally known that the severed posterior portion of an earthworm continues to wriggle, while the anterior part proceeds to move forward normally. If the two severed edges are connected by means of two strings, then the rear part will follow the front part in the typical crawling rhythm (E. v. Holst 1932, 1933).

In grasshoppers, peripheral feedback affects only the frequency of the otherwise central rhythm of the flying movements. If receptor impulses are excluded, the only result is the somewhat slower rate of rhythmic discharge of the thoracic ganglion. The resulting pattern of wing movements corresponds, however, to the normal wing beats (D. M. Wilson 1961, 1964). For the explanation of changes in the gait of leg-amputated insects it is also unnecessary to postulate adaptive reflexes (D. M. Wilson 1966; G. Wendler 1965). In walking insects—however—the system is much more influenced by peripheral reafferences: They control the phase relations between the six walking legs (E. v. Holst 1943; G. Wendler 1964, 1965; D. M.

Wilson 1968). This is clearly shown by the alternation of walking rhythms in leg-amputated arthropods, but we do not yet know the way in which reafferent signals influence the phase angle of the legs.

As we already know, walking movements of an insect's legs are caused by a system of six self-sustained, mutually coupled oscillators. Wendler in his 1968 experiments tried to find the way the oscillators work on each other to produce the observed coordination of leg movement in *Carausius* and also the alternations of walking rhythm after amputation of legs. One hypothesis is shown in Fig. 11. The system consists of six oscillators of nearly the same spontaneous frequency. Oscillators of each segment show mutual influence on each other, those of ipsilateral legs—for instance, hind and middle leg of one side—in direction caudal to front. A nonrhythmic

Figure 11. Diagram of a system of six self-sustained, mutually coupled oscillators (represented by circles). Oscillators belonging to the same segment (for example, hind legs) influence each other mutually, those of ipsilateral legs (for example, hind and middle leg of one side) in a direction caudal to the front. (Additional explanation in the text.) (After G. Wendler [1968].)

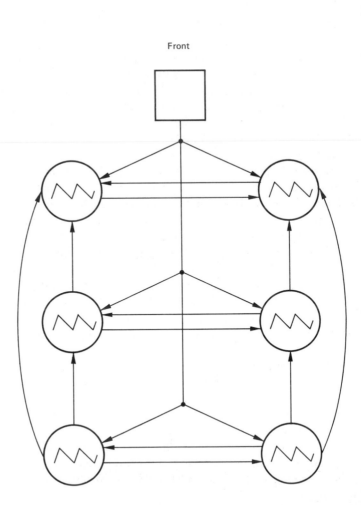

Front

signal from the central nervous system controls walking velocity. As Wendler showed by means of an analog computer model (TR 20), the system holds for all observed phase relationships of an animal's legs. It also offers the frame for several hypotheses of leg receptor influence on coordination: Leg receptors could measure the leg movement and influence directly the neighboring oscillator, thus determining the phase. An alternative system would perform central coupling of all oscillating systems without direct influence of afferent systems (receptors). In this case, receptor influence would be restricted to keeping the amplitude of the oscillator of the same leg high enough for influencing phase relations of the dependent oscillators. This hypothesis is underlined by experiments with partially leg-amputated stick insects whose leg stumps move with lesser amplitude than the intact legs (G. Wendler 1965).

K. D. Roeder (1935, 1937) concluded that an endogenous automatism forms the basis of the mating behavior and locomotion of the praying mantis. Upon removal of the supra- or subesophagal ganglion, locomotor and mating behavior were disinhibited. The two behavior patterns occurred continuously, whereas releasing stimuli were normally required. Roeder postulated the existence of endogenous, self-activating systems that are responsible for the coordinated movements, whose impulses are controlled by inhibitory centers. Additional experiments have strengthened this interpretation (K. D. Roeder 1963a).

Endogenous activity of the central nervous system has also been demonstrated by P. Weiss (1941a), who implanted a section of embryonic spinal cord and a forelimb *anlage* into intact axolotls. The developing forelimb then became innervated from the implanted spinal cord, and the motor nerves grew faster than did the sensory nerves, which reached the limb much later. However, as soon as the efferent motor connection had become established the leg began to move. Although it did not show a coordinated walking movement, an alternation between agonistic and antagonistic muscles could be observed in the irregular movement of the leg.

In the neural elements underlying spontaneous and reflex behavior only gradual differences exist as regards their threshold of excitability (K. D. Roeder 1955). In nonautomatic cells the excitation remains at a constant resting potential, and a stimulus is necessary to elevate it above the threshold. After discharge it drops to zero but increases again and the curve of excitation temporarily surpasses the level of the resting potential only to return to the resting level thereafter. For spontaneous elements, however, the readiness for discharge increases until the discharge threshold has been reached and spontaneous discharge occurs. Between these two extremes there are transitions. When resting potential and discharge threshold are close together, a simple stimulus is able to release an entire sequence of discharges because the curve reaches the discharge threshold during the phase of increased irritability.

Chick embryos move their hind limbs spontaneously and rhythmically even when all sensory ganglia have been neutralized by the removal of the entire dorsal half of the spinal cord and when all influence of the brain is ruled out by the removal of a section of the spinal cord (Figs. 12–14). Turtles and fish will move before the reflex arcs have been closed. The oysterfish (*Opsanus tau*) hatches

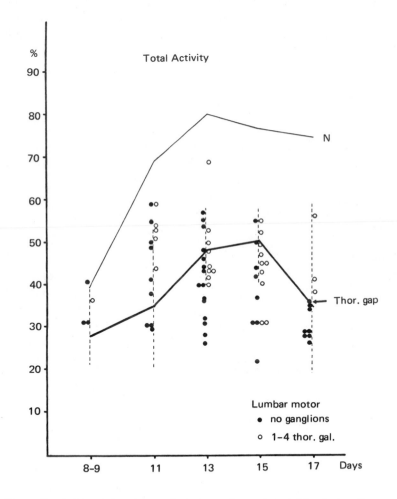

Figure 12. Activity during the standard observation period of 15 minutes. N: normal embryos. Thor. gap: embryos with thoracic section of the spinal cord (controls). The vertical dotted line indicates the range for these controls. Black dots: embryos without any ganglia. Circles indicate embryos with several small ganglia in the lumbar region which do not enervate the extremity. (From V. Hamburger, R. E. Wenger and R. Oppenheim [1966].)

Figure 13. Schema of the operation: 2-day-old embryo. Total extirpation of the neural tube in the lumbar region and extirpation of the dorsal half, including the neural crest in the lumbar region. (From V. Hamburger, R. E. Wenger, and R. Oppenheim [1966].)

Figure 14. Top: Control. Cross-cut through a lumbar piece of spinal cord of a 12-day-old embryo. The pictures below of the operated embryos are from 11- to 17-day-old embryos. DR, dorsal root; VR, ventral root; LMC, lateral motor column. (From V. Hamburger, R. E. Wenger, and R. Oppenheim [1966].)

and is able to swim in this same condition in a well-coordinated manner, before sensory input is possible (H. C. Tracy 1926).

All these examples demonstrate the spontaneous, endogenous activity of motor neurons. In response to the objection that the neurons are still in an environment capable of influencing them, and that perhaps chemical stimuli from the blood could serve as releasing stimuli, Roeder replies very aptly:

In speaking of endogenous activity of the central nervous system I have used the term in the same sense—activity (in this case detected as nerve impulses) generated by mechanisms within the central nervous system. The criterion is that it continues to take place after all afferent nerve connections with the outside have been severed. Factors in the extra-cellular medium surrounding the nerve cell play an important part in determining whether a given nerve cell will remain inactive until stimulated, or will regularly discharge impulses without stimulation. Nevertheless, the coupled regenerative system responsible for the sequence of nerve impulses must be considered to reside in the neurons themselves, and it would be misleading to think of the ambient medium bathing the cells as providing stimuli equivalent to those that normally reach it from the outside via afferent impulses (K. D. Roeder 1963b:438).

Similar statements are made by T. H. Bullock and G. A. Horridge (1965):

The term spontaneous means repetitive change of state of neurons without change of state of the effective environment—that is, activity without stimulation other than the standing conditions. Of course the activity occurs only if many aspects of the milieu remain within certain limits—for example, the temperature and the ionic balance. These could be thought of as steady-state stimuli; but unless there is evidence of physiologically significant control of milieu, the term stimulus is not appropriate (p. 314).

The fact that the oysterfish shows coordinated swimming before the reflex arcs are closed, and that a completely deafferented eel shows normal undulating movements, proves in addition to a central nervous automatism (spontaneity) that the central coordination is independent of afferent impulses. E. v. Holst (1935, 1936) explains this by postulating that within the central nervous system there are groups of cells which are spontaneously active, which send their impulses to the musculature, if not prevented by inhibiting factors. These spontaneous cell groups influence one another, which results in certain specific movement coordinations. Holst demonstrated how these mutual influences work in fish that do not show undulating movements with the whole body but swim by means of rhythmically moved fins. He transected the medulla of these fish and provided artificial respiration; the fins were connected to recording pens (Fig. 15). When the operative shock had subsided, the fins showed rhythmic movements. If only one fin moved, a regular sine wave was obtained, when several fins were moving this curve was more or less modified, which demonstrates the mutual influences of the rhythms. This must be a central influence because the passive

Figure 15. Experimental setup of E. v. Holst. The spinal lipp fish receives artificial respiration. The fins are connected with writing pens. (After E. v. Holst [1939].)

movement of a resting fin did not affect the rhythm of another fin.

The fin rhythms may influence one another equally strongly. Frequently, however, one fin or fin pair maintains its constant rhythm. While the movement of the other fin changes rhythmically. In this instance a dominant, independent rhythm is superimposed on a dependent one. The influence of the dominant rhythm on the dependent rhythm can be read from the recordings (Fig. 16). Each

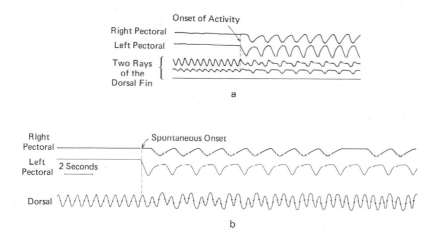

Figure 16. Records of mechanically registered fin movements of a spinal lipp fish showing superposition and magnet effects. (a) During the spontaneously appearing pectoral fin rhythms the rhythm of the dorsal fin, which until then had been regular, becomes changed. This illustrates a purely central superposition. (b) With the beginning of pectoral fin activity (dominant rhythm) the more rapid rhythm of the dorsal fin rays is slowed to coincide with the frequency of the pectoral fins. This illustrates the dependence of the dorsal fin rhythm and is an illustration of a strong magnet effect. (After E. v. Holst [1938, 1939].)

fin has the tendency to maintain its own rhythm, but the dominant rhythm becomes superimposed upon the dependent one. If the dependent rhythm accelerates too fast it is slowed down, and if it lags behind it becomes accelerated. If the dominant rhythm is of sufficient strength, then it will superimpose its own rhythm completely (*absolute coordination*). If this is not possible, then the phase relationships of the rhythms change periodically in a lawful manner (*relative coordination*). An everyday example may serve to illustrate this. If we walk with our small daughter she will try to keep in step with us. Gradually she will loose the synchronization, and the phase differential increases until the child will correct this by making a small jump that will again bring her into phase. This attraction which two rhythms have for one another E. v. Holst calls "magnet effect." During *superposition,* finally, the dominant rhythm becomes superimposed on the dependent rhythm in an arithmetic relationship. Whenever the dependent fin moves in the same direction as the dominant one, its amplitude becomes larger, and the opposite is true when their movements are opposite one another. This, then, is another way in which the independent rhythm can become imposed upon the dependent one and thus result in absolute coordination. Pure magnet effects or pure superposition are rarer than intermediate forms.

In E. v. Holst's experiments the pectoral fin rhythm was always dominant over the rhythms of the dorsal and caudal fins. Their independent rhythms appeared only when the preparation emerged from operative shock or when the animal died. In the former case "coming to" appeared from posterior to the anterior, and death occurred in the opposite way, such that the dominant rhythm dropped out first.

What is true for the mutually influencing automatisms of the various fins is also true for each individual fin, which does not beat in the manner of a stiff board. Rather, wave movements pass over it, and each fin ray has its own automatism. These automatisms influence each other in such a way that the individual rays move in regular phase intervals. We can detect a hierarchical order of the automatisms. Each automatism can be broken down into subgroups, which in turn have a magnet effect upon one another, and the coupling is usually stronger or absolute at the lower level of integration.

Theoretically there are a large number of possible interactions between various automatisms, which, however, are not realized randomly. Thus E. v. Holst found several stages of stability in the phase relationship of different automatisms, which increase with the simplicity of the reciprocal frequency relations. The most stable coordination is the absolute coordination 1:1, followed by 1:2, 1:3, and 2:3 (or 1:2:2, 1:2:3, and 2:3:4 for three rhythms). For locomotion on land absolute coordination is undoubtedly the most efficient. The adaptive character of the various forms of relative coordination in water animals is not yet understood.

When their phase relationships are stable, the reciprocating automatisms form a transposable "gestalt," because a change in the frequency of one automatism influences the other in such a way that the original phase relationship is maintained. This is also true for stereotyped, learned movement sequences. Even if a person writes at different speeds, a specific speed is maintained for each letter, and the total impression remains the same whether the person writes in capital or lowercase letters. Even these learned movements are based on automatisms. Whereas in fixed action patterns the relationship between various automatisms is genetically programmed, in learned coordinations an initially unstable relationship between automatisms becomes fixed later. The automatic groupings are led, in a manner of speaking, through success into new patterns, whereby the automatic cells seek new stable relationships. The transition from the clumsy execution of a movement into a new and stable coordination occurs suddenly, as everyone knows who has learned to dance or to ski. This explains also why practicing parts of actions is not very useful. If one first learns the components until a fixed, automatic relation is established, these relations must again be undone and again newly coordinated when a movement pattern of a higher integration is established. That this reordering can take place centrally without any aid from afferent input is shown by the experiments of E. Taub, S. J. Ellman, and A. J. Berman (1965). Their rhesus monkeys learned to grasp a cylinder from a fixed position with a deafferented hand without the aid of vision, to avoid a shock that followed an auditory stimulus. Preoperative training was not necessary.

The basic unit of the automatic movement is always the automatic-rhythmic group of cells in the central nervous system. We have discussed earlier the form constancy of the fixed action pattern. If we now ask what it actually consists of, we will quickly find the answer, because whether or not a wave passes down a dorsal fin fast or slowly, the phase distance of the muscle contractions which participate in the movement remain constant. And this is, as P. Leyhausen (1954a) emphasized, to be understood by the stereotypy of the fixed action patterns and not as an absolute unmodifiability.

A central coordination was also demonstrated by J. Gray (1950), who deafferented toads completely, with the exception of the labyrinth. In spite of this the animal swam in a coordinated fashion. The walking pattern of the toad remains well coordinated even after deafferentiation of all limbs, provided at least one spinal nerve remains intact (J. Gray and H. W. Lissmann 1946a, 1946b). There must exist then a central movement coordination, because the coordination could not come from the afferent nerves of the labyrinth. The findings of H. W. Lissmann (1946) have to be interpreted in the same way; his largely deafferented sharks were still able to swim with well-coordinated undulating movements.

Even reflex movements can be centrally organized. The coordination of the wiping reflex of a spinal frog remains if the leg that performs the movement is deafferented (E. Hering 1896).

In mammals deafferentation of a limb is usually correlated with the loss of complex movements. But it had been known for some time that the scratch reflex of the dog, in which 19 muscles cooperate in rhythmic coordination, remain well coordinated even following deafferentation (C. S. Sherrington 1931). Recently it was found that in bilaterally deafferented monkeys the function of the hands recovers almost completely to a normal level. The monkeys climbed and swung along with their hands even when their eyes were covered. They grasp and point toward an object although they cannot see their hand. After complete bilateral deafferentation (C2-S5) of the spinal cord the monkeys remain capable of performing a large number of learned and goal-directed movement patterns. Unilaterally deafferented monkeys, on the other hand, are unable to use the deafferented hand freely (E. Taub and A. J. Berman 1964). A detailed discussion of central coordination and central automatism is available in T. H. Bullock (1961, 1962), T. H. Bullock and G. A. Horridge (1965), and B. Hassenstein (1966).

At first the central coordination and automatism was used as a criterion of the fixed action pattern. In line with that criterion, behavior patterns that are coordinated by feedback from afferent stimulation would not be considered fixed action patterns, for example, reflex movements. But since the movement coordination via afferent pathways (proprioceptors) can occur as a phylogenetic adaptation, it seems useful to classify these behavior patterns as fixed action patterns as well (I. Eibl-Eibesfeldt 1963, 1966a). Unconditioned reflexes and automatic movements are extremes that are connected by means of numerous transitions.

When discussing earlier the automatisms of fins and fin rays we discovered the hierarchical organization of the fixed action patterns. This holds for all complex fixed action patterns, whether they are birdsongs or a sucking movement. They can be subdivided into elements which in turn are fixed action patterns. However it is possible, in principle, to set an upper limit where all those functional motor units are called a fixed action pattern which are activated once by external stimuli or spontaneously, and which continue in the absence of additional external stimuli. The orderly sequence does not have to rest exclusively in an automatic, central pattern of excitation. Internal sensory stimulation frequently controls the discharge of a movement pattern in the manner of a chain reflex. As long as their regulating influence has been determined genetically, and the movement pattern does not depend upon additional external stimuli, one can speak of fixed action patterns. If the concept were restricted to purely centrally coordinated automatic movements, then, in the opinion of this author, a too rigorous limitation of the concept would be the result, especially because the demonstration of fixed action patterns would then be difficult to achieve in many cases. When taxis components are added (p. 17), we speak of instinctive actions.

If, on the other hand, an ordered movement sequence comes about when the occurrence of a behavior changes the releasing stimulus situation and activates new behavior via this new stimulus situation, then we are confronted by a chain of fixed action patterns (pp. 156 ff.).

Such a chain exists, for example, if a falcon separates his victim from a flock of birds by means of a sham attack. Then, when he has been successful, he captures the single bird, plucks it, and eats it. In this case each succeeding stage presupposes a new and adequate stimulus situation, which is brought about by the activity of the animal. A fixed action pattern is always an inborn, internally coordinated sequence, which merely requires a releasing stimulus.

4   MOTIVATING FACTORS

As I have tried to make clear in Chapter 3, a behavior is not merely a response to external stimuli. The animal is not simply an automaton into which one drops a coin, for which one then receives a response. The animal is active also because of internal motivating mechanisms. This is convincingly illustrated in the study of intact animals. Animals that are maintained under constant conditions display, for example, a circadian rhythm (see p. 392), rest and activity following an endogenous periodicity that coincides approximately with the day-night rhythm. The anemone (*Metridium*) shows spontaneous rhythmic contractions in 10-minute intervals (E. J. Batham and C. F. A. Pantin 1950). Furthermore, animals that for a time have had no opportunity to perform a certain behavior pattern are in a state of specific readiness to perform precisely the behavior pattern they were unable to perform. W. Craig (1913) clearly recognized this specific drive state. The observer at first notes merely a general restlessness of the animal—one has the impression "as if it were searching for something."

That this is not merely a general motor restlessness but the expression of a specific readiness to act can be recognized by the readiness to respond to specific releasing stimuli: The thirsty animal seeks water and passes up food objects. An animal in a hunting "mood" searches for a releasing situation that permits the discharge of hunting behavior patterns, and the sexually motivated animal searches for appropriate stimuli. If it does not find adequate releasing objects, it may on occasion accept substitute objects. Female rats are so ready to retrieve during the first few days following parturi-

44

tion that they will repeatedly grasp their own tail, retrieve it in their mouth, and deposit it in their nest. Sometimes they even grasp one of their own hind legs and limp into the nest on three legs (I. Eibl-Eibesfeldt 1963; W. E. Wilsoncroft and D. U. Shupe 1965). The specific search for a releasing situation, which W. Craig (1918) called *appetitive behavior*, is variable and adaptable to changing situations. The animal must be capable of mastering detours that lie between it and a desired goal that is remembered, as when a dog in a mood to hunt proceeds toward a chicken yard known to him. Once he has found the releasing situation, the more automatically discharged fixed action patterns of prey catching run off. Their occurrence not only changes the releasing stimulus situation (p. 169) but results also, as W. Craig already stated, in a change of mood. One often speaks of a drive-reducing *consummatory act.* In cats the behavior patterns of prey catching, lying in wait, creeping, catching, jumping on, and pawing normally occur in a certain sequence, which is directed by the releasing situation. Beyond that, however, P. Leyhausen (1965a) demonstrated that each of these individual actions also has its own spontaneous motivation. If a cat has had no opportunity to perform one or the other action, this action will develop its own appetitive behavior. The animal searches for a releasing stimulus situation, merely to paw or catch. The mouse alternately becomes an object to be caught, to be intently observed, killed, eaten, or pawed. The behavior patterns leading up to each action then become appetitive behavior for the desired consummatory act in each specific case.

Appetitive behavior can also be demonstrated by electrical brain stimulation. Cats that are eating stop when stimulated in certain points of the hypothalamus and attack a rat which has until then been ignored. If they do not see prey they search, and they learn a maze when they find as a reward a rat they are allowed to attack. Rats stop eating if the appetitive behavior for gnawing is released by electrical stimulation and begin to search for objects that are suitable for gnawing. They also learn a maze under this motivation (W. W. Roberts and H. O. Kiess 1964; W. W. Roberts and R. J. Carey 1965).

The specific readiness to act or the mood of an animal is also shown in a noticeable lowering of the threshold for certain releasing stimuli. A predator in a mood to hunt will react most readily to stimuli that release hunting-behavior patterns. If prevented from hunting for some time it will even accept substitute objects, and in some instances the response may occur in vacuo following a prolonged absence of the appropriate releasing stimuli (p. 53). At the same time the thresholds for other behavior patterns, for example those belonging to the area of sexual behavior, are raised such that very strong releasing stimuli are required to detract the animal from its hunting and switch it to sexual behavior.

The observations on intact animals show that fixed action patterns often occur in sets, and then they also show a common and identical fluctuation of the releasing thresholds. This points to a common physiological mechanism. The sets of behavior patterns are to some extent mutually exclusive. In male cichlids a readiness to flee suppresses a readiness to fight as well as to court. The readiness to attack and to court, however, are positively correlated. Not so in cichlid females, where a readiness to attack generally suppresses sexual readiness (B. Oehlert 1958). In males sticklebacks the readiness to court suppresses the readiness to bite. P. Sevenster (1968) rewarded the males whenever they swam through a ring by briefly lifting an opaque screen in front of a glass pane so that the male could see a neighboring female and court it. The males learned their task slowly, because they rarely swam through rings of their own accord. Once they had learned it, however, they passed the ring several times in a session without delay in order to see the female. If the task was to bite a rod for the same reward, they learned this very fast, because they often spontaneously bite and nibble on rods. Even so, however, they never achieved a high quota of success, as they seem to be unable to repeat the biting immediately after the performance of the courting dance. Although they often stood in front of the bar, clearly intending to bite it, they could not bring themselves to do it, owing, obviously, to an inner inhibition. Biting and attacking, in contrast, do not inhibit each other. One male quickly learned to bite the bar if rewarded with the view of a rival whom he could fight through the glass pane. Whenever the opaque screen was lowered again, he would bite the rod without delay. Recognition of these kinds of relationships permits one to draw inferences about the mechanisms underlying this behavior (p. 162). During continuous observations one records the sequence of several behavior patterns, computes the correlation, and sets up models that express these relationships.

Examples for this type of *motivation analysis* are found in the work of D. Morris (1958), P. R. Wiepkema (1961), and W. Heiligenberg (1964). In male ten-spined sticklebacks (*Pygosteus pungitius*), Morris showed that 1766 dances were followed by 1232 attacks (70.4 percent). On the other hand, 208 nest-showing actions were followed by 5.3 percent attacks, 1 percent nesting, and 93.7 percent sexual behavior. One can conclude from these results that aggression is dominant during the early stages of stickleback courtship and the sexual drive later.

Appetitive behavior is always the first indication of a specific internal readiness to act; this mood is the expression of a physiological state, usually called a *drive,* which can be measured quantitatively.

By means of brain stimulation with thin electrodes, E. v. Holst and U. v. Saint-Paul (1960) activated various drives in intact chickens. Upon stimulation the chickens began, for example, to walk about restlessly. That this was not the activation of a more general activity, but a typical appetitive behavior was made clear when the animal was offered variously a rival, a female, water, or food. The chickens always responded to a particular object when a particular point was stimulated, for the duration of the brain stimulus. The strength of the activated drive was measured by E. v. Holst and U. v. Saint-Paul by the level of stimulus current necessary to release the behavior. For instance, if they activated two opposing drives, such as those for sitting down and standing up, by means of two electrodes placed at two different positions in the brain stem, they were able to measure (in volts) how strongly a chicken was motivated to sit and to stand, respectively (Fig. 17). They could measure changes in the specific readiness for each behavior. A chicken that was originally motivated to stand was brought into a sitting mood, following repeated activation of the sitting behavior; after this change a stronger stimulus was now required to release standing up.

The observations and experiments with intact animals show clearly that a living organism is not a reflex automaton waiting passively

Figure 17. Measuring a sitting and standing drive. A chicken in a mood to sit spontaneously (a) is made to stand up by stimulating a central region that controls the standing-up drive. The stimulus is strong and short; after a short latency the chicken stands up, but sits down again immediately after the cessation of the stimulus. The second standing-up stimulus is of equal strength (b) but lasts longer and the chicken stands up longer; latency and standing up are unchanged. Finally, if the duration of the stimulus is longer (c), but weaker, the animal rises more slowly but remains standing as long after the stimulus, as in (b), because stimulus strength and duration compensate each other; that is, the value remains constant. The central process that underlies these actions is designated by the middle curve. Below the zero line the physiological sitting drive is above the standing drive. The strength of these drives can be measured. As long as the animal sits, the threshold value that must be crossed to cause the animal to stand up is a measure of the strength of the sitting drive that must be overcome. (After E. v. Holst and U. v. Saint-Paul [1960].)

for stimuli from the outside to which it then responds. Rather, it is motivated from within to act in a specific manner. What are the motivating factors behind such a specific readiness to act? What are the mechanisms[1] which bring the animal into a state of specific restlessness, and what leads to the termination of the once activated behavior? Let us examine these questions with some examples.

Many investigations deal with the phenomenon of "thirst" (summary by A. V. Wolf [1958]). It was found that the appetitive behavior of searching for water is released by osmoreceptors in the hypothalamus. They report hypertonicity of the blood, and by injecting saline solution into the veins of dogs and people they can be made thirsty. On the other hand, it is possible to eliminate thirst by intravenous injection of water. If minute quantities of a hypertonic saline solution are injected directly into the hypothalamus, the animal also becomes thirsty; the same effect is achieved by electrical stimulation of this area in rats and goats.

A thirsty animal does not have to continue drinking until the normal osmotic value of its body fluids has been restored. There would be danger of too much water being taken in, because resorption takes some time. As a kind of safety measure, the amount of water in the stomach and the swallowing activity itself is taken into account (R. T. Bellows 1939; E. J. Towbin 1949).

Dogs provided with an esophageal fistula, through which all the water they drank drained outside, drank regularly and stopped after a certain time, but this satiation through the act of drinking did not last very long. If the stomach of the drinking animal was also filled with water satiation was more lasting and correspondingly less water was taken in by mouth. If a rubber balloon was inserted into the stomach and blown up, the amount of sham drinking was also significantly decreased. This drinking activity is controlled by several mechanisms; the osmoreceptors in the hypothalamus release the appetitive behavior for drinking and finally terminate it, while a short-term "satiation of thirst" is achieved by filling the stomach and by the drinking activity itself. The last activity is especially interesting for us because it often appears as if the mere performance of the movements is "drive reducing."

In this connection the observations of D. W. Ploog (1964a) and R. A. Spitz (1957) deserve special attention. Both noted a clear correlation between the degree of satiation and the amount of sucking movements in infants. If the infants had taken in a certain quantity of food during 20 minutes by sucking, they were satisfied and slept. If the openings in the nipples were too large, so that the same amount or even 50 percent was sucked within 5 minutes, they re-

[1] When speaking of motivating mechanisms, we refer to the totality of the physiological machinery involved in the specific activation of an animal.

mained dissatisfied. They continued to suck in vacuo and began to cry. If they were given the empty bottle, they continued to suck for another 10 to 15 minutes and then seemed satisfied. According to M. Mead (1937), infants of peoples who have not been influenced by Western ways and who begin to nurse on their mothers shortly after birth, do not suck their thumbs. Puppies fed from nipples with a large opening afterward sucked on their own bodies and during sleep. On the other hand, when they were nursed with nipples containing small holes they did not (D. M. Levy 1934; S. Ross 1951). Calves that are fed from buckets, and hence drink their milk too fast, develop the habit of sucking on their steel chains or on other calves. Some become stunted as a result of this; they show the so-called tongue flick, which is perhaps a vacuum-sucking stereotypy. K. Zeeb (personal communication) was able to eliminate this behavior by letting even older heifers drink only from a bucket fitted with a rubber nipple. Ducks that have been fed grain on land dabble in vacuo (K. Lorenz 1963).

There are a number of comprehensive investigations dealing with the mechanisms that regulate the intake of food in mammals (L. de Ruiter 1963; J. Mayer and D. W. Thomas 1967). The glucose level of the blood is registered by glucose receptors in the hypothalamus. The motivating systems are located in the hypothalamus: Electrical stimulation of the lateral part leads to an increased food intake, but there are also activating and inhibiting systems outside the hypothalamic region (B. G. Hoebel and P. Teitelbaum 1962; P. Teitelbaum 1961). The satiating mechanism is sensitive to an increase in the glucose level of the blood, but, as with drinking, filling of the stomach leads also to an inhibition. Mechanical receptors report the volume to the ventromedial hypothalamus, and chemical receptors report the quality of the food that was taken in.

In the blowfly the appetitive behavior for feeding depends on the amount of food in the foregut. When the foregut is filled, inhibiting impulses pass via the nervus recurrens to the central nervous system. If the nervus recurrens is cut, the feeding inhibition is removed, and the fly continues to suck up food until she becomes extremely distended and dies (V. G. Dethier and D. Bodenstein 1958).

The interaction of external and internal factors affecting the fanning drive of the stickleback was investigated by J. v. Iersel (1953). After males had fertilized three to four clutches of eggs their sexual drive waned and they began to ventilate the eggs with fanning movements of their pectoral fins. Van Iersel measured the intensity of fanning and found that the total fanning time increases from day to day until the young hatch but drops sharply shortly thereafter. The increase of fanning activity is caused by the oxygen consumption of the eggs: If the CO_2 content of the water is artificially raised, the stickleback will fan more. However, the fanning activity does not depend exclusively upon external stimuli, as is shown by the following experiment: If one presents a fanning male with a new clutch just prior to the end of the fanning cycle and shortly before the activity stops completely, one can induce a new fanning cycle that is very similar to the previous one. The peak of fanning is lower, however, and if additional cycles are induced, the peaks will successively be lower (Fig. 18). Since the releasing stimulus situation is always the same, this change must be dependent on events within the stickleback. The mechanisms of drive reduction in the sexual behavior of male sticklebacks were investigated by A. C. A. Sevenster-Bol (1962). She found that the presence of eggs in the nest and not the act of fertilization functions as an inhibiting stimulus situation for the zigzag dance and for leading by the male.

D. S. Lehrman (1961) investigated the development of the reproductive behavior of the ring dove (*Streptopelia risoria*). Males or females that were placed alone into a cage containing a nest and

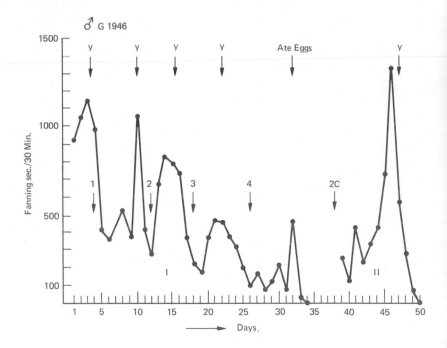

Figure 18. I. Series of fanning cycles in a stickleback male which were repeatedly induced by the exchange of spawns. Arrows 1–4 indicate the time when a new spawn was presented. The *y* symbols mark the time of hatching. The eggs that induced the fifth cycle were eaten. II. Sixth fanning cycle before which the male had courted for 5 days and then built a new nest. (From J. v. Iersel [1953].)

eggs could not be induced to incubate them. A sexual partner had to be present. When pairs were placed into a cage with a nest and eggs, they courted, built a nest, and from the fifth day on some began to incubate; by the seventh day all did. It could be that the birds would have to become used to the cage; therefore, Lehrman placed the pair into a test cage but separated the partners by an opaque partition. When he placed them together after 7 days and gave them nest and eggs, they still required 7 days before they began to incubate. Thus becoming used to a cage was not a factor. If, on the other hand, he kept the animals with their partner and nesting material in the cage and gave them eggs on the seventh day, they all incubated within 2 hours. If he did not give them nesting material first and presented them with a nest and eggs on day 7, they built nests, first; some began to incubate the same day and the others by the end of the next day.

The birds seem to pass through two stages: First, courtship induces the readiness to build a nest, and this, in turn, induces the readiness to incubate eggs. It is sufficient that females can see males through a glass partition, provided the males court. C. Erickson and D. S. Lehrman (1964) presented females with castrated males (which do not court) and this had no effect on ovarian development—an elegant demonstration of the significance of the courtship behavior. The changes induced by courtship behavior are hormonal. Lehrman injected 80 pairs of doves with progesterone 7 days prior to placing them together. If he then presented them with eggs, they incubated at once. If he injected them with estrogen instead, they began to build a nest first and incubated within 11 days. According to R. A. Hinde (1965), the courting canary male stimulates estrogen production in the female.

The interaction of various motivating factors in the courtship behavior of canaries are quite complex (Fig. 19). Changes in day length induce growth of the gonads and estrogen production via a mechanism in the hypothalamus and the hypophysis. The latter is affected by stimuli coming from the male. The female responds to the courting male while under the influence of estrogen and builds a grass nest. This, in turn, further stimulates the female. The nest, together with the estrogen, stimulates the development of the brood patch and the oviducts, which are further stimulated in their development by secondary hormones initially activated by the estrogen. Then the eggs are laid. As a result of increased sensitivity to the nest, the bird stops building with grass and selects only feathers. Contact of the brood patch with eggs and nest now induces incubation. R. A. Hinde (1965) emphasizes four points in his summary about these interactions:
1. The causes and consequences of sexual behavior are closely intertwined with nest building and cannot be understood by themselves.

Figure 19. Interrelationships of various factors in the reproductive cycle of the canary, showing relations among hormones, external stimuli, behavioral changes, development of brood patch, and so on. Solid lines indicate facilitating influences; dashed lines indicate inhibiting influences; dotted lines indicate probable but not proved facilitating influences. (After R. A. Hinde [1965, 1966].)

2. The external stimuli induce endocrine changes that are added to the immediate influences on the behavior.
3. Hormone production has multiple causes.
4. Hormones have multiple effects.

L. R. Aronson (1949) found that cichlid females (*Tilapia macro-cephala*) prepared no typical nest pits when alone. If they see a male, even through a glass in the next tank, their ovaries begin to develop and they begin to build a nest.

Similarly, other external stimuli participate, via the hormonal system, in the buildup of a drive. The increase in length of day during the spring stimulates in many songbirds the growth and activity of the gonads and hormone production and in this way induces the reproductive drive.

Common house mice build a nest shortly before parturition, for which they use up to four times as much nesting material as for their sleeping nests. This nest building can be released by the injection of the hormone progesterone, but not by prolactin. The hormonally stimulated nest-building drive normally wanes immediately following parturition, but the increased nest-building activity is maintained by the presence of the young in the nest. If the young are removed, this activity decreases. Virgin females can be stimulated to build nests if one presents them with very small young (G. Koller 1955; see also Figs. 20 and 21).

How decisively and specifically hormones are involved in the organization of drives can be seen in the numerous experiments on the sexual behavior of castrated and hormone-treated animals (F. A. Beach 1948). Female dogs urinate in a squatting position, while males stand and raise one leg. The development of this behavior depends on the male sex hormone. Young males still urinate without raising the leg and continue to do so if they are castrated before they are 4 months old. If injected with testosterone, they later raise a hind leg while urinating. Females do the same if they are spayed shortly after birth and treated with progesterone but not when they are spayed as adults (T. Martins and J. R. Valle 1948).

In the golden hamster there is a clear negative correlation between sexual receptivity and aggression, which is dependent on hormones. Golden hamsters in estrous show a decreased readiness to attack males in the rutting period (J. W. Kislack and F. A. Beach 1955).

From what has been said so far, it can be seen that a specific readiness to act can be released by many different factors and that it is activated usually through the interaction of several. We discussed external stimuli (day length, sexual partner), internal stimuli, and hormones.

The main problem in respect to the fixed action pattern, which has been little investigated to date, is concerned with the lawful fluctuation of the inner readiness to act, which cannot be explained on the basis of factors discussed so far. K. Lorenz (1937) writes about his well-fed starling, which never had the opportunity to catch live insects but would nevertheless fly up into the air from its perch, act

Figure 20. Top: effect of 1,5-I.E. corpus luteum hormone (Progesteron Boehring-Spezial) on three castrated female mice; bottom: increase in the building activity of a nonpregnant female mouse after the addition of newborn mouse pups (arrow). (Top: from G. Koller [1955]; bottom: after G. Koller [1955].)

as if it were catching something, return to its perch, perform killing movements, and finally swallow, although Lorenz assured himself that the bird had caught no prey—a good example of behavior in vacuo. H. N. Kluyver (1947) observed waxwings (*Bombycilla garrulus*) who showed insect prey-catching behavior in vacuo during strong

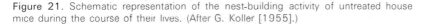

Figure 21. Schematic representation of the nest-building activity of untreated house mice during the course of their lives. (After G. Koller [1955].)

frosts when insects are absent. Young honey buzzards while still in the nest already perform the movements associated with digging out wasp nests (K. Gentz 1935). Sticklebacks also court in vacuo (N. Tinbergen 1952). Many more examples of such behavior[2] in vacuo are now known (L. Koenig 1951; P. Leyhausen 1956; and others). It has also been shown that the mere performance of a movement can be rewarding in itself. We cited the example of satisfying the sucking drive (p. 48).

In the very aggressive cichlids *Etroplus maculatus* and *Geophagus brasiliensis* the males must fight with other males before successful pairings with females can take place. If they are not given the opportunity to fight they kill their females, because they discharge their aggression on them. To avoid this, it is only necessary to separate two pairs by a glass partition; then the males fight at the glass and do not harm their females. The same occurs if other conspecifics are kept in a sufficiently large tank with them. If these "whipping boys" are removed, the male regularly attacks the female and finally kills her (K. Lorenz 1963). A. Rasa (1969) investigated this in *Etroplus maculatus* and confirmed the findings of Lorenz. Those males that could attack other fish in their tank directed few attacks against their females. Being able to attack neighbors, if only through a glass partition, they showed little aggression toward their females. If there was no stranger to attack, they fought with their females, and the number of attacks directed at them was markedly increased in com-

---

[2] M. Bastock, D. Morris, and M. Moynihan (1953) have proposed to replace the term "vacuum activity" with the term "overflow activity," because one could never be certain of the complete absence of a releasing stimulus. But the term "vacuum activity" does not imply this, so we may as well retain the original term.

parison to the number in the two other groups (Fig. 22). This increase in aggressive behavior toward their own females, which was at first not understood, may be due to the female's continued efforts to seek contact with the male in spite of his attacks. An investigation of the spontaneity of aggression in cichlids that were raised in isolation from conspecifics is underway. In view of the great theoretical significance of this phenomenon, investigations of other vertebrates are needed.

W. Heiligenberg (1964) demonstrated that the readiness to fight wanes in male cichlids (*Pelmatochromis subocellatus*) if the animals fight briefly without damaging one another. They were not generally fatigued, as was demonstrated by their readiness to perform other behavior patterns. O. Drees (1952) gave salticid spiders an opportunity to exhaust prey-catching behavior, including approach running, stalking, and creeping, long before overall physical fatigue set in. He was able to rule out avoidance conditioning and adaptation of afferent mechanisms (p. 81), so he interpreted these central damming-up and discharge processes in line with K. Lorenz's hypothesis.

If the gobbling call[3] of the male turkey is repeatedly released by a stimulus of constant amplitude and frequency, the threshold for this stimulus is raised, and the animal will no longer call. This depends primarily upon adaptive processes of afferent, releasing mechanisms (p. 81), because the gobbling calls immediately reappear in response to a tone of different frequency and amplitude, even if the new stimulus is normally less effective as a releaser than the one in the preceding stimuli series. If the two differentially effective stimuli are presented alternately in a continuing sequence, the weaker of the two is soon not responded to, which points to a central fatiguing process in addition to adaptation (M. Schleidt 1954).

In the cichlid *Pelmatochromis subocellatus kribensis* the readiness to fight is increased, as measured by the number of bites, following a short fight, but wanes increasingly as the fights continue: Short pauses lead again to an increase. The "straining" (sifting), a movement associated with feeding, suppresses attack bites. Shortly before and during straining the attack readiness is markedly decreased; afterward it is clearly higher. It appears as if the fish had "saved up" the biting behavior, or as if it had accumulated during the straining period. The digging movement of biting into the sand does not have the same influence on biting against conspecifics.

In physiology a similar accumulation phenomenon has been known since 1892 as *spinal contrast* (C. S. Sherrington). E. v. Holst (1937) investigated it in more detail in the sea horse. This fish swims about very little but usually fastens itself to some weeds with its prehensile

---

[3] A call that can be released by various noises and tones.

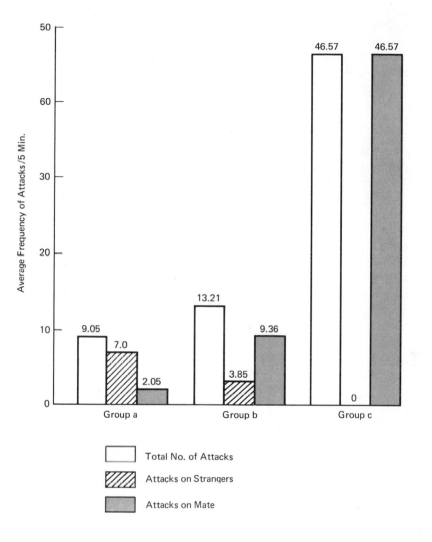

Figure 22. Average number of attacks of male cichlids (*Etroplus maculatus*) during the reproductive phase. In group a the pair lived together with several unmated adult conspecifics and several young animals. Three such pairs were observed for a total of 83 hours and 15 minutes. Two of these pairs spawned twice and one three times. In group b a glass partition separated the pair from its conspecifics. They were able to see their neighbors but could attack them only through the glass. Two pairs that raised five broods successfully were observed for a total of 62 hours and 40 minutes. In group c the pair was completely isolated from all other conspecifics. In three pairs that were kept thus the pair bond disintegrated and the female had to be removed for her own protection. In the fourth pair the bond also disintegrated, but the animals mated once more briefly before spawning. They spawned, ate the eggs, and continued to fight. One day later they again mated and remained together until the eggs hatched. Then the male drove off the female and finally killed her 3 days after the young hatched. He raised the young alone successfully. The total observation time for group c: 84 hours and 17 minutes. (After A. Rasa [1969].)

tail; its dorsal fin is collapsed. The dorsal fin is raised only when the
fish swims and it beats in an undulating movement. If the spinal
cord is cut, this does not cause, as in most fish, uninhibited, spontaneously locomoting movements (p. 33); instead the dorsal fin of
the spinal sea horse remains in a half-raised position. If the animal
is gently squeezed in the gill region, the fin is folded down completely. When the fish is released the fin is raised a littler higher than
before. If this is repeated and the fish is held longer, the dorsal fin
is raised completely and begins to undulate, having been prevented
from doing so by an inhibiting stimulus caused by the squeeze hold.
The explanation for this peculiar phenomenon is that the automatism (p. 33) for the swimming movement of the sea horse produces
only a very small amount of endogenous excitatory potential, so
that it has to accumulate sufficiently before the swimming movements can occur. This accumulation is achieved by a reflexive inhibition of the swimming movements. In the spinal sea horse this
inhibition is lacking, accumulation of excitatory potential does not
take place, and the weak action specific "excitation" is continuously
discharged, causing the half-raised position of the dorsal fin.

We are reminded in this connection that E. v. Holst (p. 33)
demonstrated the central spontaneity underlying swimming movements of some fish. The extent of the endogenous excitation production varies from species to species. In the eel, which swims a lot,
a spinal preparation will show undulating movements until it dies.
In the sea horse, which moves relatively little, the underlying automatism becomes visible only when continuous discharge is prevented
experimentally in a spinal animal, and an accumulation of excitatory
potential is achieved in this way. It is quite possible that the
relative need for movement in higher vertebrates can be understood in terms of such differences in the accumulation of central
excitatory potential. The lion, which stalks its prey, is a calm animal
that may be kept in a small cage. Weasels and wolves, which run
down their prey, have a great need to run and will continue to
discharge this drive after feeding to satiation, by running up and
down in their cages for hours.

K. Lorenz and E. v. Holst recognized the connection between
these physiological states and the observations in regard to the spontaneity of instinctive behavior of intact animals. K. Lorenz hypothe-
sized that each fixed action pattern—not only those of locomotion—
is based upon accumulation of central excitatory potential of E. v.
Holst's automatisms. With this generalization he bridged the gap
between behavior studies and physiology. The investigations of many
ethologists and physiologists have since supported this view. K. D.
Roeder (1955) noted that reflex movements possess an endogenous-
automatic base, which is not sufficient for a spontaneous discharge.

What biochemical events within the central nervous system correspond to or are correlated with the observed fluctuations of the
specific readiness to act (specific excitatory potentials) is not known.
Perhaps the key to an understanding of these phenomena is in the
catecholamine metabolism. Recent investigations indicate that animal
and human behavior is dependent in some way on the catecholamine level (noradrenaline and dopamine) as well as on indolamine
serotonin. Drugs that lower the central catecholamine level have a
calming effect; those that raise this level stimulate the motor activity
and aggressiveness. In man they have an antidepressive effect. The
manner in which these substances act is still unknown, but it has been
suspected that they are involved in aiding synaptic transmission.
Their accumulation and depletion at certain locations in the brain
might possibly explain the phenomena of central lowering as well
as raising of thresholds (G. M. Everett 1961; G. M. Everett and
R. G. Wiegand 1962; D. X. Freedman and N. J. Giarman 1963;
N. J. Giarman and D. X. Freedman 1965; J. J. Schildkraut 1965;
J. J. Schildkraut and S. S. Kety 1967).

I. J. Bak (1965) and R. Hassler and I. J. Bak (1966) were able
to demonstrate the existence of submicroscopic catecholamine stores,
which changed under the influence of drugs. Reserpine depletes these
catecholamine stores and possibly results then in a loss of spontaneous
activity. Following iprozianid addition, catecholamine stores are
repleted, and spontaneous movements increase at the same time.

The implications that follow if we postulate an accumulation of
a central nervous excitatory potential as the basis for instinctive behavior are of great importance for human ethology. Many examples
indicate that man is dependent upon an accumulation of central
excitatory potential in some areas of his behavior, which is difficult
to control because he is not conscious of it and which affects his
inner readiness to act in a specific way along with other motivating
factors. This may be true, for example, of the aggressive drive, which
in present-day human society finds very few adequate opportunities
for discharge. The constant endogenous accumulation of excitatory
potential continues to lead man to seek a discharge for this drive,
and in ignorance of the biological conditions he projects his periodically occurring "anger" outward—in his personal daily life to
those who are next to him, for instance, the spouse, and in a larger
group, possibly against minorities or neighboring peoples. Only a
clear understanding of the nature of these phenomena can help us
in the search for reasonable solutions.

5 BEHAVIOR AS A RESPONSE TO A STIMULUS

Innate releasing mechanism as a basis for innate recognition

The conception of a central accumulation of excitatory potential, discharge of which is inhibited by higher central controls, requires the assumption of a special afferent mechanism that removes these inhibitions at the biologically appropriate moment. This neurosensory *innate releasing mechanism* (IRM) allows the central impulses to proceed to the effectors only when certain key stimuli are encountered. Key stimuli are usually simple. They can be discovered by means of experiments with models that are presented to inexperienced animals. Innate releasing mechanisms, which respond unselectively to the simplest stimuli, can become more selective through individual experience. The toad, which at first snaps unselectively at moving objects, soon learns to avoid noxious prey (p. 262).

Unconditioned stimuli can also be inhibitory: The search automatism of infants—a rhythmic head movement when searching for the nipple, at once comes to an end when the child touches the nipple with the mouth (H. F. R. Prechtl 1958). Many precocial animals show an innate avoidance of a precipice, which they recognize visually before having had the adverse experience of falling off a cliff. Chicks, kids, lambs, and 4-week-old kittens which never fell off anywhere stop when they reach a cliff that is covered by the same

glass plate on which they are standing, while the less visually oriented Norway rat walks without hesitation on the glass plate above
the abyss (E. J. Gibson and R. D. Walk 1960). Parallactic changes
while moving seem to be the effective stimuli. In order to exhibit
the behavior the cats must merely have had visual pattern experience coupled while actively moving about. Cats that could walk in
a rotating striped drum, but which were prevented from seeing their
own feet, reacted to the visual cliff. Kittens that were carried passively
through the same path traversed by the other kittens, but lacking
the walking experience, did not avoid the visual cliff, although their
visual impressions had been the same for both groups of animals
(R. Held and A. Hein 1963). Day-old chicks exhibit unlearned visual
depth discrimination. The cues provided by focusing are critical for
this discrimination but not the binocular and motion parallax (P.
G. Shinkman 1963). Three-day-old chicks that were never fed but
grew up under normal lighting conditions preferred to peck at
photographs of half-spheres that were illuminated from one side,
provided they were "correctly" (that is, the bright side facing
upward) oriented. Three-dimensional objects normally are lighter
on the upper side, because the light normally comes from above.
They prefer such photographs over others that are mounted upside
down, even if they were raised in cages that were illuminated from
below, which points to an inborn capacity to utilize surface shadings
as a stimulus parameter for three-dimensional objects (R. Dawkins
1968). There undoubtedly exists, then, an ability to innately recognize releasing and inhibiting stimulus situations of a highly complex
nature, as was also pointed out by W. McDougall (1936) and J. v.
Uexküll (1921; see also Fig. 23).

Figure 23. Young and in-
experienced cat at a visual
cliff. Although the animal
had had no experiences
with a precipice, which is
here also covered by a
glass plate, it hesitates at
the edge. (After E. Gibson
and R. Walk in P. R.
Marler and W. J. Hamilton
[1966].)

The innate releasing mechanism responsible for these actions is first defined in purely functional terms. It is a stimulus filter. Its seat within the brain is unknown in most cases. The investigations of H. R. Maturana and others (1960) in the leopard frog (*Rana pipiens*) show that analysis and integration of stimuli takes place as early as the retinal level. In the retina they found five types of ganglion cells, which only respond to different stimuli. One group fires only briefly when a light is turned on and off. They also respond to each moving edge and they fire during a darkening and lighting up as well as during the passing of the leading and trailing edge of a stripe. If the object is stationary in the receptive area of the retina these cells do not respond; they only respond to changes in contrast— they are event detectors. Another group of cells does not respond to turning light on and off, only to the passing of a straight or curved edge. If it stops the frequency of discharge drops to a lower level of continuous discharge. These cells inform the frog continually about the contours of objects; they are contour detectors. One group of cells is of special interest, because it does not respond to a change of level of illumination but responds with vigorous discharges when a small object that is darker than its background passes over the receptive field. The authors call these "beetle detectors." Finally, there are special cells that measure the decrease in illumination and others that measure the light intensity. Here the selectivity of the stimulus filter already exists in the retina. In the retina of the rabbit different nerve cells have been found that process the arriving impulses prior to entry into the central nervous system. There are cells that fire only when a dark object moves across the visual field in a certain direction (H. B. Barlow and others 1964).

We find in the frog retina several overlapping receptor systems which process the retinal image. Similar events occur in the retina of the cat. Contrary to what is found in the frog, where the center of an on–off area responds at the beginning as well as at the end of a light stimulus, the corresponding receptor field of the light-adapted cat are so arranged that an on area is surrounded by a peripheral off area. The investigations of D. H. Hubel and T. N. Wiesel (1959, 1962) dealt with information processing beyond the retinal organization by recording the activity of single neurons in the corporea geniculata, in the cortex striatum, the visual center of the cat brain, along with the retinal events. The cortical cells interact with the retinal receptor areas; they are connected with specific ganglion cells in such a way that a certain retinal stimulus area is projected onto a particular cortical cell. If the retina is stimulated with a narrow long band of light instead of with a spot of light, and if one records from a cell in the striate cortex, then one obtains responses of various

Figure 24. (a) Responses of a cell in the striate cortex of a cat during stimulation of various points of the retina with a 1° light spot and (b) with a bar of light (1 × 5°) that was rotated in seven different positions around the receptive area. The receptive areas of the retina are shown on the right [above for (a) and below for (b)]; triangles denote inhibiting fields, crosses excitatory fields. (After Hubel and Wiesel [1959] from P. R. Marler and W. J. Hamilton [1966].)

amplitude dependent on the position and type of movement upon the light-stimulated area. Thus we encounter within the visual system data processing by steps that begin at the retinal level (Fig. 24).

The investigations of D. H. Hubel and T. N. Wiesel (1963) demonstrate that many of the complex physiological stimulus-processing functions within the cortex of the cat can be found in the newborn kitten. The neural connections that underly this complex data processing must be present at birth, although the newborn animals appear not to be able to utilize this functioning visual system. Visual deprivation results in degeneration within the brain which can be demonstrated histologically, but this is secondary. The lack of visual impressions does not prevent the development of neural connections but leads to the degeneration of those present at birth (T. N. Wiesel and D. H. Hubel 1963a, 1963b, 1965). Electrical recordings from the antennas of male silk moths led D. Schneider (1962) to the conclusion that the specificity for the sexual odor of the females is dependent upon the structure of the receptors.

There are a number of inborn mechanisms that process and integrate sensory data. The processing starts in the sensory receptor and we learned that the integration of the data can occur at different levels of the central nervous system. Of course not all of these data-integrating mechanisms are releasing mechanisms.

On the whole we know much less about the localization and nature of releasing mechanisms than about the way they work. This will be examined in Chapter 6 (a comprehensive discussion can be found in W. M. Schleidt [1962, 1964b]).

Key stimuli and releasers

Behavior may be activated by an internal drive but is normally released by specific stimuli from the environment. A hungry toad (*Bufo bufo* L.) moves about until it finds a worm or insect, fixates the prey, and snaps it up. What are the cues that tell the toad what prey is?

Before we can discuss this question we have to recall that each animal can perceive only a limited portion of the total environment with its sense organs. Its external environment or Umwelt, according to J. v. Uexküll, is made up of the particular stimuli that the animal perceives. Whoever studies the reactions of animals must first be acquainted with its sensory-physiological capacities, because these differ from species to species. Some examples may illustrate this: Bees can see unltraviolet light and distinguish polarized from nonpolarized light—capacities that man does not possess. Bats can hear ultrasonic sounds which we cannot hear, but they do not see very well. But because they can establish for themselves a copy (Abbild) of their environment by means of the echoes from their own calls, they are able to orient as effectively in flight as birds with good vision. Nile pike send out electrical impulses and react to small differences in the potential of the surrounding electrical field. Bees accept sugar substitutes that are tasteless to us. Pit vipers are very sensitive to infrared radiation and perceive temperature differences of 0.005°C. Red-breasted robins (*Erithacus rubecula*) utilize the earth's magnetic field for navigation.

The species differ qualitatively and quantitatively. Guinea pigs, being macrosmatic animals, are able to detect nitrobenzene in a 1/1000 dilution of the concentration we can sense. Some performances tax our comprehension. Eels respond to odorous substances in dilutions of 1:2.9 million billion. This, according to H. Autrum

(1943, 1948), is equal to 1 ml of substance dissolved in a body of water 58 times the volume of Lake Constance.

More important than absolute thresholds are differential thresholds. In minnows the difference threshold for pitch of a tone is half a note. Old World monkeys distinguish as many color nuances as we do. The critical distance between two just-distinguishable colors in the red region is 10 mμ and in the blue region is approximately 9 mμ (W. F. Grether 1939).

Finally, of utmost biological significance is the ability of an organism to locate a stimulus source with respect to distance and direction. Here the visual sense attains the greatest precision and range by means of the highly developed camera eye of the vertebrates and the compound eyes of insects. However, bats also locate their prey with the echo of their ultrasonic cries. We shall discuss the capacities of the sense organs and their functioning in space in more detail in Chapter 16.

The functioning and capacities of the sense organs is studied by sense physiologists. The methods may vary. The conditioning method has been most successful: A certain reaction, for example, eating, becomes associated with a specific stimulus, for example, a whistle. When the animal has formed an association between this stimulus and the feeding, one tries to determine, with the aid of appropriate controls, which sense mediated the relevant stimulus. For example, if one leads the whistle to the mouth without blowing it and the animal does not react, this indicates that the animal heard and reacted to the sound and has been conditioned to it. In this way K. v. Frisch (1923) demonstrated hearing in the dwarf sheatfish. The same author used the conditioned technique to demonstrate color vision in the honeybee. C. v. Hess (1913) had placed bees into a darkened room with two different-colored lights of differential brightness. Bees approached the brighter of the lights regardless of whether it was red or green. They oriented to brightness, and v. Hess concluded that bees are color blind. K. v. Frisch (1914), who was unwilling to accept that an insect which searches out flowers should be color blind, offered food to bees on yellow paper which was placed between papers of various shades of gray. The bees did not confuse the yellow with a single shade of gray—hence they could see color. This indicates that the animals could behave differently in different functional systems. The investigations of N. Tinbergen and his collaborators (1943) give an additional, impressive illustration. Male grayling butterflies (*Eumenis semele*) fly randomly toward models of females of various colors as if they were color blind. When approaching blossoms, however, they prefer certain colors over others and distinguish them from shades of gray of equal brightness. In this functional system they do demonstrate color vision.

The capacities of sense organs can be elegantly investigated electrophysiologically by the recording of action potentials (H. Autrum 1958; J. Schwartzkopff 1962). The structure and functioning of sense organs are discussed in detail by H. Autrum (1952), W. v. Buddenbrock (1952), R. Granit (1955), D. Burkhardt (1960, 1961), D. Burkhardt, W. Schleidt, and H. Altner (1966), L. J. and M. Milne (1963), and H. Heran (1966).

Of all the sensory stimuli perceived by an animal, only relatively few innately release reactions. In dogs only stimuli from food objects initially release salivary secretion, and an appropriately conditioned dog will show this reaction later to a bell or light stimulus (p. 251).

We must therefore distinguish between the *perceived* and the *effective* stimuli (N. Tinbergen 1951). The former are the subjects of sensory physiology; the latter are studied by ethologists.

It has been shown in numerous investigations that "unconditioned" stimuli and stimulus patterns exist to which an animal will react with appropriate actions prior to any experience with them. It has been demonstrated that the sense organs may serve quite different functional systems. The carnivorous water beetle (*Dytiscus marginalis*) reacts with prey-catching behavior, not to a moving tadpole in a glass vial, which it normally readily attacks, but to the odor of meat extract (Fig. 25).

Figure 25. (a) Head of a *Dytiscus* water beetle with well-developed complex eyes; (b) the beetle was alarmed by the meat extract and is caught in a "physiological trap" (the odor of the meat). (From N. Tinbergen [1955].)

Compound Eye

0,1mm

1   2mm

a

b

Which specific stimuli or stimulus combinations release a specific reaction is determined by the use of models. Observation alone will often tell the ethologist something of their nature. To return to the example of the prey-catching toad: Whenever a toad spies a moving insect, it fixates upon it, moves toward it, or even pursues it. As soon as the insect remains motionless, nothing further happens. The toad continues to stare at the spot where the insect was last moving, and will rarely snap at the motionless prey. After some time the toad seems to loose interest. Thus it appears as if the movement of a prey is a strong releasing stimulus for the prey-catching response. Experiments with models support this. If we move objects such as stones or paper pieces with a string, the toad will fixate and snap at them. If the object surpasses a certain size it will release escape or defensive behavior. Toads react this way shortly after metamorphosis. In the clawed frog (*Xenopus laevis*) visual stimuli and vibrations in the water will release the prey-catching behavior, and will do this also in the inexperienced animal, which can be readily demonstrated. Tadpoles are plankton filterers that never snap at objects with an oriented movement. If they are isolated shortly before metamorphosis in clear water, so that they have never snapped at prey, and one then projects a light spot against the background of the container, the clawed frog will at once swim toward it, make the specific fanning movements with the forelimbs, and snap at it. A small jet of water from a pipette or slight touch of the legs also releases turning toward the stimulus and the prey-catching movements. The animal snaps at the stream of water.

This behavior will normally lead a frog to its prey, because in general small moving objects in its environment are prey animals. Toads and frogs can learn something new very quickly. If a toad has snapped at a leaf repeatedly without success it will refrain from doing so in the future. The same is true if she caught a bad-tasting or even stinging insect. For avoidance to take place only one bad experience is necessary. The innate knowledge is limited to snapping at small, moving objects. These characteristics sufficiently identify the natural prey objects. Large objects release escape behavior (I. Eibl-Eibesfeldt 1951a, 1962a). Similarly, unselective is the response of a dragonfly larva to small moving objects; it snaps at small objects and flees from larger ones (M. Hoppenheit 1964). When deprived of food for some time, larger objects are taken than before. Pike innately snap at moving objects. Whitefish (*Coregonus wartmanni*) fixate and snap even at floating, nonmoving particles (E. Braum 1963).

Such releasing cues are called key stimuli. In analogy to a key that opens a lock, the key stimuli act upon a mechanism (the IRM) that normally prevents the release of central impulses when it is not appropriate and will only open the way to the musculature when the appropriate key stimuli are received. Each functional cycle has its own key stimuli, and the animal reacts to correspondingly different key stimuli. For a herring gull to roll an egg into its nest it must be spotted; when robbing eggs this characteristic is of no importance (J. P. Kruijt 1958). The grayling butterfly reacts to colors when visiting blossoms; during courtship it acts as if color blind (N. Tinbergen and others 1943).

Key stimuli exist for almost all senses. The night moths of the family Noctuidae and Geometridae show escape reactions of dropping, flying downward, or other evasive maneuvers when they hear the ultrasonic cries of bats or artificially produced ultrasonic stimuli (K. D. Roeder and E. A. Treat 1961). Crickets and grasshoppers react to their own specific song in a predictable fashion (J. Regen 1924; A. S. Weih 1951; A. Faber 1953; W. Jacobs 1953; A. C. Perdeck 1958). The males of the mosquito *Aëdes aegypti* react selectively to the whirring sound of the females' wings (L. M. Roth 1948; H. Risler 1953, 1955), and many frogs respond to the calls of their own species (C. M. Bogert 1961).

The female turkey recognizes her chick only by its calls and will brood a stuffed polecat that is fitted with a loudspeaker uttering the call of newborn turkeys. She will kill her own young if she is deaf and cannot hear their calls (W. M. Schleidt, M. Schleidt, and M. Magg 1960). Mallard ducklings (*Anas platyrhynchos*) and wood ducks (*Aix sponsa*) each prefer the calls of their own mothers (G. Gottlieb 1965a; see also p. 226). It might be that ducklings and chicks which were incubated could hear their own calls while still in the egg and would generalize from them to their mothers. If this were so, newly hatched chicks should prefer their own or calls from other chicks over the somewhat different calls of their mother. This, however, is not the case, as G. Gottlieb (1966) found in recent experiments. The chicks always preferred the specific call of their mother, even if prior to hatching they were exposed to additional calls of other chicks. In this case they will later follow their mother's call even better with a shorter latency and a higher proportion of followers, than those chicks which could only hear themselves in the egg. The ability to recognize the call note of their own species is therefore acquired as a phylogenetic adaptation. It is quite possible that the innate releasing mechanism which underlies this capacity to respond is facilitated in its embryonic development by auditory stimuli. A summary of investigations of acoustical key stimuli and releasers can be found in R. G. Busnel (1964).

Spiders react to slight vibrations of their net with prey-catching behavior. The ant lion throws sand at the ant if loose sand comes tumbling down into the funnel that it has constructed. Water boatmen

react to vibrations of the water surface. Disturbances in the surrounding water release prey catching or search in many fishes and
clawed frogs (G. Kramer 1933).

Special alarm substances warn fish swarms and the tadpoles of
the toad; they all escape when they detect substances secreted by an
injured member of their species (K. v. Frisch 1941; I. Eibl-Eibesfeldt
1949; F. Schutz 1956; W. Pfeiffer 1963). Alarmed honeybees secrete
an odorous substance through their widely exposed cloaca which
excites the others and makes them aggressive (K. v. Frisch 1965).

Some gastropods of the tidal zone escape when they perceive
substances which predatory starfish secrete from their feet, but they
do not flee when a plant-eating starfish approaches (T. H. Bullock
1953).

The male silk moth is very sensitive to the sexual odor of the
females (I. Schwink 1955), which was analyzed by A. Butenandt
(1955) and R. Beckmann, D. Stamm, and E. Hecker (1959). In man,
too, there are sex-specific reactions. Men, and women before sexual
maturity and past their menopause, can hardly detect the odor of
certain arousing substances derived from musk glands, whereas young
women can smell these substances, which are widely used in per-
fumes, especially about 2 weeks after menstruation. Men can detect
this substance following estrogen injection (J. Le Magnen 1952).
Additional examples of communication via chemical substances can
be found in E. O. Wilson (1963, 1965) and G. Cavill and P. Robertson (1965; see also p. 438). Substances that are effective in intraspecific communication are called *pheromones.*

Chemical stimuli often aid in the search for food. The moray eel
hunts at night with the aid of its olfactory sense. The cuttlefish is
able to neutralize the olfactory sense of this predator by secreting
its "ink" as part of its defensive reaction (I. Bardach and others 1959).
Many parasites find their hosts by means of olfaction, and they
selectively react to the specific odor of the host species (D. Davenport 1955; G. Osche 1963; M. Lindauer 1963). Many marine polychaetes that live on shrimp and starfish react to the odorous substances that are diffused in the water around the host species. The
ichneumon fly (*Pimpla bicolor*), which parasitizes the larvae of
the South African night moth (*Euprotis terminalis*), is attracted by the
latter's odor. The ichneumon fly (*Alysia manductor*), which parasitizes fly larvae, approaches only the odor of fresh flesh. *Nasonia
vitripennis*, which only attacks pupae of flies, is only attracted by the
odor of decaying flesh. The coddling moth (*Carpocapsa pomonella*)
is only attracted by the odor of apples. Newly hatched garter snakes
(*Thampnophis species*) react prior to all feeding experiences selectively
to extracts of certain prey (G. M. Burghardt 1966).

In sharks the odor of blood releases search for prey. They are able to locate bait with their olfactory sense (I. Eibl-Eibesfeldt and H. Hass 1959; see also Fig. 26). For mosquitos, bedbugs, and mites, the key stimulus that attracts them is the heat radiated from the warm-blooded animals which normally leads them to food. L. J. Milne and M. Milne (1963) reported that an electric clock which radiated heat attracted mites each night, as did the chickens in whose house the clock was mounted. The clock, which did not work for this reason, began to run again when the mites left in the morning. We shall discuss additional examples of visual key stimuli. It can be demonstrated that innate behavior is activated by key stimuli, so there must have evolved corresponding releasing mechanisms that

Figure 26. Sharks (*Carcharhinus menisorrah*) alarmed by odors from a bait that was placed above the reef (Maledive Islands). In the lower picture one shark has grasped the bait. (Photograph: I. Eibl-Eibesfeldt.)

serve as stimulus filters. Such an adaptation may be a unilateral one, wherein only the recipient is adapted to a specific relevant environmental situation. This is true for the perceptual mechanisms of a predator, while its prey will not develop special signals that would make it more recognizable; to the contrary, the prey develops characteristics that make it as hard as possible for the predator to recognize.

It is quite different where the contact between two organisms is of selective advantage for both, as in the interrelationships between mates, between the mother and young, or in a symbiotic relationship. In these instances receiver and sender of a signal are mutually adapted to one another. A fish is not apt to evolve a signal for its predator, but a female is apt to evolve a distinguishing characteristic for the male. These signals may consist of special morphological structures, odorous substances, calls, or conspicuous movements or postures. Such highly differentiated structures and behavior patterns which serve special signal functions have been called *Auslöser* (social releasers) by K. Lorenz (1935).

As man is especially attuned to the perception of visual signals, most of the investigations have been concerned with visual key stimuli and releasers, especially in fish and birds. For signaling we find bodily structures such as plumage patterns, color patches, and manes, as well as special behavior patterns which have been called expressive movements (p. 91). The latter we shall discuss separately.

Various methods of analysis are available for the investigation of releasing stimuli. G. K. Noble and B. Curtis (1939) offered jewelfish females a choice between males exhibiting courtship coloration and others with inconspicuous coloration, by presenting one each on either side of an aquarium containing the female. The females always spawned on the side of the conspicuously colored males. If, however, a colorful male was blinded with eye cups, which resulted in lack of movements, the female spawned next to the colorless but active male.

Releasing stimuli can also be studied by making changes on the living animal. The schooling fish *Pristella riddlei* has a conspicuous dorsal fin with a black mark. A group of such fish with an amputated dorsal fin are less attractive to an isolated fish than a group of intact fish, so one may assume that the black dorsal fin is a visual following signal (M. H. Keenleyside 1955; see also Fig. 27). G. Noble

Figure 27. The fish *Pristella riddlei,* which shows the dark spot on the dorsal fin that serves as a releaser for the following reaction by conspecifics. (After M. H. A. Keenleyside [1955].)

(1934) investigated the stimuli that released fighting in the lizard *Sceloporus undulatus*. Only the males of this species possess a blue stripe at the border of the belly and a blue patch on the throat. When Noble painted such stripes on a female it was attacked. On the other hand, males courted males when he had painted over their markings. If one paints the black bills of juvenile zebra finches (*Taeniopygia castanotis*) red so that they are like those of the adults, they are not fed by the adults despite their intense begging behavior (K. Immelmann 1959).

Employing Skinner's instrumental conditioning method, G. P. Sackett (1966) demonstrated innate recognition of threat expressions in rhesus monkeys. Four male and four female monkeys were raised in isolation from conspecifics from birth until 9 months. Controlled visual experience consisted of slides projected against the cage wall which showed monkeys and neutral objects (sunset, landscape with trees, geometric figures, and so on). After each presentation the monkeys could self-project the slide they had just seen by pressing a lever. They could do this repeatedly during a 5-minute period and in each projection the picture was visible for 15 seconds. The monkeys soon learned the method of self-projecting, and they viewed the pictures they preferred more often than others. It was found that they preferred pictures of conspecifics, especially the picture of a young monkey and that of a threatening adult. These two pictures released also the most frequent social responses (vocalization, invitation to play, climbing about, and visual and manipulatory exploration of the pictures). At $2\frac{1}{2}$ months of age the young monkeys suddenly reacted with fear to the picture of the threatening adult. They withdrew before the picture, crouched, clasped themselves, and showed fearful facial expressions. At the same time lever presses for this picture were markedly reduced (Fig. 28). Not until 2 months later was an increase noted for this picture. The animals recognized the threatening expression at $2\frac{1}{2}$ months, although they had never seen a conspecific or their own mirror image. There must exist then an innate releasing mechanism that matures in the absence of social experience. Habituation may account for the increasing interest at a later time.

For the experimental analysis of the releasing stimuli the technique of adding components to a model is frequently used. One tries to release the behavior in question by the simplest of all possible stimuli in inexperienced animals. Prior observation aids in making the initial selection. In this way D. Lack (1943) released the most intense fighting behavior in the redbreast (*Erithacus rubecula*) by presenting a bundle of red breast feathers in the territory of the male. A stuffed juvenile without red feathers was ignored (Fig. 29). This allows the conclusion that the behavior of territorial defense

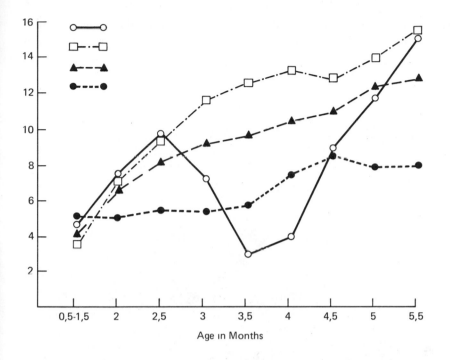

Figure 28. Frequency of bar presses for viewing of pictures by rhesus monkeys raised in isolation. The ordinate shows the average number of bar presses per 5-minute test. The abscissa shows the age (in months). The curves show the reactions of the monkeys to the picture of: open circles, a threatening conspecific; squares, a young monkey; triangles, other monkeys; and filled circles, control pictures. (After G. P. Sackett [1966].)

Figure 29. Two robin models. Left: a stuffed young bird without red breast feathers; right: a bundle of red feathers. (Explanation in the text.) (After D. Lack [1943].)

is released in the redbreast by the red breast feathers. Similar results were obtained by V. A. Peiponen (1960) in the bluethroat, where the blue breast feathers are the releaser.

A male toad approaches all moving objects during the breeding season and tries to clasp them. The male releases its grip only if the clasped object gives the defensive call of another male. Females remain silent and continue to be clasped, as are carps, the human hand, and so on (Fig. 30). The innate releasing mechanism is quite unselective in this case, but suffices, because during the breeding season only toads of this species are encountered in ponds (see also p. 158).

In comparing the effectiveness of various models they are usually presented consecutively. Only a few experiments are possible with one individual because learning is quite rapid. This shortcoming of the method of successive presentation of stimuli can be overcome by presenting two different models simultaneously to the animal.

a

Figure 30. (a) The male clasps the female; (b) two fingers as an acceptable model of a female; (c) the male clasps the tip of a rubber boot. (Photographs: (a) I. Eibl-Eibesfeldt; (b) and (c) H. Sielmann in I. Eibl-Eibesfeldt [1954].)

b

c

D. Franck (1966) investigated the pecking reaction of the chicks of the common gull to beak models, where he compared the methods of simultaneous versus successive presentation. Both showed agreement; however, by simultaneous comparison Franck was able to measure differential preferences that could no longer be detected with the method of successive presentation.

An exemplary analysis of innate releasing mechanisms we owe to E. Kuenzer and P. Kuenzer (1962). The following reaction of the young of cichlids which spawn on the substratum is released by the movement and coloration of the mother. Form and size have no effect. The key stimuli for the following reactions are species specific and correspond to the reproductive colors of the females. Young of *Apistogramma reitzigi* approach yellow models, while the young of *A. borelli* approach models that are painted contrastingly with black and yellow. In a similar manner, in *Nannacara anomala* the selectivity with which the young react to various models with the following reaction fits exactly the behavior and appearance of the parents. The young in these experiments were without experience, so the adaptation of the innate releasing mechanism and releaser must have originated during phylogenesis (P. Kuenzer 1968).

In the stickleback the red belly releases fighting; a plump wax model with a red underside, but lacking all other fish characteristics, is attacked at once, while models resembling a stickleback but lacking the red markings do not release fighting (Fig. 31). It is important, however, that the underside is red; if the model is turned upsidedown, it looses its fight-releasing qualities. The cues release fighting only when they are presented in a particular relation to others, in this case "red below" (N. Tinbergen 1948).

A male stickleback recognizes the female by her swollen abdomen, which she also presents in a definite manner. One can imitate the swollen abdomen and the posture with a simple dummy and elicit mating behavior. Even animals raised in absolute isolation react correctly to the signals of the same or opposite sex of their own species. No differences can be observed between the latter and animals that grew up normally (E. Cullen 1960).

A relationship between two characteristics (cues) constitutes a *configurational stimulus*. This kind of stimulus also releases the food-begging response of the herring gull, for example, a red dot at the tip of the bill (N. Tinbergen and A. C. Perdeck 1950). The blackbird (*Turdus merula*) gapes toward the simple model of an adult where head and body are represented by two black disks of different size. The smaller of two cardboard disks is gaped at as if it were the head of the adult; here the key stimulus is the size of the head in proportion to the body. If two-headed models are offered, the birds will prefer one of them and orient to a certain size of the

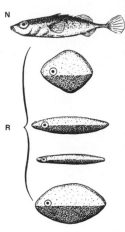

Figure 31. Stickleback models. N, a carefully made form- and color-true imitation of a stickleback without a red belly. It is less frequently attacked than the four simple red-bellied models of series R. (After N. Tinbergen [1951].)

head. The two pictures of the models show in each case heads of the same size only the bodies are of different size. In Figure 32a the smaller head is responded to; in b, the larger is gaped at. The animals respond to a certain size relationship between head and body which is the key stimulus. As in gestalt perception, relations between stimuli are attended to and this seems to be true for perception in general: If a bird is trained to the lighter of two gray stimuli, but then one exchanges the darker of the two with one still lighter than the stimulus which was positive before, the bird will prefer this new, lighter stimulus. This is also true of key stimuli. In the mouthbreeding cichlid *Haplochromis multicolor,* the young disappear in the mouth cavity of the mother when danger approaches. They also try to enter a simple model of the mother's head and orient to the position of the eyes and a point between these. If the

Figure 32. Two-headed models that release the gaping reaction of young black birds: (a) and (b) are models; (c) experimental situation. (Explanation in the text.) (After N. Tinbergen and D. J. Kuenen [1939].)

eye spots are positioned in a horizontal plane, the model is more effective than when one eye is above and one below (H. Peters 1937). In *Tilapia mossambica* the young gather predominantly near the underside of disk-shaped models. They also approach dark spots and try to dig into any depression. This response would normally lead them into the mouth of the mother (G. P. Baerends 1957; see also Fig. 33).

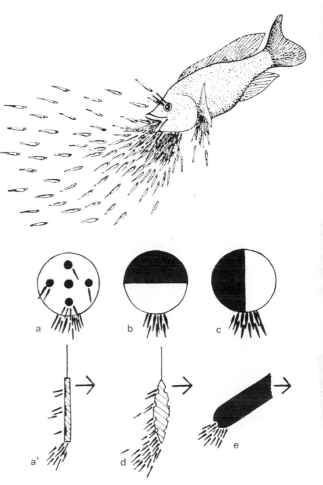

Figure 33. Model experiments with *Tilapia mossambica*, a mouthbreeding cichlid. Above, mother with young, which move toward her mouth. Below are models in the form of a flat disk: (a–c and a'), a disk with indentations; (d) a disk with pits; (e) a dark-colored test tube with an opening at the bottom. When the young fish are startled they approach these models and seek to enter them at the underside. They are also attracted by dark spots. (From G. P. Baerends [1957].)

In response to a silhouette of a model pulled across the sky, turkeys, geese, and ducks show specific escape reactions. K. Lorenz (1939) found that in geese and ducks the important cue is the "relative speed" of the moving silhouette (expressed in diameters per time unit) and that the form of the object is without relevance. However, in his experiments with turkeys he found that a hawk-shaped model was more effective than that of a goose. A model, shaped like a cross with two arms of even length, with a long post and a short head end, frightened the turkeys when it was pulled so that the short end pointed forward; the same model, pulled in the opposite direction so that the long post pointed forward like an outstretched neck of a goose, proved rather ineffective. Lorenz concluded that the "short-neckness" of the escape-releasing model was an innate characteristic. N. Tinbergen (1948, 1951) generalized this opinion to "gallinaceous birds, ducks, and geese." W. M. Schleidt (1961a, 1961b), in an attempt to repeat the original experiments with turkeys, found that the relative speed is also of importance in this species. However, his turkeys, unexperienced with flying objects up to that point, responded equally well to shapes with long or short necks and to silhouettes of birds of prey as well as to simple disks. Presenting such models at different frequencies (for example, the "longneck" 10 times as frequent as the "shortneck"), the model shown more often soon decreased in effectiveness, but the rare ones lost little of their frightening effect. Lorenz in his pilot study had used turkeys that were frequently exposed to ducks and geese flying overhead, so we can expect that they were already habituated to long-necked flying objects before the experiments were started, and therefore only the short-necked models appeared sufficiently different to elicit escape. D. Mueller's (1961) investigations of the escape-releasing stimuli in unexperienced capercailles also failed to show differences in response to various shapes and provide evidence for an optimal effective range of "relative speed." Recent experiments by Schleidt indicate that turkeys, exposed to a variety of shapes of equal size but all shown at the same frequency, habituate more to some than to others. This would indicate some "unlearned" preference for certain shapes over others; however, there is no clear correspondence to the contour of a bird of prey.

The male lightning bug (*Lampyris noctiluca*) reacts specifically to the configuration of the species-specific stimulus pattern of the female's light organ, which consists of two parallel bars, one behind the other, and two dots. Stencils of this pattern illuminated with a flashlight from behind elicit approach by the males in preference over others. The males of *Phausis splendidula* possess a less selective innate releasing mechanism in that they always approach models of larger area (up to four times), even when it shows great departures from the species-specific pattern (F. Schaller and H. Schwalb 1961; see also Fig. 34).

The same behavior is often elicited by several key stimuli. These stimuli, which can also be presented separately become additive in their effectiveness if they are combined. We already mentioned that the red belly of a stickleback is a strong fight-releasing stimulus. If we now observe sticklebacks threatening one another, we will

Figure 34. (a) *Phausis* female in the typical lighting position; (b) light organ of *Lampyris noctiluca;* (c) light organ of *Phausis splendidula* (both females). Below in three columns each: models ordered as to the effectiveness on the species signal shown above each. On the left the most effective model, on the right the least effective one. (After F. Schaller and H. Schwalb [1961]; in W. Wickler [1967a]; H. Kacher, artist.)

notice that the rivals assume a head down position before each other. If we imitate this threat posture with models they release fighting—even those that are placed horizontally will not release fighting at all. The head-down position, a behavioral cue, is therefore also a fight-releasing stimulus. If a stickleback is presented with a red-bellied model, which already releases fighting in the horizontal position, then, if shown head down it will release much more intensive fighting behavior (N. Tinbergen 1951). This *law of heterogeneous summation* was first seen and described by A. Seitz (1940).

The males of the cichlid *Astatotilapia strigigena*[1] are blue with black marks on the dorsal and ventral fins. These characteristics are displayed during threat and constitute at the same time the lowest intensity of fighting behavior. The displaying fish shows his lateral side with erected fins to the opponent. What happens next depends on the behavior of the other. If the other responds likewise and shows reproductive coloration, then the opponents will stand parallel

---

[1] The species has not been clearly identified. It is probable that this is a species of *Hemihaplochromis.*

to one another and spread the skin that covers the branchiostegal rays. This leads to the second stage of the hostile encounter. The males exchange tail beats. They stand parallel to each other facing in opposite directions and hit the opponent in its face. This is followed by the third stage, the actual fight, in which the males open their mouths and ram the other fish wherever they can reach him. The other fish will always try to escape and attempts in turn to ram the other fish himself, so a circling ("merry-go-round") results.

Scitz found that the blue coloration, the black marks at the fins, as well as the behavior patterns of lateral position, spreading fins, tail beats, and ramming thrusts, each alone released threat behavior of varied intensity. The stimulus components are exchangeable up to a certain degree. The tail beating by a model without reproductive coloration is as effective as a model that shows only spreading of fins and the reproductive coloration. If all these characteristics are combined one will obtain a stronger response.

U. Weidmann (1959) investigated the stimulus summation phenomenon quantitatively. He released the peck reaction in young black-headed gulls by simple cardboard models and counted the number of pecks directed at them. In this way he found that a gray, round cardboard disk, for example, received y number of pecks as opposed to y' number of pecks for a square model. If he painted both models red,[2] the number of pecks increased by the same amount x for each model. The effects of various releasing stimuli can be additive, but they do not add up in such a simple manner in all cases, as was shown by E. Curio (1961, 1963).

A behavior is not only dependent upon the strength of the releasing stimuli, but, as we have said earlier, also upon the internal readiness to respond. This has to be considered in experiments with models. When the readiness to respond is high, even a weak stimulus can release the complete behavior with full intensity. On the other hand, an animal with low motivation may show strong reactions only when the releasing situation is especially effective. The interaction of internal and external factors has been studied by G. P. Baerends, R. Brower, and H. Waterbolk (1955). First, they determined how the various conspicuous markings of male guppies are correlated with their specific sexual motivation. This gave them good indicators of the inner readiness to respond with sexual behavior. They were able to present releasing stimuli of varying degrees of sexual motivation and found that both factors compensate one another in a lawful relationship (Fig. 35).

In order to assess the effectiveness of a model one has to know whether or not the animal is ready to react at all. This is done by

---

[2] Adult gulls have red bills and red releases pecking.

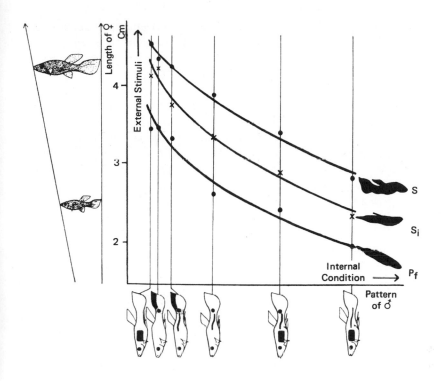

Figure 35. Interaction of internal and external factors, illustrated in the courting behavior of *Lebistes*. The strength of the external stimulus is indicated by the size of the
female (ordinate). The internal state can be read from the melanophore pattern
(abscissa). The measure, according to which the patterns were entered in the
abscissa, was derived from the recording of the readiness to perform sexual behavior
patterns. The curves show the combinations that release the same behavior with
respect to the behavior patterns of swimming after (Pf) and S form of the male (two
intensities, S and Si). (From G. P. Baerends [1956].)

testing the animal with the normal releasing object following the
experiment with models—but not before, because the behavior may
wane in intensity. This is the method of double quantification.

Experiments with models are made more difficult by a phenomenon called *afferent throttling* (M. Schleidt 1954). If a behavior
is repeatedly released by the same stimulus, the animal reacts less
and less until there is no longer a response. This decrease in responsiveness does not have to be due to a central fatigue of the
motor mechanisms. The gaping reaction of 5- to 7-day-old chaffinches can be released by vibrations of the nest, imitated calls of
the parents, and visual stimuli. If the gaping reaction has been
released by one kind of stimulus, the young stop responding after
10 to 13 elicitations. However, they gape with full intensity if
shortly afterward one presents a new releasing stimulus. If the stimuli

are exchanged in this manner one can release gaping in such a bird up to 46 times. H. F. R. Prechtl (1953a) spoke of an adaptation of afferent mechanisms which must be beyond the sense organ because it was demonstrated that the sense organ continued to respond to stimulation: Following repeated, visually released gaping the birds then pressed down into the nest in response to the same stimulus, demonstrating that they still perceived it.

In the turkey the gobbling call can be released by sound stimuli of a certain frequency. When at last the animal no longer responds, it still is ready to respond fully if stimulated with a new frequency.

It is more remarkable that artificial stimulus situations can be set up which surpass naturally releasing objects in their effectiveness. This was discovered by O. Koehler and A. Zagarus (1937). The ringed plover, for example, prefers to roll white eggs with black spots into his nest in preference over his own, which have dark-brown spots. Even more surprising is its preference for large eggs. A large egg four times the size of its own is preferred, although the bird is unable to sit and incubate it properly (Fig. 36). Male grayling butterflies (*Eumenis semele* L.) approach black models more frequently than those with natural colors (N. Tinbergen and others 1943). Male butterflies (*Argynnis paphia* L.) prefer models that have the species-typical brown coloration, but the illumination, size of the colored area, and number of stimulus changes per unit of time can be exaggerated. A horizontal, rotating cylinder with brown horizontal bars was preferred by males over an actual female. The highest number of approach flights was obtained when the cylinder rotated so quickly that the males could just barely perceive the change between brown and dark stripes (D. Magnus 1954, 1958).

Figure 36. Supernormal models: The oyster catcher tries to roll a giant egg into its nest. The bird prefers it to its own normal-sized one. (After N. Tinbergen [1951].)

Firefly males (*Lampyris noctiluca*) always prefer the pattern of light of their own species, but they prefer, as reported earlier, a model with a larger illuminated area. They also prefer a model that contains a larger amount of yellow than is contained in the light produced by their own females (F. Schaller and H. Schwalb 1961).

This responsiveness to "supernormal" releasers is exploited by some parasites. O. Heinroth referred to the European cuckoo as the scourge of songbirds because the gaping mouth of the young cuckoo releases in the foster parents stronger reactions than do their own young.

This exaggeration of the releasing stimuli also shows that the evolution of the existing releasers is not necessarily completed. This may be due to counteracting selection pressures. A signal should be as conspicuous and unique as possible; that is, it should not be confused with others and thus lead to errors. Thus from the receiver comes a selection pressure in the direction of conspicuousness and uniqueness with a corresponding lack of confusion for the sender of the signal. But whatever is conspicuous is also more readily seen by a predator; hence a selection pressure in the opposite direction exists. The result frequently is a compromise. Many bony fishes, for example, carry their releasers on fins that can be folded. During courtship they spread these fins and wave them in a manner which exposes their signals. Other fish can quickly change their colors. The unicorn fish *Naso tapeinosoma* Bleeker, which lives above coral reefs in the open water, normally has an inconspicuous, darkish coloration. However, when a male courts a female he develops within seconds a light-blue saddlelike spot on its back, similar vertical stripes on the sides, blue lips, and a blue caudal fin (Fig. 37). As quickly as these brilliant colors appear, the former darkish

a

b

Figure 37. (a) Swarm of *Naso tapeinosoma* Bleeker (Maledive Islands), with inconspicuous swarm coloration; (b) courtship coloration of the courting male (light-blue lips, light-blue saddle spot, cross-stripes, and light-blue tailfin). This is an example of color changes for the purpose of signaling. (Photographs: I. Eibl-Eibesfeldt.)

colors reappear when the fish ceases to court (I. Eibl-Eibesfeldt 1962). Many cichlids, which normally possess stripes that camouflage the outlines of their bodies, are also capable of these sudden physiological color changes. During fights and courtship they acquire very conspicuous patterns and colors. They are even capable of displaying several such colorful dresses and are thus able to present several signals (Fig. 38 and Plate I).

That these various colorations and patterns are properly understood by other fish was shown by H. Albrecht (1966) in *Haplochromis wingatii,* which, when of a disposition to escape shows horizontal stripes, and vertical stripes in an aggressive mood. A mother with young does not attack her young when they have horizontal stripes but will do so when one establishes a territory and shows vertical stripes. She also attacks models with vertical but not with horizontal stripes.

In the territorial fish of the coral reef, which in their adult state are rarely captured by predators, releasers were developed fre-

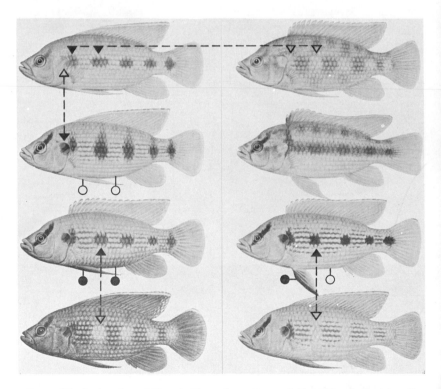

Figure 38. Explanation of Plate I. The color patterns of *Hemichromis fasciatus.* The symbols denote pattern components that vary independently of one another. They can be light (empty circles) or dark (filled circles). This illustration is still quite simplified; all phases are connected by smooth intergradations. (From W. Wickler [1965e].)

PLATE I

Color patterns of *Hemichromis fasciatus* in eight different moods. Top left: without territory (neutral mood); below: three stages of increasing territoriality and readiness to fight; bottom left shows the greatest readiness to fight. Right bottom: spawning animal; above that, while caring for young; the two top right pictures show the fright coloration of a fish that would like to hide. The top picture shows a fish that is hiding in a plant; the one below is one that has no hiding place and must remain in the open. Note how the melanophores on various locations of the body sometimes function together and sometimes in opposite ways. The coloration of the very aggressive fish (bottom left) is actually a negative of the neutral mood. (From W. Wickler 1964a; H. Kacher, artist.)

PLATE II

Top: Pygoplites diacanthus (Red Sea) as an illustration of a reef fish that is marked like a poster. (Photograph: I. Eibl-Eibesfeldt.) *Center:* courting frigate bird *(Fregata minor)* (Galápagos Islands). (Photograph: I. Eibl-Eibesfeldt.) *Bottom:* gape releaser of a young Gouldian finch. (Photograph: I. Eibl-Eibesfeldt.)

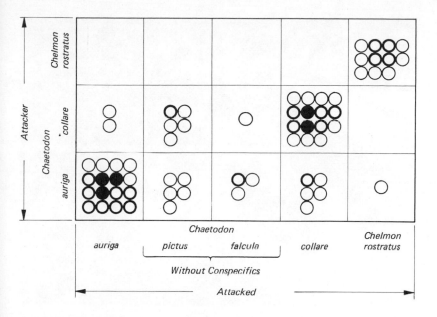

Figure 39. Distribution of intra- and interspecific fights within a group of butterfly fish (*Chaetodontidae*) that were kept together: light open circles, fights that lasted 2 seconds; heavy open circles, 3 to 9 seconds; filled circles, 10 seconds and more. Total time of observation 13 hours. To compensate for the uneven number of animals of the various species, the average number of fights that each animal would have fought with another of a different species was computed. For example, six *Chaetodon collare* had 36 fights with three individuals of *C. auriga*. One *C. collare* had on the average only 2 fights with one *C. auriga*, *C. pictus* and *C. falcula* had no species members in their tank and fought only occasionally with very similar species. (From D. Zumpe [1965].)

quently without apparent compromise. Many coral fish exhibit their optical signals continuously. They appear to be moving advertising signs. In other words, they are quite conspicuous, and they share with posters the fact that they are not readily confused with others despite their simplicity (Plate II). The patterns possess a high degree of improbability; that is, it would be highly unlikely that another fish would evolve the same pattern, unless it were a mimic (p. 151). Fish primarily fight with conspecifics or, if they have no opportunity to do so, species similar in appearance (D. Zumpe 1965; K. Lorenz 1962; see also Fig. 39). If a mirror is placed on a coral reef many fish will fight their mirror image (Fig. 40). The gaping mouths of many altricial birds present conspicuous visual signals. The young of the cave-nesting gouldian finch developed light-reflecting papillae (Plate II). Many grass finches recognize their young by their species-specific gape mark-

Figure 40. Two parrot
fish fight their mirror
image on a coral reef.
(Photograph: H. Hass
[1957].)

ings in the mouth. The whydah birds, which parasitize these species by laying eggs into the nests of the grass finches, imitate these gape markings exactly (p. 154). Finally, there is the example of a confusing signal that has the function of detracting predators. In the tropical seas there exist several species of saber-toothed blennies (*Runula, Aspidontus,* and so on) which specialize in biting off pieces of skin and fins from the fishes for their food (I. Eibl-Eibesfeldt 1955a, 1959). They attack the eyes preferentially, and in many coral fish these are camouflaged by a dark eyeband. In addition, some species developed an eyespot elsewhere on their bodies, and these are reported to divert attacks away from the eyes (W. Wickler 1961b; see also Fig. 41).

Figure 41. Example of a deceptive signal (Maledive Islands). The butterfly fish (*Chaetodon auriga*) protects itself from attacking saber-toothed blennies (*Runula*) by camouflaging its eyes with a black band and by an additional eye spot near the end of the dorsal fin. (Photograph: I. Eibl-Eibesfeldt.)

T. C. Schneirla (1965) advanced the hypothesis that all reactions can be explained by a simple principle of approach and withdrawal. Weak stimuli or stimuli decreasing in intensity would activate a system that results in approach, whereas strong stimuli or stimuli increasing in intensity result in withdrawal. This is sometimes true. With increasing voltage stimulation E. v. Holst and U. v. Saint-Paul (1960) observed a change from attack to escape in their brain-stimulated chickens. It is also known that toads approach small worms with prey-catching responses but flee from larger ones, which could be explained by Schneirla's hypothesis. But on the basis of the observed facts we cannot accept his view that there are no a priori differential reactions to effective key stimuli but only the principle of approach and withdrawal responses. The fact that a grasshopper responds to the song of its own species but not to another species song, that a chick within the egg, before it could hear or follow a hen's call, responds to a call of a hen instead of the call of a duck, whereas ducklings respond the opposite way, that firefly males prefer the signal pattern of their females—this can only be explained by assuming the existence of specific releasing mechanisms.

After all we have said it has become obvious that the animal does not possess a "picture" of the conspecific or of nonspecies members. The conspecific seems to be an object that emits various releasing stimuli for various responses. However, there are instances where the conspecific has few species-specific markings, and in this case the partner learns to recognize the other individually. K. Lorenz (1935) showed this in mallard ducks, which display a well-developed sexual dimorphism. The males are conspicuously colored (for example, head green with a white neckband); the females, on the other hand, are cryptically marked. Lorenz raised one male and one female with pintail ducks, in the absence of any contact with conspecifics until sexual maturity. At this time the female mallard did not react at all to the courting pintail males with which she had been raised. But she responded at once to the courting behavior of a male mallard which she saw for the first time through a crack in the wall of her cage. The male mallard indiscriminately courted the male and female pintails with which it had been raised. More recent investigations by F. Schutz (1963, 1965) support the view that male mallard ducks cannot distinguish conspecific females from those of other species. They courted all ducks, with the exception of male mallards, which they also recognize innately. Male mallards raised with other species congregate in groups for communal courting with their own kind.

A. Seitz (1940) tried in vain to release the courtship behavior of the cichlid *Astatotilapia*. The same animals that responded with fighting to the crude dummies of other males refused to court detailed models of females. All animals had already had experiences with real females. When he was able to raise males in complete isolation they courted, much to his surprise, even very simple models, for example, a glass rod with a round wax ball. Apparently these males normally learn the appearance of their females, and henceforth are not deceived even with a good imitation of the real fish. This phenomenon has been known to gestalt psychologists for a long time. The complex quality of a learned gestalt is made up of many details. A small change of one detail can so change the appearance of the whole that it is no longer recognized. Thus a tame redbreast became very frightened by Lorenz when he wore glasses for the first time. K. Lorenz (1943) discussed at length questions of this kind, pointing out that it is also possible for an animal to abstract secondarily, as it were, simple cues from a complex gestalt, to which an animal then reacts as to a releasing signal. He referred to an example of a sea lion which in the investigations of M. Spindler and E. Bluhm (1934) responded in one instance to the cockade on the cap of the caretaker, and another time to the feeding bucket as a releasing characteristic of the total feeding situation.

For a long time the gestalt principle was seen as the criterion of the acquired releasing situation, while the stimulus-summation phenomenon was seen as incompatible with it and as characteristic for an innate releasing situation. According to a new investigation by G. P. Baerends, K. Brill, and P. Bult (1965) the stimulus-summation phenomenon is now also valid in acquired releasing situations, so that a difference in principle between innate and acquired integration of stimulus situations is perhaps not warranted. It appears, however, that an innate situation is usually simpler than an acquired one, and as a rule of thumb we may use K. Lorenz's (1954b:29) statement: "Where an animal can be 'tricked' into responding to simple models, we have a response by an innate releasing mechanism; where it cannot be thus confused, we have an acquired recognition of a gestalt."

By this criterion K. Lorenz (1943) concludes that man, too, responds on the basis of innate releasing mechanisms to certain stimulus configurations (p. 431). Releasing signals evolve from already existing structures, the course of evolution being determined by the receiver of the signal. Some examples may serve to illustrate this. Mouthbreeding cichlids pick up their eggs after spawning. The eggs act as releasing stimulus upon a releasing mechanism of the adult fish. A strong selection pressure, caused by predators, favors a fast uptaking of the eggs. In some species of the genus *Haplochromis,* and *Tilapia* the females pick their eggs up immediately after laying, thus leaving their males no time to fertilize the eggs on the ground. Fertilization is achieved by a special adaptation. After the female has taken the eggs in their mouth, the male presents its anal fin, which bears yellow egg spots. The female tries to pick these egg dummies up and thereby inhales the sperm. Comparative studies revealed that the egg spots evolved from much less conspicuous perl spots of the vertical fins, which have no signaling function. It can be shown, however, that even such perl spots can attract the attention of females. When a female *Haplochromis wingatii* was spawning with a male that accidentally had no anal fin, she picked at the perl spots on his caudal fin. The existence of perl spots in the ancestral form of the egg-spot cichlids was, however, just one of the prerequisites necessary for the evolution of the egg dummies. Another precondition was a behavioral one. Before the signal evolved, the males were presenting their side to the females during courtship and spawning. This is still done in other species of *Haplochromis* which still fertilize their eggs on the ground (W. Wickler 1962).

That this behavioral characteristic was decisive for the evolution of egg dummies on the anal fin can be deducted from another development that took place within the genus *Tilapia*. Here, too, the females of some species take up their eggs immediately after spawning. Although the males have perl spots on their vertical fins, no single *Tilapia* species has evolved spots simulating eggs. During courtship the males guide the females to the spawning site, swimming in front of her, with the head bent down and the ventral side exposed to the female's view. In those species in which the females take the eggs in their mouth immediately after laying, the males evolved signals around their cloaca which release a snapping response in the female (p. 153).

To evolve the egg spots as a signal, the following preadaptations had to exist:

1. The readiness of the females to pick up eggs.
2. The existence of perl spots on the vertical fins of the males.
3. A lateral display of the males during courtship and spawning.

The evolution of signals in progress can be studied in some species of gulls (N. Griffith Smith 1966). The arctic gulls *Larus argentatus, L. hyperboreus, L. glaucoides kumlieni,* and *L. thayeri* are very similar to each other. Nonetheless, cross-breeding is rare, even where the ranges of several of the species overlap. Experiments revealed that the gulls recognize their species mate mainly by the coloration of a small ring around the eye (yellow to orange) and by the color of the iris. If one paints the eye ring of a male with the color of another species, he gets no female of his own species. If one changes the eye-ring color of a female that is already mated, no copulation takes place and the pair finally breaks up. Thus the eye-ring coloration is a signal for both males and females.

Only in the Kumlien's gull does the coloration of the iris vary considerably. Where this species overlaps in its distribution with other species there is less variation. At the south coast of Baffin Island, where Kumlien's gull occurs together with the light-eyed herring gull, dark-eyed Kumlien's gulls are preponderant. In areas where this species is sympatric with the dark-eyed *Larus thayeri*, bright-eyed Kumlien's gull prevail.

Assortative mating takes place only where this gull lives with other gull species. Light-eyed females pick light-eyed males and dark-eyed females pick dark-eyed males. The assortative mating system may be maintained by a fixation of the chicks on the iris type of their parents. The evolution of distinguishing signals of these gull species is evidently still in progress.

6 RELEASERS (EXPRESSIVE MOVEMENTS AND OTHER SOCIAL SIGNALS)

Origin of expressive movements and other releasers

A courting bird behaves conspicuously. It spreads its feathers, assumes certain postures, sings, and frequently offers the female food and other gifts. One dog greets another, wagging its tail, or growls and bares its teeth at a stranger. A threatening cat humps its back and hisses but purrs when in a friendly mood.

Behavior patterns of this kind have a communicative function. Their effectiveness is often enhanced by conspicuous, morphological structures (feathers, manes). The behavior patterns that have become differentiated into signals are called *expressive movements*. They evolved in the service of coordinating social behavior and are therefore releasers like the morphological structures that evolved as signals.

Certainly one animal can understand many behavior patterns of another. If someone shivers, he communicates something. It is best to distinguish such undifferentiated expressive behavior from differentiated expressive movements that have become signals, although the former may become transformed into expressive movements. Expressive movements may be innate or learned. They may be quite simple movements or postures. Frequently several expressions are superimposed on one another, which leads to a multiplicity and seeming variability of expressive behavior; nevertheless, they can be traced back to a few invariables (fixed action patterns).

This has misled some investigators. R. Schenkel (1947) writes that the richness of expressions and the variability of the facial expressions in the wolf argues against the validity of the concept of fixed action patterns in mammals. In a reply to this statement K. Lorenz (1952) showed that in the dog's facial expressions the combination of the intention movements to flee with those of fighting leads to a great variety of expressions. The intention to flee is characterized by pulling back the corners of the mouth, retracting the upper lip, and wrinkling the muzzle and forehead. Both kinds of expressions can be superposed in varying degrees. Fighting and fleeing are often activated simultaneously, so one usually sees a combination of the two expressions, rarely a pure expression of one or the other (K. Lorenz 1953; see also Fig. 42). P. Leyhausen (1956b)

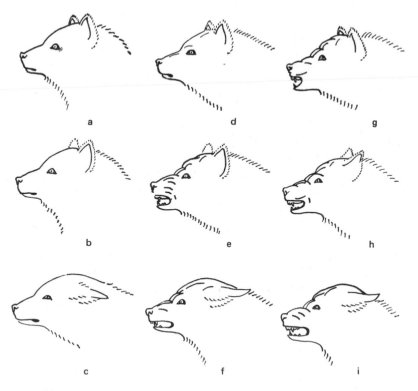

Figure 42. Various facial expressions of the dog which result from a superposition of various intensities of fighting and flight intentions. (a)–(c) Increasing readiness to flee; (a)–(g) increasing aggression and the corresponding superpositions. (From K. Lorenz [1953].)

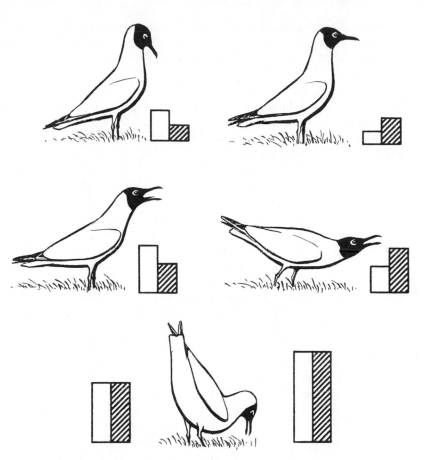

Figure 43. Various display postures of the black-headed gull, in each case with the appropriate strength of the simultaneously activated attack and escape drive. The block diagrams next to the figures: crosshatched, attack drive; open, escape drive. (After M. Moynihan [1955].)

confirmed this in the facial expressions of the cat. Depending on the relative strength of the simultaneously activated drive to attack or to flee, the black-headed gull shows various display postures (Fig. 43), which can be interpreted as superpositions or combinations of various intention movements (M. Moynihan 1955).

Species members understand such combinations, and some mammals communicate complex messages through the combination of various expressive movements. Pregnant polecats threaten the male by defensive biting, but simultaneously pacify him by uttering a call that indicates readiness for contact (Muckern) (I. Eibl-Eibesfeldt 1956c). In Indian macaques M. R. Chance (1963) observed how females aroused the interest of the males by threatening them, but

at the same time neutralized the released aggression by also showing submissive behavior at the same time. The night monkey (*Aotes*)
produces complex messages by arranging stereotyped elements into
a sequence (M. Moynihan 1964).

When two Hamadryas baboons fight, one often flees to a higher-
ranking animal. He appeases the latter by conspicuously presenting
to him at the same time threatening his opponent, thus directing the
aggression of the high-ranking individual against his opponent ("protected threat" H. Kummer 1957).

The richness of expressive movements is quite different from
species to species even in closely related ones. The wolf possesses a
much more varied expressive repertoire than the fox (G. Tembrock
1954). This is correlated with the fact that wolves hunt in packs and
need to communicate more while coordinating this activity, while the
fox is a solitary hunter.

By comparing the behavior of related species it has been possible
in some instances to reconstruct the phylogenetic history of various
expressive movements. Expressive movements are often derived
from other behavior patterns when these accompanied a certain
state of arousal or activity of the partner frequently enough so that
they can serve as cues to others. Social grooming behavior, for
example, is always an expression of readiness for social contact and
peaceful intentions. A dog in a friendly mood greets by licking and
nibbling the same way as does a tame badger (I. Eibl-Eibesfeldt
1950a) and many other mammals. The companion does indeed
understand the "friendly" meaning of this gesture, which is also
shown by the mother in caring for the young. Such a gesture can
calm an aggressive animal; O. Antonius (1947) cites an impressive
example. He kept a wild onager stallion (*Equus hemionus*) which
was very aggressive and always attacked him whenever they met.
Because the stallion could not reach Antonius, he redirected his
aggression by biting into the fence or against animals in adjacent
enclosures. Once, while trying to attack his neighbor, he had his
back to the fence so that Antonius could reach the animal with a
bundle of keys. The result of this contact was dramatic. The stallion
acted as though he had received an electric shock, he stood still,
turned around for a second, showing intention to bite, but refraining
from doing so. Instead, he continued to permit this obviously pleasing stimulation. From that moment on the animal was tame. Whenever Antonius appeared, the animal no longer threatened, but
approached, showed greeting behavior, turned around, and presented his croup in order to be scratched. In a similar manner I
was able to tame a galago (*Galago crassicaudatus*). The animal
seemed to like being scratched behind the ears and in the armpits.
Soon it showed that it wanted to be scratched there by raising an

arm. Since the behavior patterns of social care of skin and fur already expresses contact willingness, it is understandable that they sometimes became ritualized into expressive movements. The lemur (*Lemur mongoz*) greets others with a movement that is used to comb the fur, a behavior that is common to this group. This combing movement with the lower mandible is made into space, accompanied by rhythmic calls and even licking the air at high intensity. *Macaca speciosa* will make similar licking movements and rapidly open and close the lips. By imitating these movements one can pacify aggressive animals. Many lower monkeys will groom only after performing the intention movements of licking (R. J. Andrew 1963a, 1963b). Vervet monkeys smack their lips before cleaning one another and also gnash their teeth. This generally expresses a peaceful mood (T. T. Struhsaker 1967). During the courtship of many birds and mammals preening and grooming behavior play a great role. They aid in pacifying the aggressiveness of the partner. Even the behavior of an attacker can be transformed into "friendly" grooming. In herons, cormorants, guillemots, and other birds the bird that is attacked pacifies an attacker by presenting its head. The attacking behavior then leads into preening behavior (C. J. O. Harrison 1965). If one rat bites another accidentally during play, the one that was bitten will squeak and is at once groomed by the one that bit it (I. Eibl-Eibesfeldt 1957a).

Frequently behavior patterns that led to attacks evolved into threat gestures. Thus the opening of the mouth preceding biting as an intention movement evolved into the baring of the teeth in many mammals (carnivores, rodents, and so on).

In crabs, threats with the main weapon, the claws are ritualized in several ways. In only a few instances are the claws raised and lowered in a slow rhythm. This is done by the shore crab (*Grapsus grapsus*) in threats against conspecifics as well as against others (H. Schöne and I. Eibl-Eibesfeldt 1965). The mangrove crab (*Goniopsis cruentata*) threatens similarly, and in a slightly modified form this claw movement is used in a waving motion during courtship (H. Schöne and H. Schöne 1963). Ritualization has evolved farthest in the fiddler crabs (*Uca*): one enlarged claw of the male is waved, and each species has evolved its own mode of waving (H. Hediger 1933; J. Crane 1943, 1957; R. Altevogt 1955, 1957; see also Figs. 44 and 45). Not all movements are derived from threats. In a number of fiddler crabs the feeding movements of the claws have become ritualized into the waving movement (J. Crane 1966).

Many threat postures seem to have evolved out of movements of preparing to jump at another. We may see only an intention of raising, or the animal may actually jump toward the opponent, and stop short in an exaggerated manner, thumping the feet hard

Figure 44. Various types of claw waving in crabs: (a) *Uca rhizophorae* (vertical waving); (b) *Uca annulipes* (waving sideways); (c) *Uca pugilator* (waving type with the claw stretched far out); (d) *Dotilla blanfordi*; (e) *Goniopsis cruentata*. (After H. Schöne and H. Schöne [1963].)

Figure 45. Various types of claw waving in the genus *Uca*: Top: *Uca lactea* (Fiji Islands), lateral waving type. The claw, which is initially pulled in (a), is stretched out in a sideways movement (b); then it is raised (c) and returned in an arc into the original position. e–h, Examples of vertical waving; e and f, *U. rhizophorae* (Malaya); g and h, *U. signata* (Philippines). (After J. Crane [1957].)

onto the ground, frequently erecting the hair at the same time (for example, in badgers and squirrels as a "putting-on-the-brakes" display). In man, stamping the feet in anger seems also to be a ritualized attack movement (I. Eibl-Eibesfeldt 1967a). Movements to protect parts of the body, such as pulling back of the ears to protect the pinnal and the inner ear, have evolved into expressive movements. So have epiphenomena accompanying general arousal, such as displacement activities (N. Tinbergen 1952) or autonomic events (blushing, paling, gland secretions), but only if they characterized the physiological state of the animal unambiguously (D. Morris 1956; I. Eibl-Eibelsfeldt 1956b, 1957a). Thus many snakes vibrate the ends of their tail. In some species this has become a threatening gesture and special rattling devices have evolved (rattlesnakes). In many rodents and in other mammals as well, tail movements and ear movements became ritualized into expressive movements (H. A. Freye and H. Geissler 1966; R. Schenkel 1947; P. Bopp 1954).

Movements of embarrassment and other epiphenomena of excitation do not have other functions that could counteract selection in a certain direction, so they seem to be especially suitable for modification into signals.

Porcupines possess spines on their tails that have become modified into sound-producing organs, as was noted by Darwin (Fig. 46). Similarly, the raising of hair and feathers may have led to the evolution of manes and conspicuous feathers. Blushing may have led to the development of bare skin areas that are strongly vascularized, which can be exhibited as tumescent bodies, and so on. The habit of marking with urine, found in so many mammals, may have been derived from urination when the animal was frightened (I. Eibl-Eibesfeldt 1957a). In a strange environment rats leave trails by regularly secreting drops of urine. They also do this when crawling over conspecifics, thus marking them. In porcupines (*Erethizon*), agoutis (*Dasyprocta*), and maras (*Dolichotis*) the males raise up and spray urine with an erected penis on females as well as against male rivals (R. Kirchshofer 1960). In maras the mere presentation of an erected penis serves as a threat. W. Wickler (1966c) suspects that the genital presentations of male primates must have evolved in a similar manner (p. 430).

Whenever it is of advantage for an animal that some of its incidental behavior be understood by another, selection operates to transform the behavior pattern in question into a conspicuous signal. This modification of a behavior pattern to serve communicative function is called *ritualization* (J. S. Huxley 1923; see also Fig. 47).

Figure 46. Tail quills of a porcupine that have been modified into sounding organs. (From C. Darwin [1872].)

Figure 47. *Emblemaria pandionis* courting before the cave in which it lives by waving with the dorsal fin, which has become enlarged as it assumed the signal function. (From W. Wickler [1966a].)

On the basis of homologous patterns such as grooming or parental feeding, expression movements evolved often independently and along similar lines in different animal groups as homologies (p. 190), especially in instances where it is not paramount that the species are clearly differentiated in their expression movements from one another. At times similarities come about in that certain characteristics are required for the effectiveness of a signal and then selection pressure leads to converging evolution. This is true for many threat vocalizations, such as hissing, roaring and spitting, as well as those which can be heard above the roar of breakers at the seacoast and signals that are difficult to detect by predators (P. R. Marler 1956b; I. Eibl-Eibesfeldt 1957a).

Figure 48 shows several quite similar calls which five songbird species give when a predator passes overhead. The high-pitched, thin, and long-drawn-out calls ("siit" note of the chaffinch) can almost never be sufficiently localized to orient a predator. They are too high to be useful in detecting binaural phase differences, according to P. R. Marler (1956b), and are too low for the detection of appreciable differences in intensity. In addition, they begin and end unnoticeably, so that binaural comparisons with respect to time differences are also of no help. The call serves to warn conspecifics of danger; they, in turn, seek the nearest hiding place without trying to locate the caller. Chaffinches and many other songbirds also react to the quite similar warning calls of other species.

Figure 48. Calls of five different species of songbirds while a diurnal raptor flies overhead. The calls range between 6000 and 9000 cycles per second. Note the narrow frequency range and the lack of irregularities. (From P. R. Marler [1956b].)

On the other hand, when the signals distinguish species they are substantially different, especially in closely related species, where the chance of hybridization exists (P. R. Marler 1957a). Thus the territorial and courtship songs of the three warbler species, the willow warbler, chiffchaff, and wood warbler (*Phylloscopus trochilus, P. collybita, and P. sibilatrix*) are very well differentiated (Fig. 49).

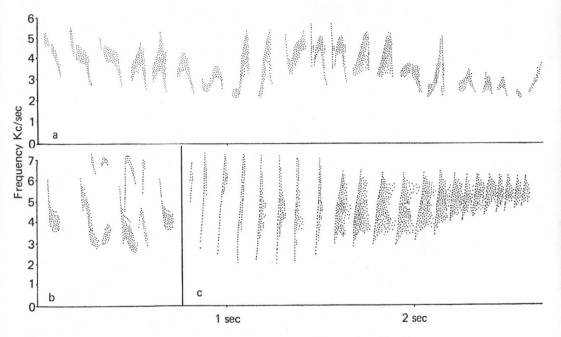

Figure 49. Songs of the willow warbler (a), chiffchaff (b), and the wood warbler (c). (b) consists of four parts: tsip-tsap-tsap-tsip. Note how much the songs of these closely related species differ with respect to length, tempo, and structure. (From P. R. Marler [1956b].)

They can also be easily located because of the break between frequencies in their song patterns. On small islands where fewer birds live together, the calls are much more variable within a species in comparison to the calls of conspecifics that live on the continent. This is true for the calls of the blue tit of Tenerife Island and for the goldcrest on the Azores (P. R. Marler and D. J. Boatsman 1951). Closely related species of grasshoppers which live sympatrically have very distinct calls (Fig. 50) and sympatric lightning bugs have different blink signals (F. A. McDermott 1917). Frog species with partially overlapping ranges have calls more clearly distinguishable in areas where they occur side by side than where only one species is found (W. F. Blair 1958). Contrast is emphasized only where it seems important. This phenomenon is called *character displacement.*

Figure 50. Songs of five field grasshoppers of the genus *Chorthippus* (after oscillo-grams by A. Faber): (1) *C. biguttullus.* (2) *C. brunneus;* these species are closely related to each other, occur side by side, and differ greatly in their songs. (3) *C. montanus.* (4) *C. longicornis.* (5) *C. dorsatus*; the close relationship is reflected in a certain similarity of the songs; (3) lives in very wet, (5) on moderately wet, (4) on fairly dry ground. These three species do not live sympatrically, so no selection pressure favored the development of divergent songs. (From W. Jacobs [1966].)

Such selection pressure toward uniqueness with the resulting lack of confusion of signals exists also when expressive movements have evolved in opposite directions within a species (see Darwin's principle of antithesis, p. 324).

All the changes that accompany ritualization seem directed to meet the requirements of signaling to an optimal degree. This means that in most instances the signal is conspicuous, precise, and not easily confused. Selection takes place via the receiver, who selectively perceives and responds to the signal. From many comparative studies we know that the understanding of a certain behavior pattern preceded its development into a signal. This is valid for the attracting and mood-inducing effect of feeding movements of chickenlike birds on conspecifics, which in mother hens—in its ritualized form—serves to attract hens. Within the phasianidae this food calling has become ritualized into an important courtship movement, as will be shown below.

The following specific changes can take place during ritualization:
1. The behavior undergoes a *change of function*. The original food enticing in various phasianids became a courtship behavior (p. 103). The same holds for the inciting of ducks (p. 108).
2. The ritualized movement can become independent from its original motivation and develop its own motivating mechanisms. Such a *change in motivation* was illustrated by W. Wickler (1966c) with the example of the "presenting behavior" of female baboons, which has become a "greeting gesture." A similar autonomy in respect to a specific innate movement can be traced during the course of the increasing ritualization of the inciting behavior of ducks (K. Lorenz 1941).

3. The movements are frequently *exaggerated* in respect to frequency and amplitude, but they are at the same time simplified, by the dropping out of some components while others become exaggerated. We have already mentioned the waving of claws in fiddler crabs as an example. The visual effectiveness of this movement was not only strengthened through the exaggeration of the *amplitude of movement* but also by the frequent *rhythmic repetition* (K. Lorenz 1941; A. Daanje 1950; W. Wickler 1963).

4. The *threshold values* for releasing stimuli often change to such a degree that the more ritualized behavior pattern, in general, is more easily released (A. Daanje 1950; B. Oehlert 1958).

5. Movements frequently "freeze" into *postures*. In this way many threat postures developed out of the opposing motivations of attack and fleeing, which are usually activated simultaneously during encounters with enemies (K. Lorenz 1951).

6. Components of *orientation* are changed. (An example is "inciting" in ducks (p. 108).

7. A behavior pattern that previously varied in respect to the intensity of motivation and stimulus can become changed in that it occurs stereotyped with constant intensity (frequency and amplitude), even if the animal is strongly motivated (*typical intensity*, D. Morris [1957]). In this manner the behavior becomes unambiguous (B. Dane and W. G. van der Kloot 1964). As one example we may cite drumming in woodpeckers (p. 111).

8. Variable movement sequences can become compressed into stereotyped and simpler ones (see zigzag dance of the stickleback and inciting of the duck, pp. 108 and 157).

9. Along with these behavioral changes there frequently occurs the development of very conspicuous body structures, such as ornamental plumes, enlarged claws for waving, manes, sailfins, tumescent bodies, and so on.

All these changes can occur during phylogenesis as well as during ontogenesis, because even a learned expressive movement can be improved in effectiveness as a signal. This can be observed in the development of begging behavior in zoo animals (p. 112). This is learning by reward, in that the zoo visitor rewards behavior or postures that are most attractive or pleasing to him, so that the animals exaggerate the desired postures and leave out the unnecessary ones.

The "polishing" of ceremonies, which at times occur between two married partners, is not included under the process of ontogenetic ritualization.

The cultural ritualizations of man follow the pattern of phylogenetic ritualization. Lifting the hat as a greeting developed from removal of the helmet. The military form of greeting, the salute, in which the hand is brought to the edge of the cap, developed from the movement of raising the visor. Both are gestures expressing confidence. These traditional greeting forms are widely ritualized, and hardly anyone using the behavior knows the origin of these gestures. This will be discussed further later (p. 454).

The similarities in the development of individually acquired expressive movements with the phylogenetic process of ritualization have at times led to Lamarckian interpretations. However, the principle of selection alone is sufficient to achieve a directed evolution or development. On the whole it appears unreasonable, only upon a superficial consideration, that nature should provide the raw material for selection only through "blind" mutation rather than striving towards adaptation with the aid of directed mutations. Upon further reflection one can see that organisms would be in danger of running into evolutionary culs-de-sac in this case. Only through blindly random mutations is it possible to try out all possibilities that could be of use in meeting the challenges of a changing environment. This method of evolution, which at first appears so unwieldly, actually turns out to be the most appropriate one, that is, the most advantageous in terms of selection. Random changes in the environment can only be successfully matched by a random exploration of all possibilities.

As can already be seen from the summary, the concept of ritualization refers to the improvement of a signal—to the development of releasers. Behavior patterns and changes in structures in respect to signal function, as well, fall under this concept. The direction of evolution is determined by the perceptual mechanism of the receiver. We also discussed ontogenetic ritualization and contrasted it with phylogenetic ritualization. The term "stylization" has been proposed for the concept of ontogenetic ritualization (D. Morris 1956), but it is possible to avoid adding yet another concept. It is important to recognize that the elucidation of a signal does not always mean becoming more conspicuous. Many animals have ritualized behavior which make them less conspicuous. We have in mind the waving motions of leaf-dwelling insects with many examples of convergencies. These are signals, too, but we deal with deceptive signals, which hide the animals in the moving leaves. Frequently animals developed conspicuous patterns which serve to facilitate specific learning processes. One is reminded of the wasp pattern, whose meaning has to be learned by many song birds. Therefore, not all releasers also have their counterpart in an innate releasing mechanism.

Not included in the concept of ritualization are all those changes which are concerned with the improvement of the reception of signals, that is, the development of innate or acquired releasing mechanisms (W. M. Schleidt 1962). W. Wickler (1967b) has proposed that all processes which lead to an improvement of communication be collectively termed *semantization*. This can take place unilaterally from the sender (semantization from the sender—German, senderseitige Semantisierung) and is then called *ritualization*. Often sender and receiver develop together by mutual adaptation, and finally a semantization on the part of the receiver can take place. This concerns all development of releasing mechanisms. This is the case when a night moth evolves an innate releasing mechanism responding to the ultra sounds of a bat which releases escape responses, or when a toad develops an innate releasing mechanism for a prey that "does not want to be noticed." The imparting of meaning, so to speak, occurs from the receiver's direction. For this reason the development of the sender does not mirror exactly the development of the receiver.

After this discussion we now return to the concept of ritualization, illustrating the process with some examples.

R. Schenkel (1956, 1958) was able to trace the progressive ritualization of food enticing during courtship by a comparison of various phasianid birds. The least ritualized form is shown by the male chicken (*Gallus*). It scratches several times with its feet, steps back, and pecks at the ground while uttering calls to attract, even if there is no food present. It will, however, pick up small stones, as if they were food objects; the hen comes running (Fig. 51a). If the rooster actually found food, she will eat it; if not, she will seek in vain.

The ring-necked pheasant (*Phasianus colchicus*) attracts hens in a similar manner (Fig. 51b). The courting impeyan pheasant (*Lophophorus impejanus*) bows low with a slightly spread tail before the hen and pecks vigorously at the ground.[1] The hen approaches, searches in front of him, but now he maximally spreads his wings and tail feathers and keeps his lowered head still. Only his spread-out tail bows rhythmically up and down during this stage of "ecstasy" (Fig. 51c).

The peacock pheasant (*Polyplectron bicalcaratum*) behaves similarly. After scratching on the ground in the manner of a food-enticing male domestic chicken, he will bow with raised wings and spread tail feathers (Fig. 51d). If the female approaches he will

Figure 51. Evolution of courtship in the phasianids out of the food call: (a) domestic rooster food calling; (b) the calling phase in the ring-necked pheasant; (c) "ecstacy" in the impeyan pheasant; (d) "ecstacy" in the peacock pheasant; (e) "ecstacy" in the peacock. (From R. Schenkel [1956].)

[1] The impeyan pheasant does not scratch for its food but pecks it free with its beak.

quickly move his head back and forth in her direction. If one gives him food, he will offer it to the female, thus revealing the original motivation. The movements seem to be ritualized feeding movements. Normally the male does not feed the female at this time, although she pecks on the ground in front of him. The courtship movements of the peacock (*Pavo*) are already so ritualized that one cannot recognize their origin without a knowledge of the intermediary forms. The male peacock spreads the tail feathers, shakes them, and moves back several steps. Then he bends the spread-out tail forward and points downward with his beak, while his head is still upright (Fig. 51e). As a result, the female runs in front of him and pecks in a searching manner on the ground in the focal point of the concave mirrorlike shape of the fanned tail. The male peacock points, so to speak, with his fanned-out tail toward imaginary food. Young male peacocks, incidentally, food entice still in the original form with scratching and pecking. Thus there is a recapitulation of phylogenesis during ontogenesis.

This food enticing (food calling) also appears in the functional cycle of care of young. A hen attracts her chicks by scratching, pecking, and giving special calls, and this is undoubtedly the less ritualized form of the behavior. We can often observe that expressive movements which indicate a readiness for contact are derived from actions performed during the care of young. W. Wickler (1967c) and I. Eibl-Eibesfeldt (1966a) recently compiled a series of examples. Among many songbirds the adults feed one another during courtship, as if the partner were a young bird, while the other "begs" with the infantile "wing trembling" (Fig. 52). In the cuckoo (*Clamator jacobinus*), which no longer feeds its own young, we can only observe the derived feeding of the pair during courtship. Many carnivores feed their young. Young of the jackal (*Thos mesomelas*)

Figure 52. Left: raven feeding its young; right: raven pair during courtship feeding. (Photographs: E. Gwinner.)

of Africa cause their parents to regurgitate food by pushing their parents with their muzzle (Fig. 53). Between adults this pushing into the corners of the mouth is part of a greeting ceremony (W. Wickler 1966c). The "affectionate mouthing" of wolves probably developed in a similar manner (R. Schenkel 1947). In some seals, which no

Figure 53. Jackal (*Thos mesomelas*) feeding young (Tanzania). (Photograph: W. Wickler.)

Figure 54. Top: seal mother (*Arctocephalus galapagoensis*) greets her young by rubbing snouts; bottom: sea lion male (*Zalophus wollebaeki*) greeting a female (Galápagos Islands). (Photographs: I. Eibl-Eibesfeldt.)

a   b

Figure 55. (a) The 1-year-old gorilla child Jambo is fed a cherry by her mother. The mother holds a piece between her lips and offers it to her young; (b) chimpanzees greeting one another by kissing. (Photographs: (a) P. Steinemann (Zoological Garden, Basel) from E. M. Lang [1964]; (b) Baron and Baroness H. van Lawick (Tanzania), with permission of *National Geographic.*)

longer regurgitate food, pushing and rubbing with muzzles is used exclusively as a greeting between mother and child and among adults as well (Fig. 54). Sea lion bulls also use this greeting ceremony to stop fights between their females (I. Eibl-Eibesfeldt 1955b). In anthropoid apes and man mouth-to-mouth feeding of young by their mothers has become known (Fig. 55a),[2] but adult chimpanzees also greet one another with a kiss (J. v. Lawick-Goodall 1968), and as early as 1915 M. Rothmann and E. Teuber (1915) thought that kissing in man is ritualized feeding. This will be discussed further (Fig. 55b). S. Freud saw the similarity between care behavior for young and sexual behavior in man, but misread the direction of development in the opposite direction by saying that the mother responded with feelings from the sexual realm, that is, when stroking, kissing, and rocking the child, which then became a substitute for the sexual object (S. Freud 1950).

In analogy, we find similar ritualization in insects, in which mutual feeding also is an important part in keeping the group together. E. Roubaud (1916) proposed the hypothesis that in wasps

[2] A photograph of a Pupuan mother mouth-to-mouth feeding its baby was published by A. Dupeyrat (1963:128).

the larval saliva is the means of establishing a bond between mother and larvae and thus resulted in the raising of many young. The feeding is mutual. W. M. Wheeler (1928) coined the term *trophallaxis* for this phenomenon. Exchange of food is also believed to be the basic cohesive factor among ants and termites, where behavior in respect to care of young is continued into adulthood (T. C. Schneirla 1946; W. Wickler 1967b). Bees that want to enter a strange hive appease the guards by offering them food.

By comparing the courting behavior of many species of ducks, K. Lorenz (1941) was able to reconstruct the evolution of some highly specialized courtship movements (Fig. 56). In some species,

Figure 56. Courtship movements of the mallard drake. Two top rows, the basic movements of courtship: (1) bill shake, (2) shake and stretch, (3) tail shake, (4) grunt whistle, (5) head-up tail-up, (6) looking toward the female, (7) nod swimming, (8) showing the back of the head, (9) pull up, (10) up-down movement. The movement patterns 1–4 and 10 appear during group courtship of the drakes; 5–9, on the other hand, appear during the sexual courtship before the female in a coupled sequence from 5 up. In the lower four rows complete movement protocols are represented. (After K. Lorenz [1958].)

for example, the up-down (auf-ab) movement is lacking. In its place in the sequence there is a "drinking toward" (antrinken) the partner, which has the same appeasing function. The similarity in form, the position within the sequence, and, in this instance, the similarity in function as well, indicate that the up-down movement is probably a ritualized form of the drinking movement.

Many drakes preen their wings during courtship, which has been interpreted as conflict behavior (displacement activity, p. 177). The primary feathers are only touched in passing with their bill, as if they were pointing at their colorful wing speculum. In the mandarin drake this movement became an actual demonstrative movement. In the mandarin drake special conspicuous feathers evolved, which are exposed during this apparent preening (Fig. 57). Other comfort movements, such as shaking or bathing, have also become ritualized within the duck family. An extensive presentation and discussion we owe to F. McKinney (1965).

Another behavior pattern that allows the step-by-step tracing of its evolution is the inciting behavior in ducks. This movement patterns belongs to the behavior repertoire of females and serves the function of separating a particular male out of the group of communally courting males, by becoming aggressive against other females. In sheld ducks this behavior is still found in its least ritualized form. If one pair meets another, the female attacks with threatening gestures. But as soon as she approaches the other pair,

Figure 57. (a) Sham preening of the garganey drake. It preens the exposed side of the wings, exposing the blue shoulder feathers in the process. (b) Sham preening of the mandarin drake. It touches a sail-like arm feather which is turned upward. This movement is firmly coupled with the "drinking toward" (c and d). (After K. Lorenz [1941].)

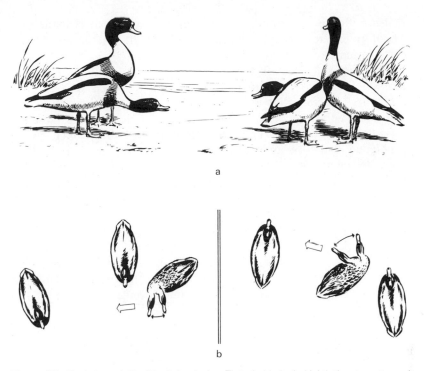

a

b

Figure 58. Evolution of "inciting" in ducks. The sheld duck (right) threatens toward a neighboring pair. The mallard duck (left) always threatens back over her shoulder. When the threatened opponent is positioned to her side and behind, her neck is bent back farther than when the opponent is in front of her. (Additional explanation in the text.) (From K. Lorenz [1963].)

her escape drive is activated and the female returns to her drake. There her aggressive impulses become stronger again. She stops and threatens back toward the strange pair, but she does not turn directly toward them. Her body remains turned in the direction of her mate, and she threatens with her head over her shoulder at the other pair (Fig. 58). This is not a fixed position but is the result of the conflict between attack and escape. In mallard ducks this movement sequence became a fixed action pattern: The female always threatens over her shoulder, even if the bird toward which she threatens is in front of her and she then points away from him. She continues to look at him and does not turn the head back as far as if the opponent were actually behind her (Fig. 58). In low intensity inciting she may point her head directly toward him, but with increasing excitement her head is forced back.

The oscillation between two antagonistic intention movements became ritualized into an expressive movement in the stickleback. The dance of the courting male consists of movement components that are directed alternately toward the female and toward the nest. The former are activated by a drive to fight, the latter from a drive to mate (J. v. Iersel 1953).

W. Wickler (1963, 1964b) showed that the nod swimming of the mimic (*Aspidontus taeniatus*) of the cleaner wrasse is derived from a conflict behavior that is widespread in the Blennidae. He filmed and observed a large number of species and noted that all these fish raise their head when emerging from their hiding place. When frightened and moving backward, however, they press their head down to the ground. When both tendencies are equally strong, they nod. They also nod while swimming when they are in a conflict between continuing to swim on and to turn back. If the disturbance is caused by a conspecific, this nodding occurs in a rhythmic pattern, that is, is clearly ritualized, but at other times the pattern is irregular (Fig. 59). In the mimic of the cleaner fish the nod swimming is very similar to the ritualized nodding and nod swimming of its relatives, except that this species exhibits the ritualized form of the behavior in intra- as well as in interspecific interactions. This

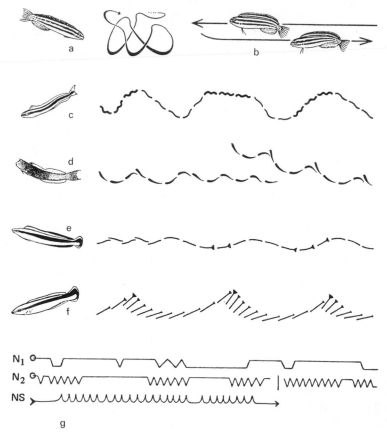

Figure 59. Forms of nod swimming. (a) Courtship of a *Petroscirtes* male. (b) Threatening *Petroscirtes* female. When swimming forward the fish slightly raises its head; when swimming backward it lowers the head. Out of these intention movements nodding can be derived; (c) *Runula*, threat swimming. (d) *Ecsenius* male, courtship. (e) *Aspidontus*, nod swimming. (f) *Labroides*. dance. (g) Comparison of movement frequencies in *Ecsenius*. N_I, irregular nodding when disturbed (unritualized); N_{II}, ritualized nodding toward conspecifics; NS, ritualized nod swimming (blennies; f, cleaner wrasse). (From W. Wickler [1964b].)

comparative investigation not only explains for the first time the evolution of mimic behavior patterns but documents it completely with motion picture film. According to U. Weidmann (1955) the head flagging of the black-headed gull should be interpreted as intention to flee. If the two partners of a pair are acquainted with one another, the female will beg from the male and is fed. Out of nest building and nest repair come many expressive movements of birds—for example, in several species of pelicaniformes where they have greeting and threat function (G. P. v. Tets 1965; see also Fig. 60).

Figure 60. Probable evolutionary course of postures or movements with slight threat functions toward strangers and greetings toward the partner, which probably were derived from nest-building or repairing movements in the *Pelicaniformes*. (a) *Pelecanus erythrorhynchos*, reduced grasping of nest material; (b) *Phalacrocorax auritus*, nest repair; (c) *Morus bassanus*, bowing with lowering of the bend of wing; (d) *Sula leucogaster*, trembling bow; (e) *Phalacrocorax aristotelis*, bowing forward; (f) *Sula sula*, bowing forward with raising of the wings; (g) *Anhinga anhinga*, bowing with bill clapping (in vacuo or toward branches). (After G. P. v. Tets [1965] from W. Wickler [1967].)

In the black woodpecker two expressive movements are derived from chipping out a nest cavity (H. Sielmann 1956). Drumming against dry branches signals to other males that the territory is occupied. Translated it would mean "someone is working (chipping) here." Females are attracted. This movement is sharply modified by quick rhythmic repetitions. A second expressive movement is less ritualized and therefore more easily recognized as derived from the original chipping behavior. If a woodpecker who builds a nest cavity wants to be relieved, it will fly to the entrance of the nest cavity, where it will peck with deliberate slowness at the edge of the entrance hole. The partner approaches to relieve the bird and continues the work. This "relief drumming" is also shown by birds that incubate eggs when the partner is asked, so to speak, to take over. Then the gentle pecking is directed against the wall of the nest cavity. This behavior then seems to be a general signal for relieving from other tasks as well as chipping. In one of our films adult woodpecker finches (*Cactospiza pallida*) can be seen using the food-begging behavior shown by the young as gestures during pair formation. The females actually beg from the males and are fed. The male begs with the same behavior patterns at the entrance to the nest and invites the female to follow (I. Eibl-Eibesfeldt and H. Sielmann 1965). This shows that expressive movements in the course of evolution experience changes (generalization) similar to those traceable in the evolution of language. (K. Lorenz 1964).

Cichlids generally lead their young with "follow-me" signals, which can be thought of as ritualized swimming movements. Normally, a fish that swims away will collapse its dorsal fin and make undulating body movements. The cichlid *Aequidens* swims a short distance with exaggerated undulating movements and collapsed fins and waits for its young. In the dwarf cichlids, on the other hand, we observe these follow-me signs only in the form of an exaggerated head shaking as a last remnant of the swimming movement. In the cichlid (*Herichthys cyanoguttatus*) this follow-me movement became a warning signal during danger; the same movement in the jewel fish (*Hemichromis bimaculatus*), in which, in addition a quick, repeated raising and collapsing of the dorsal fin serves as a signal to call the young to the nest pit—a movement that K. Lorenz (1951) called a releaser for "putting the young to bed." In *Hemichromis,* then, two expressive movements with separate meaning have been derived from the movement of leading young.

Several interesting examples of a change of meaning in the course of evolution have been described by W. Wickler (1965b, 1965e). The sexual presenting behavior of female hamadryas baboons (*Papio hamadryas*) also has an appeasing effect and is used with this intention by males, which in further assimilation to the females also imitated their red tumescent bodies in the anogenital region. In the spotted hyena (*Crocuta crocuta*) the presenting of the male genital region is an appeasement greeting ceremony. The sexes cannot be externally distinguished; the females have penislike organs capable of erection, with which they greet the males.

Innate expressive behavior may be brought under willful control in higher animals and man. Everyone knows that children who have been only slightly injured will only begin to cry loudly upon arriving near home. A female gorilla in the Basel Zoo, whimpered only when she was sure that she could be heard from the time she was 4 months old (R. Schenkel 1964).

Of the well-established *learned expressive movements* we mention first the begging movements of many domestic and zoo animals, which usually are derived from intention movements of grasping and approach and sometimes by imitation of gestures (K. H. Winkelsträter 1960). The animals learn by trial and error, and the movements become stereotyped, rhythmic, and frequently exaggerated, like the innate expressive movements.

Higher mammals can also be deceptive. A chow chow female that disliked to accompany her owner by following his bicycle limped badly when called to come along. On the return leg of the trip, however, she bounded ahead in a lively fashion (K. Lorenz 1950c). A gorilla female in the Basel Zoo, which disliked being alone, sought to call the keeper to her by a number of staged accidents. A new woman caretaker fell for this ruse, when the gorilla pretended to have caught her hand between the cage bars. When the woman entered the cage, the contact-starved animal rushed toward her and held her in her arms throughout the night (E. M. Lang 1961).

Dogs use the learned movement of "shaking hands" as an appeasement gesture, which comes close to an ability to talk, as K. Lorenz (1950c:178) emphasized.

Who does not know the dog who has done some mischief and now approaches his master on its belly, sits up in front of him, ears back, and with a most convincing "don't-hit-me" face attempts to shake hands? I once saw a poodle perform this movement before another dog of whom he was afraid.

Dogs also are able to make themselves understood through simple gestures appropriate to the situation at hand.

If your dog bumps you with his nose, whines, runs to the door, scratches there, or places its paws on the edge of the sink and looks around in a questioning manner, then he does something which is much more like human language than anything a jackdaw or graylag goose can ever say . . . (K. Lorenz 1949:125).

The female gorilla child Goma began at age 2 to communicate her desires to her keeper by pointing at what she wanted. If she wanted to have the door opened she put her finger into the keyhole. Later she pulled people to the door by their hand (R. Schenkel 1964).

Even closer to human speech is the behavior of a tame raven (*Corvus corax*) called Roa which was hand raised by K. Lorenz. Ravens have a special way of inviting conspecifics to fly along with them. They fly from behind closely above the other raven, wag their folded tail, and call crack-crack-crack. Roa did this with his caretaker, especially when he noticed the latter at a place where he had been frightened at a previous time. He then approached the caretaker flying from behind as he would have done with another raven. Instead of the innate call he now called roa, roa, roa, in imitation of the human voice. At the same time, however, the bird retained its innate warning call and used it appropriately with his conspecific sexual partner.

E. Gwinner (1964) had a raven that was called to the wire of its cage with the German word "komm." Later he called his female in the same manner. B. Grzimek's (1951) raven called all children Gregor, after the first child he had come to know. The learned sounds are used in communication between ravens who are friends, but the sounds used in immediate, species-preserving situations such as begging and position calls of young, food, attack and threat calls, and calls preceding copulation cannot be replaced by learned ones (E. Gwinner 1964).

O. zur Strassen (1952) reported on a young grey parrot that had learned to say "bitte" (please) when given something to eat. Once the bird said "bitte" continuously when its teacup was empty and the bird was obviously thirsty. After receiving its tea the bird talked continuously, but only other words. Since then E. Gwinner and J. Kneutgen (1962) found that paired ravens and shama thrushes (*Copsychus malabaricus*) call one another with the song strophe they have learned from their partner. They name, so to speak, the partner with the song motive that is characteristic of the other and use this solely to call it back.

Crows of the United States have alarm, fright, and collective call notes that differ from those in France. If tape recordings of American crows are played to French crows, they either do not react or interpret them wrongly. For example, they gather instead of flying away when they hear the American alarm call. French herring gulls do not react at all to the calls of American conspecifics (H. Frings and M. Frings 1959). They do not understand their "language."

The highest level is reached by learned expressive movements in human language. What in animals is commonly referred to as "language" is with the exception of the two last examples, nothing more than interjection, noninsightful sound production.

Human language can be used as a definition per se. Interestingly enough, human language does not lack an innate basis (E. H. Lenneberg 1964). In the first 3 months of life the children of deaf and dumb parents cannot be distinguished from those of normal parents (E. H. Lenneberg and others 1965). Children born deaf begin to babble but cease to do so after a time, apparently because feedback, necessary for future development, is lacking. Based on this motor ability for language it is conceivable that children would develop a language of their own. O. J. Jespersen (1925) describes the case of two Danish children who grew up neglected and were cared for by their deaf-mute grandmother. They conversed fluently in a language no one else could understand and which had no similarity to Danish.

The chimpanzee that was raised by C. Hayes (1951) babbled the syllables pu, pwa, bra, bu, wa, io, aho, baho, and gurgled k-k with saliva until the fourth month of its life. With great difficulty and repeated coercion of the movements it was possible to teach the animal the four words Mama, Papa, cup, and up. However, the animal did not always use them appropriately, although it was able to obey 50 verbal orders correctly. The chimpanzee failed when presented with new combinations ("kiss the cup," "kiss the dog").

R. A. Gardner and B. T. Gardner (1967 and 1968) are teaching a young female chimpanzee a kind of sign language. She has already learned several signs, such as those for come, come and hug, and come swing. She correctly reacts to pointing and the signs for look, stay, no, more, and sweet. She is also able to signal some of these herself. Thus she begs with an outstretched hand, palm up, when, for example, seeking help. This is probably an inborn behavior, because it has been observed in wild-living chimpanzees (p. 128). This view is further strengthened by the observation that this gesture occurs spontaneously in the third week of life. From this gesture handshaking developed. The female chimp points with an extended index finger and places the caretaker's hand on that spot of her body which she likes to have cleaned. Behaviors expressing anger are probably innate, such as stomping with a foot, waving of a raised clenched fist, and the shaking of large objects. The chimp demands that the ball be thrown to her by waving the right hand toward her own body, and more recently she uses the arbitrarily acquired signs for more and sweet.

By the end of the training the female had learned 30 signs and was able to use these spontaneously and correctly. The first signs were simple requests such as "come," "tickle," and "go away." Later object names were added, such as "flower," "blanket," and "dog." The names were used as requests and answers to questions. Subject names were also used to denote pictures of known objects. Once she was familiar with 8 to 10 signs, she began to use these in sequences of two or more, many of them in a free and original combination. "Open flower" stood for the garden gate and "listen eat" for the alarm clock that signaled the meal. These results show that chimpanzees can learn spontaneously, and spontaneously string together numbers of combinations of these signs (see also W. H. Kellogg [1968]).

According to A. Kortlandt (1967), chimpanzees communicate by simple gestures in nature. These gestures show local peculiarities in form and meaning, which are presumably local conventions. The intention movement of lifting a baby onto the mother's back is generally a signal for the young to mount the mother (J. van Lawick-Goodall 1968). However, Kortlandt observed it in one area functioning as a warning signal between communicating males.

Classification of releasing signals according to function

The expressive movements can be grouped according to their function. It seems appropriate to me to make a distinction between those expressive movements that are used in intraspecific communication and those used between species. R. A. Stamm (1964) raised the valid objection that certain contacts (threat) with other species are equal to those with members of their own species. Frequently, however, they are not. The predator is often threatened in a different manner than the conspecific, something frequently overlooked, and this is the reason for emphasizing these differences. In both categories we can distinguish expressive movements signifying a willingness for social contact from those indicating avoidance. Often one and the same signal has a different effect on different perceivers. The "long call" of the black-headed gull attracts unmated females but repels rivals. The "head-to-ground" posture repels neighbors and attracts the female (G. Manley 1960; see also drumming in woodpeckers, p. 111).

Releasers for intraspecific behavior

Signals that promote group cohesion

Introductory remarks

Between members of one species there exist attractive as well as repelling forces. Frequently, the conspecific is the bearer of aggression-releasing signals, which erect a barrier, in a manner of speaking, that discourages approach. At certain times, however, this barrier must be overcome, for example, when males and females have to come together for mating, or when aggressive animals of a species are to live temporarily or permanently in a group. In such instances a multitude of behavior patterns and signals play the role of buffers against aggression, when new contacts are first attempted as well as during continued contacts. The signals that regulate intraspecific interaction and serve to isolate species from one another as a rule are so specific that they are only understood by members of the same species. This is especially true where closely related species occur sympatrically. Such species are clearly distinguishable in the calls with which they attract the sexual partner. This is true for grasshoppers (A. C. Perdeck 1958a) as well as for frogs (C. M.

Bogert 1961), songbirds (P. R. Marler 1957a), and other animals. We have already given some examples of this (p. 99). Here is an additional example. Male fence lizards exhibit a specific head-nodding pattern that differs from species to species (Fig. 61). D. Hunsaker (1962) imitated such patterns of head nodding with plastic models of lizards and presented them simultaneously to females of *Sceloporus torquatus* and *S. mucronatus*. The females turned toward that model, which nodded with the species-specific rhythm.

Figure 61. Nod patterns drawn after motion pictures in various species of *Sceloporus*. The curves show how the body is raised and lowered. The amplitude is given in inches. (After D. Hunsaker [1962].)

Courtship behavior

Many courtship ceremonies serve to establish contact between animals. The sexual partner is first attracted by special behavior patterns and signals. Its hesitancy to make contact then is reduced, and the behavior patterns of the partners become synchronized to one another so that fertilization is possible. Frequently males and females court, but one partner is usually more active; as a rule this is the male. For example, the female pipefish courts the male (K. Fiedler 1954).

In order to attract partners at great distances, we find the use of odorous substances by female insects as well as the courtship songs of bird and locust males. In addition, male birds frequently make themselves conspicuous by special display behavior. Male frigate birds (*Fregata*) inflate their reddish throat pouch and wait for passing females (Plate II). Birds of paradise display their magnificent feathers, often accompanied with unusual postures. The bluebird of paradise (*Paradisaea rudolphi*) and the white bird of paradise (*P. guilielmi*) hang upside down on branches during courtship (Plate III). In the latter species, two males, which court in groups, form a symmetrical figure: One sits on a branch and the other hangs below with its back to the ground (H. O. Wagner 1938). The superb lyre bird (*Menura novaehollandiae*) cleans a courting place. There he courts, calling loudly, placing his fanned out tail on his back so that the patterned underside becomes visible (Fig. 62).

Figure 62. Courtship of the superb lyrebird (*Menura novaehollandia*): (a) seen from the side; (b) seen from the front. The bird folds its spread tail over its back, exposing the conspicuously marked underside. (Photographs: H. Sielmann.)

a

b

PLATE III

Top: bluebird of paradise *(Paradisornis rudolphi),* courting with head down (New Guinea). (Photograph: H. Sielmann.) *Center:* Lauterbach's bowerbird *(Chlamydera lauterbachi)* (New Guinea). The male courts with a red fruit in its bill. (Photograph: H. Sielmann.) *Bottom:* Satin bowerbird *Ptilonoryhnchus violaceus)* building its bower (New Guinea). (Photograph: H. Sielmann.)

PLATE IV

Top right: Automeris acutissima exhibiting the eye spots when frightened. (Photograph: E. Priesner.) *Top left:* larvae of *Dicranura vinula* Top, resting position; below, frightened. The larvae has ejected two red threads from the raised abdominal pair of legs and exposed its face mask. (Photograph: I. Eibl-Eibesfeldt.) *Center left:* hornets *(Vespa crabo)* and the moth mimic *(Aegeria apiformis).* (H. Kacher, artist.) *Center and bottom right:* top row, poisonous Malayan butterflies whose larvae feed on poisonous aristolochiacees; left to right: *Atrophanura varuna, A. coon, A. nox.* Bottom row, three different morphotypes of the nonpoisonous butterfly *Papilio memmon;* these are females of the same species; each morphotype mimics another model. In this plate the various morphotypes of *Papilio memmon* were placed under the models they mimic. (After examples by M. Tweedie [1966]; H. Kacher, artist.)

The most unusual way of courting is undoubtedly that of the bowerbirds of Australia and New Guinea. They clean special areas for courting and build bowers and decorate them, so to speak, with a "removable courtship plumage" (E. T. Gilliard 1963). The crested gardner bowerbird (*Amblyornis macgregoriae*) selects a straight young tree and decorates it with twigs and lichens (Fig. 63). He dances around this "maypole" displaying his erected nape feathers.

Figure 63. Pair of crested gardner bower-birds (*Amblyornis macgregoria*) before a tower made of branches and covered by lichens. (Photograph: H. Sielmann.)

The orange-crested gardner bowerbird (*Amblyornis subalaris*) builds a tentlike stage which he decorates with fruit, flowers, and snail shells and colorful beetles (Fig. 64). H. Sielmann (1967) observed these birds more closely and for the first time filmed them in the wild. He found that the males are very selective about the objects they choose. They put a flower into the wall of moss that they have erected, stand back, and look it over. It frequently happens that they remove it and place it elsewhere.

Figure 64. Male of the orange-crested gardner bowerbird (*Amblyornis subalaris*) inserting a flower into a wall of moss. (Photograph: H. Sielmann.)

Lauterbach's bowerbird (*Chlamydera lauterbachi*) builds a bower with a central basketlike cabin. On either side of this cabin he erects two more walls, creating two more passages. The central cabin is decorated with bluish ficus fruits. In front of the bower he keeps red fruits. Should a female approach, he picks one up in his bill and presents it to her (Plate III).

Another bowerbird (*Chlamydera nuchalis*) builds a bower that is decorated with bones and other bright objects (Fig. 65). This bird also brings one of these objects as a kind of symbolic present to the female, then turns his nape, which has conspicuous feathers when they are spread, toward her. Some bowerbirds, for example, the satin bowerbird (*Ptilonorhynchus violaceus*), paint the inside of the bower with saliva mixed with crushed berries and charcoal; frequently they use a leaf or piece of bark as if it were a brush.

An interesting parallel to the bowers of these birds is found in the males of the cichlid *Tilapia macrochir,* which decorates his spawning pits with furrows radiating outward from the center so that the entire structure appears like a star. Thus the spawning pit becomes a signal that attracts females (M. Huet 1952).

Once the sexual partner has been called or approached, its fear of close contact must be overcome. Even gregarious animals, and more so solitary ones, often maintain a certain distance to conspecifics and react defensively toward transgressions of their individual distance. This barrier of aggressiveness must be overcome and is done with special appeasing gestures. Terns (*Sterna hirundo*)

a   b

Figure 65. Courting bowerbird (*Chlamydera nuchalis*): (a) presenting the object. The area before the bower is decorated with bones and other bright objects; (b) the male displays the colored feathers of its nape to the female whom he has enticed into the bower. He still holds an object that he has presented to her earlier in his bill. (Photographs: H. Sielmann.)

offer fish to females which they court (H. Rittinghaus 1963; see also Fig. 66).

Figure 66. Courting terns. The male courts with a fish. (Photograph: N. Tinbergen.)

Some insects behave similarly, for example, the predatory flies (*Empididae*), where the males are in danger of being devoured by the females during mating. The females of *Empis trigramma*, for example, will attempt to do this if they have nothing else to eat. Males of *Empis borealis* and *Empis tessellata* avoid this danger by catching a prey before mating which they pass to the female. This is the beginning of a very interesting sequence of ritualizations. The males of *Empis poplita* and *Hilaria quadrivittata* spin a cocoon around the prey object with the aid of spinning glands on the fore legs, before presenting it, so that the females are kept busy longer. In *Hilaria maura* the present is merely a symbol: The males spin a cocoon around an inedible object, for example, a leaf. At the end of this sequence we find *Hilaria sartor*, whose males spin balloonlike structures that serve as visual attracting signals (O. M. Reuter 1913; see also Fig. 67).

Figure 67. Ritualized passing of food during the courtship of *Empis*. The male presents a balloonlike spinning package that contains a fly. (From J. Meisenheimer [1921].)

This giving of food may be of different origin despite similar appearance. The "bill flirt" and courtship feeding of the bullfinch (J. Nicolai 1956) and many other birds is ritualized feeding derived from the functional system of parental care (see p. 104). This may also be true for the mutual feeding of chimpanzees, who pass food to one another with their mouths (M. Rothmann and E. Teuber 1915). An orangutan mother nursed her infant with pablum by mouth feeding, as did a gorilla mother in the Basel Zoo. Mouth-to-mouth feeding was also practiced until recently in the German province of Holstein (D. W. Ploog 1964a). The children react appropriately when approached by protruding their lips and making licking movements upon contact with the mouth. Kissing in man could be derived from this. Upon closer observation one can see pushing movements with the tongue which remind us of movements used in passing food, and the noticeable sucking movements may be interpreted as associated with receiving food. In the flies discussed above, food passing almost certainly is not derived from the brood-care system.

Other gifts besides food are brought. Diamond dove males give nesting material (Fig. 68) to their females. Courting exotic finches (p. 196) and many other birds (P. Kunkel 1959) behave similarly. Infantile behavior inhibits aggression just as effectively. Behavior patterns usually shown by the young, most frequently used by males, seem to have this effect (I. Eibl-Eibesfeldt 1957a; D. Burkhardt 1958). The courting bearded titmouse shows begging movements with the wings in the manner of a young bird (O. Koenig 1951). The woodpecker finch male (*Cactospiza pallida*) entices the female to the nest with this behavior. At the same time she begs for food from him with the same movements (I. Eibl-Eibesfeldt and H. Sielmann 1965). Male hamsters call like nestling young when they court.

Figure 68. Courting diamond doves (*Geopelia cuneata*). The male presents a twig to the female. (Photograph: J. Nicolai.)

When a man is courting he often speaks tenderly with childlike words that emphasize diminutives.

The yellow babuin (*Papio cynocephalus*) greet with "lip smacking." These movements, which are made into the air, can be explained as rapidly performed sucking movements and also during ontogeny as being derived from the sucking movements of the young. Pink parts of the body, such as the nipple, the penis, the female reproductive region, and the face of the child all act as strong releasers of smacking. As these parts are also attractive to other baboons; they contribute most likely to the cohesion of the group (T. R. Anthoney 1968). With this the female breast received an additional signal function that aids the cohesion of the group. This is interesting, because we can prove a certain parallel development in humans (page 436).

Finally, many animals appease during courtship by covering up signals or weapons that ordinarily release aggression. According to N. Tinbergen (1959) head flagging in the black-headed gull serves to hide the black face mask that releases aggression (Fig. 69). Terns hide the tops of their heads with a stretch posture, and the raven appeases by looking away and raising the bill (E. Gwinner 1964).

Figure 69. Presenting the back of the head (head flagging), an appeasement gesture of the black-headed gull. (Photograph: N. Tinbergen.)

In a particular courtship sequence several of these expressive movements occur often. In courting albatrosses we can observe behavior patterns that can be interpreted as ritualized food begging. appeasement gestures, showing-of-nest, preening behavior, as well as others that are not yet understood (Fig. 70). The courting ritual, which is repeated many times by one pair, begins with a dance.

The male walks around the female with his neck pulled in and wobbles conspicuously and in synchrony with the steps from one side to the other. Maintaining the rhythm to his steps both birds turn their heads alternatingly to the side so that the bill touches their raised shoulder. The dance is then followed by fencing with the bills (Fig. 70a). The birds stand facing one another, stretch their

Figure 70. Courtship of the Galápagos albatross (*Diomedea irrorata*): (Hood, Galápagos Islands). (a) bill fencing; (b) clapping the beaks; (c) preening the shoulder by the left and clapping the beak by the right animal; (d) "presenting" the beak by the right bird; (e) showing the nest site; (f) social preening. (Photographs: I. Eibl-Eibesfeldt.)

necks forward, and hit their bills together with rapid sideways motions of the head; at the same time they nibble with their bills. Young birds beg for food in this manner. Perhaps this behavior is ritualized begging. Other behavior patterns may follow as time goes on, for example, clapping the bill, which is done by standing up, opening the bill widely, and closing it with a loud noise. This is frequently done by both birds at the same time (Fig. 70b). This can also be observed when albatrosses threaten one another. More bill fencing or display movements may follow, the birds (Fig. 70d) raising their bills straight up and calling. This reminds one of the appeasement gesture of boobies that N. Tinbergen (1959) described. Sometimes one bird clappers its bill—stretched forward (Fig. 70c)—and at this time the partner always preens its shoulder feathers. At the moment the bird stops clacking and raises himself up, the other throws up his bill in display, clapping once loudly. These behavior patterns may follow one another in various sequences. At the end of such a courtship sequence the birds bow to each other, pointing their bills to the ground (Fig. 70e); while doing this they utter two-syllable calls. This could be symbolic pointing at the nest site. Usually they both sit down and begin preening each other's neck feathers (Fig. 70f). After a short interval this entire sequence may be repeated.

Some mammals take symbolic possession of the courted female by marking her with scents, creating an odor bond. For example, porcupines (*Erethizon dorsatus*), agouties (*Dasyprocta aguti*), and maras (*Dolichotis*) approach the females on their hind legs and spray them with urine. Similar urine ceremonies are known in guinea pigs and rabbits (ref. in I. Eibl-Eibesfeldt [1958]). A gesture of contact willingness in females is the special movement (p. 112) presenting the genital region, frequently conspicuously altered. The partner can also become attached to an animal by activating his aggression against a third, which is the case with the inciting behavior of ducks (p. 108). The "triumph" calls of geese, which are a greeting gesture in the wider sense (p. 129), may have a similar origin.

Submissive gestures, greeting and other appeasement behavior

Many of the discussed courtship behaviors are appeasement gestures that also play an important role in other situations. Frequently the loser of a fight appeases the victor by so-called submissive gestures that are usually the opposite of threat postures. The marine iguana submits by prostrating itself before the winner, who then ceases fighting but waits in a threat position until the vanquished retreats from the area (p. 319). The appeasing function of behavior has been shown in chaffinches by P. R. Marler (1956). Animals that behave submissively are permitted to come closer to conspecifics than those who show a threat posture. Some vocalizations have appeasing effects, such as the muttering (Muckern) calls of polecats (p. 93) and the squealing of young rats. If one rat bites another too hard during play, the latter will squeal upon which the former will gently groom its fur. Many postcopulatory displays of birds can be interpreted as appeasement ceremonies, and many, but not all, greeting ceremonies have this function. In general, contact between conspecifics, frequently of different sexes, is established and maintained by greeting ceremonies.

When the flightless cormorant (*Nannopterum harrisi*) returns to its nest and mate, it will bring a gift of a sea star or a bundle of

Figure 71. Appeasing gesture: passing of nest material to the mate while exchanging places at the nest by the flightless cormorant (*Nannopterum harrisi*) (Narborough, Galápagos Islands). (Photographs: I. Eibl-Eibesfeldt.)

seaweed which is presented to the bird on the nest. The latter often pulls it away aggressively (Fig. 71). One can recognize from the vehemence of the behavior that aggressiveness is directed toward the gift. A simple experiment shows that this is indeed the case. If one takes the gift away from a bird returning to its nest, which is possible because the birds are quite tame, the bird is driven away from the nest by its mate (I. Eibl-Eibesfeldt 1965b).

The male sea lion employs appeasing greeting gestures to keep his herd together. If two females fight, the ruling bull approaches at once and pushes himself between the two combatants extending greeting toward both, which has a calming effect (I. Eibl-Eibesfeldt 1955b).

Storks greet their mates by placing their heads over their backs and clapping their bills. This can be interpreted as a pronounced turning away of the weapon, because during a threat the tip of the bill points at the other bird. In principle the head flagging of gulls is also a turning away of organs used in fighting. Here it is primarily a threat signal, while in storks the weapon is demonstratively turned away. This is also known from other birds.

In the so-called contact species, where individuals know one another well, touching the body serves greeting function. Cats greet by "presenting the head" (O. Antonius 1939; P. Leyhausen 1956). Grooming behavior patterns clearly express a readiness for social contact, and they frequently evolved into greeting ceremonies. We remember the lemur *Lemur mongoz* discussed earlier. According to J. v. Lawick-Goodall (1965, 1968) chimpanzees have several greeting gestures. They embrace one another and kiss with a touching of lips when they meet someone they know (p. 106). The embrace can probably be derived from infantile clasping, which now serves the function of maintaining group cohesion. The gesture calms both animals. Goodall reports that even large males, when frightened, clasp young chimpanzees and calm down as a result. Another appeasing greeting behavior is the sexual presenting of females (Fig. 72a–c), whereby the greeting animal turns its posterior toward the other, a behavior found in baboons and other apes.

Males also use this originally female behavior (p. 112). Chimpanzees also shake hands the way people do. The initiative is taken by the lower-ranking animal, who reaches toward the higher-ranking animal with the hand palm up in a kind of begging gesture (Fig. 73). In response to this gesture, probably derived from the infantile search for contact (p. 346), the higher-ranking animal gives his hand, which in turn calms the other. Lower-ranking animals solicit approval in this way from higher-ranking members when they attempt to obtain food at a common feeding place. They also bow when greeting others (J. Goodall 1968). The very aggressive cichlids (*Tro-*

a

b

c

Figure 72. Greeting chimpanzees (Tanzania): (a) arriving female presents and is touched at the genital region by the male; after she has turned around (b) he touches her face; (c) she bows "laughingly" and the male begins with social grooming, which remains, however, a mere symbolic gesture. (Photographs: I. Eibl-Eibesfeldt.)

Figure 73. Chimpanzees greeting by shaking hands. The lower-ranking animal presents the hand to the higher-ranking one by holding the palm up in a begging fashion. (Photographs: Baron and Baroness van Lawick-Goodall, with permission of *National Geographic*.)

Figure 74. Quiver display of *Tropheus moorii*, an appeasement movement probably of sexual origin. (Photograph: W. Wickler.)

Figure 75. (a) Threatening zebra colt; (b) "greeting face" in the same zebra; the expression is as in the threat but the ears are erected. (From E. Trumler [1959].)

pheus moorii), which live in groups, appease others by presenting a yellow band to the attacker, which is also shown during spawning and courting (W. Wickler 1965e; see also Fig. 74).

Interestingly enough, some greeting gestures can also contain elements of threat behavior. According to E. Trumler (1959), this is true for the "greeting face" of horses; opening the mouth and exposing the corners of the mouth are clearly aggressive. But this is "canceled out" by also raising the ears (Fig. 75). The facial expression of a mare in estrous (Rossigkeitsgesicht) originated out of the mimic expression of threat. Redirected threat movements can be observed in graylag geese, which threaten with outstretched necks past one another as if they were confronting a common enemy (K. Lorenz 1963a). The "triumph ceremony" (Triumphgeschrei) plays a special role during pair formation. The male at first makes sham attacks toward objects that are normally avoided. Following such an attack he "triumphantly" returns to his intended mate and threatens beyond her (Fig. 76). If she joins the "triumph ceremony," a defensive alliance has been formed, which is a prerequisite for the successful rearing of a brood (K. Lorenz 1943). This behavior continues to maintain the pair bond by functioning as a greeting

Figure 76. Sequence of behavior patterns of the "triumph ceremony" in the graylag goose. The male threatens toward the opponent (F) (1 and 2) and drives him off (3), turns around and returns in an impressive display posture to the female (4), who approaches him with rolling calls. They roll and cackle together (5 and 6). (From H. Fischer [1965].)

ceremony. The neck movements of a greeting goose have the same form as the threat movement and undoubtedly were derived from it. The "cackling" (schnattern) of the "triumph ceremony" has developed from the contact call of the young (H. Fischer 1965). For this reason and by the orientation of the threatening neck posture past the other, this gesture becomes neutralized, similarly to the threat expressions of horses, which become "greeting faces" by the erection of the ears. Although the "triumph ceremony" of the graylag geese is undoubtedly derived in part from aggressive behavior patterns, it has its own motivation that is independent of aggression (H. Fischer 1965). The "räb-räb" palaver of ducks, as well as the greeting ceremony of the female bullfinch, still contains much aggression to this day (K. Lorenz 1941; J. Nicolai 1956).

Two threat postures, the upright and forward, are a part of the greeting ceremony of the mated black-headed gull pair (Fig. 77). At first these postures have a distinctively aggressive function. The female is attacked as an intruder and chased off. Soon this picture changes; she is allowed to remain, although he continues to threaten, but now with a new orientation. When alighting, he greets with the "long call"; she approaches with her neck stretched forward, and he reacts likewise. Finally, they no longer point at each other, but stand parallel, similar to geese during the "triumph ceremony." This ritual is reduced to a bare minimum as the birds get to know each other better (G. Manley 1960).

In rhesus monkeys the mounting by the males of others of the same sex does not only mean aggressive threat or assertion of rank but is an expression of an accepted order within the group, strengthening the bond between the two individuals. The higher-ranking animal usually mounts first, but is frequently mounted in turn by the lower-ranking one. C. B. Koford (1963a) compares this kind of greeting with a military salute.

In the raven, the ceremony of mutual feeding between members of a pair becomes increasingly more superficial (p. 140). Initially both partners actually feed one another; finally this is done only rarely, consisting mostly of a brief grasping of the bill. This has been interpreted as an ontogenetic ritualization but does not seem to be correct in my opinion. For this to be true one would have to demonstrate that better communication is achieved by simplifying the behavior, that is, that the signal has become more understandable and effective. This kind of ritualization does not seem to have occurred in this case. Instead it appears as if the ceremony has become more superficial, deritualized—perhaps because the animals know one another better and the need to appease aggressive behavior has diminished.

The smile of man is an important buffer against aggression. A

Figure 77. Upright and forward (standing parallel) postures in the threatening greeting of the black-headed gull. (Photographs: N. Tinbergen.)

smile can be disarming, and reports from warfronts contain examples of how aggression may be inhibited by a smile (see p. 104). Every traveler has experienced the release of tension between strangers by a smile. People usually smile politely when making a refusal to someone, and they smile when excusing themselves. However, a smile not only inhibits the aggression of another person, it also frequently brings about a friendly reply. Infants smile, and this increases the bond with the parents. In adults a smile will build a bridge to total strangers, and people smile at each other when flirting as well as during friendly greeting (pp. 411, 417–422).

A smile frequently changes into laughing, which in some ways can be seen as an increased level of the smile, but not exclusively so. People will open the mouth and make rhythmic sounds. J. A. Ambrose (1963) interprets smiling as an ambivalent behavior that is derived from the simultaneously present tendencies of turning to and turning away from someone. Light tickling releases turning toward in human babies, and strong tickling releases turning away. This is also true for other stimuli, such as sudden surprise, booh calling, and so on, which have to be presented in just the right intensity and frequency if they are to be successful in releasing laughter in a child. Ambrose sees certain similarities between smiling and crying—a rejecting gesture. K. Lorenz (1963a) interprets laughing as a greeting ceremony derived from a threat movement, a view made plausible by the exposure of the teeth. N. Bolwig (1964) interprets it as ritualized biting (a playful intention to bite). It is quite certain that there is some aggression involved in laughing. The rhythmic vocalizations remind one of similar sounds made by primate groups when they threaten in unison against an enemy. Such a combined threatening unites the members of a group, and it has been observed during investigations of laughing behavior that a strong bond is established between members of a group in a similar way. Outsiders of such a group are quite uncomfortable in the presence of such laughter, especially when this has the character of "laughing at," when it is definitely aggressive and challenging. In its original form laughing seems to unite *against* a third party. In the smile, however, the aggressive component (the sound utterance) is lacking. It furthermore must be emphasized that the baring of teeth—which strengthens the idea that there might be a phylogenetic connection between smiling and threat behavior—differs markedly from the teeth exposure during furor. In furor the corners of the mouth are opened and drawn downwards (p. 422). In some primates (Gelada baboon) this leads to the full exposure of the upper canines. During a smile only the front teeth are exposed, and these are used in social grooming and play. It is more plausible to assume that these social situations gave origin to the smile as a friendly showing of teeth.

A large number of greeting ceremonies in man have their functional analogies in animals, which serve in the establishment and maintenance of a bond between individuals who are acquainted with one another. The reader may convince himself of their appeasing function by not greeting his closest relatives and friends for one week. It is surprising to find out how quickly aggression that is not buffered by these appeasing gestures is turned against him. Human greeting gestures include, besides the smile, gestures of symbolic submission. People bow or nod their heads, in Europe in the same way as in Japan. One bares the head and removes weapons, thus demonstrating trust by giving up one's protection.

The greeting with the raised, open right hand is widespread. The Chom Pen of the Great Nicobar Island in the Indian Ocean, who had had no contact with Europeans, greeted us with this gesture in the same way as the Karamojong tribesmen in East Africa (p. 438).

Sometimes weapons are used in greeting. They are demonstratively turned away, to a position which is not dangerous, as when presenting arms. When someone greets us with a spear he does not point its tip against our stomach. The ever-present cultural differences do not affect the basic principles. Greeting ceremonies have become changed in characteristic fashion depending on rank and sex, a problem that needs still further investigation.

Quite often presents are given and this seems especially appropriate if one enters another's territory, such as a gift of flowers when we visit someone's apartment or house. This seems to require stronger, more effective gestures. If this formality is omitted in cultures where it is customary, the omission is interpreted as impolite and the unappeased aggression is experienced as annoyance.

In man threat gestures were sometimes changed into greetings. As in the inciting of the duck (p. 108), smiling in its *original function* seems to be a common and thus uniting threat against an enemy (p. 129). One demonstrates the willingness to attack together. The greeting with a raised fist is another example. The uniting function of communal threats becomes obvious in all military parades.

Finally, we must include under the heading of greeting the *farewell*, which also needs further investigation. Its function seems to lie in the strengthening of the bond for the future. Another component may possibly play a part. If one departs from another person there exists the potential danger that aggression, hitherto inhibited, could become released. A person who leaves the room backward and bowing continuously is probably afraid as well. We find a functional parallel to this in many of the complicated "postcopulatory" displays in many birds, where besides elements of display we find

Figure 78. Postcopulatory display in the striped goose (*Anser indicus*).   (Photograph: H. Kacher.)

many behavior patterns of appeasement (Fig. 78). According to O. Heinroth (1924) eider ducks tend to become aggressive following copulation. Aggression, which until then was suppressed by the sexual drive, has probably become released (see also p. 126) and needs to be appeased if the pair bond is not to be endangered.

Signals that establish and maintain contact

Animals within a group frequently maintain vocal contact, for example, members of a jackdaw or bearded titmouse flock, as do male and female of many species that are mated, as well as the maternal family of the squirrel. These calls, which serve to maintain cohesion of groups, are called *contact calls*. Many animals learn to recognize the voice of their partner individually. We already listed ravens and shama thrushes. Sea lions and sheep recognize the calls of their young individually, and only this individual familiarity keeps them together. Sea lions attack young other than their own (I. Eibl-Eibesfeldt 1955b). Young guillemots (*Uria aalge*) recognize their parents by their voice. They react only to the call notes and not to other vocalizations of the parents. The call notes of other parent birds had no effect on the young. The parents respond with calls to the vocalizations of the young while it is still in the egg. Whether or not they also recognize the young at this stage could not be determined because they recognize the individual egg and no longer accept strange eggs (B. Tschanz 1965, 1968).

Some birds have developed alternating or duet songs, which maintain the pair bond. This is found in some Australian honeyeaters (*Meliphaginae*) and some African shrikes. In the honeyeater (*Acanthagenys rufogularis*), both members of a pair sit next to one another; one sings the strophe, and as soon as it ends, the other continues. Other species sing their duets at the same time and in surprising synchrony. The highest level of development is reached in the duets of antiphonal songs of African shrikes. In *Laniarius aethiopicus* both birds have one song which each is able to sing alone. Members of a pair, however, sing frequently only to certain parts of a strophe in alternation, but so perfectly adjusted to one another that one does not recognize at once that two birds sing one melody. These duets are found primarily in birds that live in dense forests (K. Immelmann 1961; W. H. Thorpe and M. E. W. North 1965).[1] There exists then one group of common signals that keep a group of animals together but that does not presuppose individual recognition, and another group that depends on it. This also holds for olfactory signals, which in many mammals serve as cues that maintain group cohesion.

[1] In convergence, duet singing evolved in the genus *Trachyphonus*, a monogamous bird that does not belong to the song birds (H. Albrecht and W. Wickler 1968).

Norway rats, bees, and many other gregarious animals recognize one another by odors common to the group or hive, without, however, recognizing one another individually (p. 351). Sea lions, on the other hand, recognize each other individually.

If a group is to be kept together, it is advantageous that its members do the same things at the same time. A flock of birds could never keep together if each bird did something different; one would sleep, another eat, and others might want to fly. We can often observe that feeding is contagious: If one bird eats, others follow. Frequently specific expressive movements have evolved which facilitate the synchronization of moods. Graylag geese ready to fly off begin to walk, shake their heads with stretched-out neck, and call. If some members of a flock begin with this activity, others follow, and within a short time, they all leave together. In man yawning seems to have a similar contagious effect: it makes everyone sleepy.

This was described among others by K. v. Steinen, who was the first European to come into contact with the Bakairi of Central Brasilia.

If they seemed to have had enough of all the talk, they began to yawn unabashedly and without placing their hands before their mouths. That the pleasant reflex was contagious could not be denied. One after the other got up and left until I remained with my dujour (K. von den Steinen 1894, new ed. 1917:183).

The maintenance of contact is also achieved by all those signals which attract a conspecific to flee, such as the conspicuous spot on the rear end of deer and antelopes.

Communication about the external environment

Warning and distress calls

Many warning calls have evolved which alert conspecifics about the presence of a predator. The ground squirrel (*Citellus citellus*) and the marmot (*Marmota marmota*) utter a call before fleeing from a predator, and many birds do the same. It has been shown to be an effective warning to conspecifics.

Chicks fall silent while still in the egg and cease scratching movements when they hear the warning call of chickens (E. Baeumer 1955), but herring gull chicks eagerly respond prior to hatching to all calls from the outside, including the gull's own distress call (F. Goethe 1955). The chicks of some wading birds (Scolopacidae) also do not know the meaning of the alarm call and must first learn to associate it with the silhouette of the aerial predator (O. v. Frisch 1958).

Gregarious aquatic animals warn conspecifics chemically by alarm substances. For example, if the snail *Heliosoma nigricans* perceives the body juice of an injured conspecific, it will bury itself in the mud (W. Kempendorff 1942). Minnows and many other schooling fishes flee when they perceive a substance that is released from the skin of injured conspecifics (K. v. Frisch 1941; W. Pfeiffer 1960, 1963; F. Schutz 1956), as is true for tadpoles of the common toad (*Bufo bufo*) that move about in swarms (I. Eibl-Eibesfeldt 1949; E. Kulzer 1954). The death cry of many animals may be analogous to these warning signals and bring about an association with the dangerous situation. Systematic studies are not available to my knowledge. I have personally caught many rats, one after another, in traps of the type that, when tripped, breaks the rat's back. The animals stepped on the trigger and even ate from the dead conspecifics, until one rat was merely injured and did not die at once. From that time on, no more rats were caught (I. Eibl-Eibesfeldt 1953b).

Often animals react to the distress call of a conspecific that has been caught by an enemy. Many apes and monkeys who know their keeper well will attack blindly when a conspecific, after being grasped by the keeper, calls (W. Köhler 1921; S. Zuckermann 1932). This happens so automatically that, for example, terns will come to the aid of a strange chick, which they will then attack themselves (K. S. Lashley 1915).

Language of bees

We know from the pioneering and careful investigations of K. v. Frisch (last summary of his work in 1965) that honeybees communicate the direction and distance of a food source to their fellow hive members by means of special dances.

The returning forager bee begins to dance on the comb in a very specific manner. When the feeding place is near the hive, the bee performs a round dance which contains no directional information. New foragers become alarmed by this dance and then search around the hive in all directions, searching for the odor which the dancer brought back from the food source. When the food source is more than 25 meters away from the hive, the bee waggle dances. Wagging her abdomen, she runs straight for a short distance while accentuating this wagging distance with a rasping sound produced with her wings. Then she turns to one side and returns to the original position but without wagging. She then repeats the wagging dance over the same route but turns to the other side to return again to the point of origin, and so on (Fig. 79). A number of bees in the hive become excited by this dance and follow the dancer. They perceive the odor of the blossoms visited by the forager and they learn in which distance and direction they have to search. If the food source is close to the hive, then the straight wagging run is short, and the

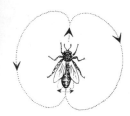

Figure 79. Waggle dance of the honeybee. (After K. v. Frisch [1959].)

wagging dances follow one another more rapidly. This enables the bees to compute the distance to the feeding place. An experienced observer can do this with a stopwatch.

Wind velocity and direction are also reflected in the dancing rhythm. When head winds prevail the bees dance more slowly; thus they report a greater distance. They do this also when they have flown to the food up a steep slope. Thus the dance does not indicate either actual distance or the flight time but refers to the expended energy that the bees require to reach their goal. This information is passed on by the amount of time they wag, which is emphasized by the rasping sounds mentioned above.

More recent investigations by H. Esch (1967) indicate that these sounds are of more importance than had been assumed earlier. Sometimes returning foragers dance on the comb without making sounds. Esch observed 15,000 of these dances, and in no instance did this silent dance lead other bees to the food source.

The direction is given by reference to the position of the sun. If a bee dances in front of the hive, which rarely occurs, then the method of transmission of information can be observed. The straight line of the wagging dance runs in the same angle to the sun which the bee maintained on the straight-line course to the food source (Fig. 80). The bee also does this when the combs of the hive are in a horizontal position and when they can see the sun. Then the straight portion of the traversed path leads also directly toward the target. If one covers the sun the bees become disoriented and the alarmed new foragers find the feeding place only by chance. When v. Frisch placed four odor plates in all four compass directions, then they were visited equally often by the newcomers in this experiment. However, when the foragers danced in the sun light on the horizontal combs they could indicate the direction and the new foragers visited one of the odor plates in preference over the others. Normally the bee dances on vertical combs within the dark hive. In this case, the

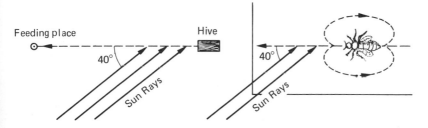

Figure 80. Indication of direction with respect to the position of the sun during the dance on a horizontal plane: left: broken line, direction of flight to the feeding place; right: the waggle dance on a horizontal plane. (From K. v. Frisch [1959].)

angle to the sun is translated into an angle with respect to gravity (Fig. 81). If the feeding place lies in the direction of the sun, the straight part of the wagging dance points straight up. If the location is 50° left of the sun, the wagging dance deflects 50° from the vertical to the left. If the food source is away from the sun, the wagging dance is executed straight down. This ability to transpose is also shown by the dung beetle, which does not dance. If allowed to walk on a plane, it will maintain a certain angle to an artificial light source. If one then raises the plane into a vertical position and illuminates the area from above with diffuse light, the beetle transposes the angle of the direction he was running before with respect to gravity just as the bees (G. Birukow 1953). It is not known of

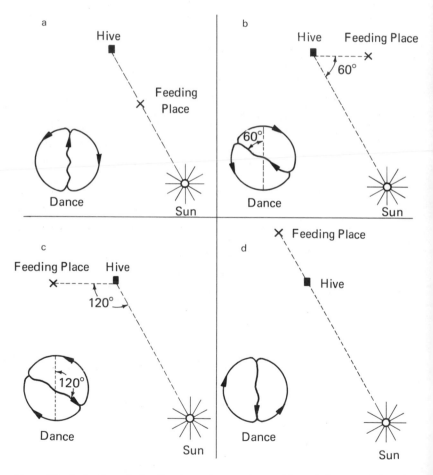

Figure 81. Indication of direction according to the position of the sun during the dance on the vertical plane of the comb. On the left is shown in each case how the bee dances on the comb at a given location of the feeding place. (Additional explanation in the text.) (From K. v. Frisch [1959].)

what use this ability is to the dung beetle. Other insects also show this capacity to transpose.

Ants also transpose the directional angle to the light source into an angle with respect to gravity when the horizontal experimental table is tipped vertically and the light is turned off at the same time. However, ants are not as precise as dung beetles or bees. If the angle of direction was 20° to the right of the sun, they will maintain 20° to the perpendicular line, but it may be either to the right or the left, and they may walk up or down with respect to gravity. The lady bug, which equates the direction to the light with a vertical up direction may, however, deviate right or left; the dung beetle transposes in the manner of the bee except that the direction toward the light coincides with a downward direction. G. Birukow (1956) attempted to explain these phenomena. An integration of light and gravity orientation can be found in bumblebees. When they leave their hive, they are positively phototactic in respect to their orientation to the sun. If one places them on a vertical plane in the dark when they are on their way to leave the hive, they will go upward, the same way they would within their dark nest in the ground when they were ready to leave. On the other hand, they are negatively phototactic on the return flight and crawl into darkness. If placed on a vertical plane in darkness they move downward. On the ground they will find their way to their nest in this way (U. F. Jacobs-Jessen 1959).

If the bees were to indicate the angle to the sun when they leave the hive, this would ultimately lead to errors, because the position of the sun changes and with it the angle to the sun, although in a lawful manner. v. Frisch's experiments have shown that the bees take this movement of the sun into account, although it is not known how they accomplish this. A prerequisite for this ability is that a bee experiences this movement of the sun herself, and from this they learn what is necessary. Bees that had seen the movement of the sun for only a few afternoons in their lives, spending the remaining time in the hive in a dark basement, acquired the knowledge about the daily course of the sun. If they were subsequently tested in the morning during daylight, they did not make errors in the direction in which they had been trained.

When making detours bees indicate the distance in a straight line, but they also communicate the length of the detour. Dancing is innate in bees, and there are several dialects. The Egyptian honeybee begins wagging dances when the food is more than 10 meters from the hive; the Krainer race only beyond 50 to 100 meters. The latter also has the fastest wagging dance.

D. L. Johnson (1957) and A. M. Wenner (1967) recently proposed that bees are able to communicate direction and distance of a goal by means of the dance, but that this is not of any use to other worker bees. These are interested in the dancers but do not utilize their information. Instead, the odors of food, hive, and other bees is said to aid them in finding their goals.

These statements cannot be brought into agreement with the experiments we have discussed so far. During the detour experiments returning foragers danced in a straight line to the food source, and the recruited bees flew in a straight line toward the goal and over the obstacles, while the previous foragers made the detour. Thus it is inconceivable that the recruits could have followed an odor trail.

If one now raises the question as to how Johnson and Wenner could come to their different opinions, one discovers several errors of experimental design which the authors overlooked. K. v. Frisch set up his fan experiments in such a manner that he trained marked bees to a food source 250 meters from the hive. The food source consisted of a weak sugar solution, which did not result in much dancing in the hive and which only barely excited other bees. During the actual tests seven odor plates were placed 15° apart at a distance of 200 meters but without food. At the same time the original feeding place was marked with the same odor as that on the other odor plates, and a strong sugar solution was offered. Now the trained bees danced and the number of new recruits on the various odor plates indicated in which direction the recruiters were searching. The number of arrivals at the food source was not counted, because the marked bees also made use of their odor glands, which are an additional attractant.

Johnson performed similar experiments, but he then added a "control experiment." To create identical conditions at all locations he placed another bee hive containing light-colored bees next to the first hive. These lighter bees visited three locations that were closer to the hive, while the darker bees continued to visit the food source more distant from the hive, but which was located on an extension of the central of the three but closer feeding places. Only marked bees of both hives were fed; other arrivals were killed until the actual tests began. Johnson began counting recruits from a certain time on. As one would have predicted from the findings of K. v. Frisch, the recruited dark bees preferably visited the central feeding stations. Alas, they did so even when Johnson shifted their feeding place to the north again on an extension of one of the closer feeding places. From this finding he concluded that the bees were not guided by the dance but merely visited the geometric center of all the feeding stations. However, during the experiment a strong southeast wind prevailed, and because the bees on the nearer hive used their odor glands, the bees on their way to the northern food location were detracted by the more intense odor reaching them from the southern feeding place. Just such detracting effects had already been described by v. Frisch, and Johnson might have avoided this error by a more careful reading of the literature. Furthermore, Johnson tested in only three directions, while v. Frisch covered a much larger area with seven odor plates. He also tested bees over greater distances (600 and 1250 meters).

Wenner's experiments are just as uncritical as are those of Johnson. He fed bees, using Frisch's gradient-experiment procedure, 400 meters from the hive. The nearest and farthest feeding places (200, 300, and 500 meters) were visited by only a few new bees. At the original food source 74 percent of all new bees arrived, although all feeding places were supplied equally well with food. Then Wenner placed another hive with lighter bees directly north of the line of flight and fed one group each of these bees at locations 200 and 300 meters, while the trained dark bees continued to fly to the 400-meter location. The recruits of the darker bees no longer responded in accordance with the distance message received from the trained bees, since 18 percent alighted at the 200-meter place, 48 percent at the 300-meter place of the light bees, and only 33 percent arrived at the 400-meter food source of their own bees. One percent visited the feeding place at 500 meters. These results could also be predicted on the basis of v. Frisch's experiments. In a strong wind, as prevailed during Wenner's experiments, bees fly close to the ground. Under these conditions many of them become attracted to the closer food sources where the learned odor and the odor secretions of other bees attract them (K. v. Frisch 1968).

Observations of other insects hint at likely ways along which dancing might have evolved. V. G. Dethier (1957) fed sugar to blowflies (*Phormia regina*). If he withheld it from them, they performed a kind of round dance, running in circles to the right and left. They also did this when they were taken to another location. They "danced" more rapidly when the sugar solution was of higher concentration. These dances were oriented in respect to light and gravity. The flies approached the light source and ran up and down on a vertical plane, without, however, indicating direction.

These flies also regurgitate food when they meet another fly, and the latter then begins to search. The "dance" here is always a searching, but one can imagine that the "round dance" of the bees might have evolved from such initial stages, that is, that it is a ritualized searching that stimulates others to search also. With increasing delays, the searching behavior of the fly decreases in intensity.

A parallel to the waggle dance was discovered by A. D. Blest (1960) in some New World saturnid moths. These moths shake their body after landing by alternately bending and straightening their legs. The duration of this shaking increases linearly with the length of flight time.

A similar physiological mechanism may have been possessed by the ancestors of bees, and which could then have become useful in indicating distance. The indication of direction can be interpreted as derived from an intention movement of flying toward the feeding place. This interpretation is probable when we consider the behavior of primitive bees. The stingless bee *Trigona postica* leads her fellow hive members. On the way home she marks various landmarks with secretions of her mandibular glands. Upon returning to the hive she runs about on the comb with fluttering wings and bumps other bees. She dispenses food samples and when a number of bees have gathered around she flies off along the odor-marked trail that she prepared earlier (M. Lindauer 1961). This leading behavior is shortened in the more advanced stingless bee species. Several species of the genus *Melipona* perform a kind of Morse code on the comb to indicate distances: Short bursts of sound indicate close goals; longer bursts signal larger distances (H. Esch 1967). They indicate the direction by flying toward the goal first in a zigzag course and then in a straight line. After repeated indications of direction in this way, some worker bees fly off in the same direction, as if they had understood the message. The dwarf honeybee (*Apis florea*) shows a complete dance, but only in sunlight and on a horizontal plane. Here the wagging dance can be recognized as a repeated intention to fly off. This fits with the observation of the kinds of sounds made during wagging dances in the advanced bees. In this way it becomes plausible that the waggle dance evolved from a more general invitation to search to the wagging dance, which indicates distance and direction of a goal. The ability to transpose orientations from the sun to gravity was probably present as a preadaptation, because it also exists in many insects that do not dance.

The dance language shows some similarities to human language. It is a means of communication between conspecifics, and relations between things are communicated. In contrast to human language the system is a stereotyped, innate coding system. Human language is also based on an inborn potential for specific sound production and perhaps on the drive to speak (p. 114), but the language symbols are individually learned and passed on by tradition. Individual experiences can be described by words and passed on, and abstract thinking permits communication about relationships between relationships. The bee dance is similar to human language in so far as it is also a symbolic language by which inexperienced individuals acquire knowledge without the object in question being present. The transfer of knowledge, however, is only done by animals that experienced the location of the feeding place in a previous flight. No bee will communicate a message just received by another bee to a third animal unless it has visited the feeding place. This language is "rumorproof" in a manner of speaking (W. Wickler 1967). O. Koehler (1949, 1952, 1954b, 1955) has repeatedly commented on this subject, most recently in a review of the work of C. F. Hockett (1960). He presents one of Hockett's tables amended by him in which he compares six methods of communication among animals (including bee dances) with human language and instrumental music. This he does in respect to which of 13 characteristics of human language are present or absent.

Intraspecific threat signals

By this term we refer to all those behavioral and morphological characteristics that serve to reject a conspecific. Colorful plumage, for example, has this effect. K. Immelmann (1959) showed that zebra finches with colorful plumage maintain a certain distance from one another while all white birds of the same species sit much closer together. Many sounds and vocalizations produced by insects and vertebrates have this same function. Many fishes threaten one another with calls, for example, cichlids (A. A. Myrberg 1965) and damselfish (I. Eibl-Eibesfeldt 1960a; H. Schneider 1963). The territorial song of birds is well known and is directed against rivals. Male sea-lions roar at the owner of the neighboring territory. Many rodents utter ultrasonic threat calls. These are only a few examples. Threatening sounds may also be produced by other means. Many monkeys display anger by shaking the branches of their trees. Japanese monkeys, in addition, beat against resounding objects, behavior that is considered to be derived from tree shaking (S. Kawamura 1963).

Gorillas and sometimes chimpanzees pound their chests (Fig. 82); chimpanzees beat against "drumming trees" and in captivity against other resonating objects (G. B. Schaller 1963; J. Goodall 1963, 1965; B. Grzimek 1954). Human beings do the same. Macrosmatic animals try to intimidate their opponents with odors.

Figure 82. Play fighting young gorillas. After rough-housing, the victor drums with the hands on his breast while the loser climbs away. (Photographs: Johnston, from the film "African Apes," F.96 of the former R. St. f. d. Unterrichtsfilm.)

In addition, special postures and movements serve to keep others at bay; these often consist of ritualized elements of attack behavior (for example, biting and rush attacks). Generally the threatening animal will make itself larger and more impressive, and may display its weapons (Figs. 83–86). The animals may raise up high and spread manes, skinfolds, fins, and feathers, which frequently exhibit conspicuous patterns or colors. On occasion it is possible to distinguish aggressive from defensive threats. An aggressive squirrel pulls its ears back and chatters with its front teeth, but if cornered and defensive, it threatens by raising up its ears, which appear still larger because of tufts of hair on their tips; at the same time it squeals (I. Eibl-Eibesfeldt 1957a; see also Fig. 87). Threatening calls are very widespread. Many species threaten predators in the same way as against conspecifics, for example by displaying their weapons.

Figure 83. Lateral enlargement of the threatening Andes anolis: top, a threatening male; below, normal posture. (After W. Kastle [1963].)

Figure 84. Threatening marine iguana (bite threat and lateral display) (Narborough, Galápagos Islands). (Photograph: I. Eibl-Eibesfeldt.)

Figure 85. Galápagos sea lion, (*Zalophus wollebaeki*) threatening near the border of his territory. (Photograph: I. Eibl-Eibesfeldt.)

Figure 86. By raising the opened claws toward the photographer a coconut crab (*Birgus latro*) threatens. In intraspecific disputes crabs often threaten with their claws and this has led to a widespread ritualization of this gesture (p. 96). (Photograph: I. Eibl-Eibesfeldt.)

Figure 87. (a) Defensive, (b) aggressive threat in the European red squirrel. (From I. Eibl-Eibesfeldt [1957a].)

Signals effective in interspecific communication

Signals indicating contact readiness between species

Animals belonging to different species often come together to their mutual advantage. Some shrimps (*Alpheus*) live together with gobies on the sea bottom, which is bare of all cover. The shrimps dig a hole in which they live; the gobies, which do not dig, profit well from this, and in turn warn the shrimp of approaching danger (W. Luther 1958; W. Klausewitz 1961; see also p. 300). Anemone fish (*Amphiprion*) live in certain specific anemones without being injured by their poisonous tentacles (I. Eibl-Eibesfeldt 1960a; E. Abel 1960a). In symbiotic relationships of this kind a problem of communication between species exists. This has been investigated in more detail in the "cleaning symbioses" (I. Eibl-Eibesfeldt 1955, 1959).

A number of marine fishes are specialists in freeing other fish of parasites (I. Eibl-Eibesfeldt 1955a, 1959; J. E. Randall 1958; C. Limbaugh 1961), for instance, the cleaner fish (*Labroides dimidiatus;* Fig. 88). This fish entices its hosts to permit themselves to be cleaned by means of a signal consisting of a special nod swimming (cleaner dance). It butts with its snout against their fins or operculae so that they spread them, and against their mouths so that they open them, so that it can get inside. While the cleaner fish inspects its host, it continuously fibrillates with its ventral fins against its body so that it knows at all times where it is being cleaned. It can be clearly seen that the host reacts to this procedure, because it stops moving the fins that have been so touched.

a   b

Figure 88. (a) Cleaner wrasse cleaning *Plectorhynchus diagrammus* (Maledive Islands). The cleaner fish swims above the mouth of the larger fish; others wait nearby to be cleaned. (b) The cleaner fish is just disappearing in the mouth of the larger fish. (Photographs: I. Eibl-Eibesfeldt.)

On the other hand, host fish invite the cleaners to do their job by opening their mouths, and they signal them when to leave by closing their mouths half-way with a jerky motion and opening them again. The cleaners leave the mouth cavity in which they are cleaning following this signal. The host signals to the fish cleaning it on the outside its intention to swim on by shaking its entire body. In this way the cleaner and the fish being cleaned communicate with a few expressive movements. The saber-toothed blenny (*Aspidontus taeniatus*) imitates the cleaner fish and sneaks up on its victims (p. 151). It looks like the cleaner and imitates the nod-swimming behavior in all its details, although this style of swimming is not typical for this group of fish. In this way the host is deceived and the mimic bites chunks from the fins and gills.

The honey guide, a bird of the savannahs of Africa south of the Sahara Desert (*Indicator indicator* and *Indicator variegatus*) leads honey badgers and man to bee hives which they themselves cannot open but on whose honeycombs they feed. They produce conspicuous calls, spread their tails when one approaches them exposing a white pattern, and fly off some distance.

The begging movements of young brood parasites such as the European cuckoo and the learned begging movements of zoo and domesticated animals are all directed toward members of other species.

Threat postures and other signals for warding off nonspecies members

A large number of threat postures and gestures, which are often similar to those used in intraspecific disputes, are used to repel members of other species. Many animals threaten each other with their weapons. Carnivores show their teeth; crabs threaten with raised, open claws (p. 145). Specific postures adapted for use against predators are found in many butterfly larvae. The caterpillar of *Dicranura vinula* raises itself up when touched or when the leaf it sits on vibrates and displays a conspicuous face mask, which if touched is turned toward the stimulus (Plate IV). At the same time the animal projects two long, red threads, which emanate from the last pair of modified legs that are raised into the air, and which are turned into a spiral repeatedly before they are again withdrawn. In general, an animal defending itself makes itself larger and more conspicuous. Some species imitate stronger species (see mimicry, p. 149). One category of defensive behavior consists of the so-called mobbing reactions with which many songbirds attack birds of prey. They have special mobbing calls and often make sham attacks from all sides against the enemy, who usually departs. In addition to being bothersome, the advantage seems to be that the detected predator is unable to surprise a prey (E. Curio 1963). Some fish species also mob predators. While diving in the waters near the Maledive Islands I observed fusiliers (*Caesio*) which repeatedly swam toward a moray eel and passed closely overhead until it disappeared (I. Eibl-Eibesfeldt 1964c).

A *distraction display* is behavior that serves to mislead a predator. Many incubating and brooding birds flutter to the ground and run away on the ground limping as if they were injured when a mammalian predator appears. This behavior is also shown by the Galápagos dove, which lives in an environment totally lacking predatory mammals. This behavior seems to be a remnant from those times when the ancestors of this dove had encounters with mammalian predators (I. Eibl-Eibesfeldt 1964b).

7 NATURAL MODELS AND MIMICRY

The simplicity of key stimuli permits model making not only by the ethologist and the fisherman but also by many animals. They, too, are able to imitate specific releasing stimuli with which to release behavior patterns in others to their own advantage. This is true first of all for a number of predators. The alligator snapping turtle (*Macroclemys temminckii*) rests at the bottom of rivers with its mouth open; the inside of the mouth and tongue are darkly colored. At the tip of the tongue are two thin, red processes; these are dangled in the water and move like small worms. They attract fish, which are caught as soon as they begin to nibble at the protuberances. The large-mouthed catfish (*Chaca chaca*) has two small moving barbels which serve the same function (H. Schifter 1956). The anglerfish (*Antennarius* and others) has a movable first ray on its dorsal fin which has skin attachments at its tip that serve as lures. These animals are camouflaged and lie quietly on the bottom with only their lures moving (W. Wickler 1964a, 1964c). The same author has recently shown that different species of anglerfish possess different lures which are adapted to specific prey animals (Fig. 89).

The snail (*Succinea*) is the intermediary host for the sporocysts of the liver fluke (*Leucochloridium*), which parasitizes song birds. The sporocysts develop extensions which penetrate into the tentacles of the snails and attract birds with their pulsating movements. The conspicuousness is enhanced by yellow-green rings and the strong swelling of the antennae. The bird is deceived by this model of an insect larvae and eats the antenna, including its contents thus becoming infested with the parasite (C. Wesenberg-Lund 1939).

Figure 89. Top left: various anglers: the alligator snapping turtle (*Macroclemys temminckii*) fishing with its tongue; below it the large-mouthed catfish (*Chaca chaca*) angling with its barbels; top right: the angler fish (*Phrynelox scaber*) fishing with its wormlike lure. The lure is not only moved by the stem on which it is fastened, but it wiggles by itself; below it *Ogcocephalus*, which fishes prey hidden in the sand with a lure that points downward. (From W. Wickler, [1967a]; H. Kacher, artist.)

The small swordtail characin (*Corynopoma riisei*) from Venezuela attracts the female with a model of a daphnia when he wants to copulate. The male's gill operculum is modified into a long extension with a dark knob at its end, which is moved in a trembling fashion before the female. She may actually bite at it (K. Nelson 1964). The male then takes advantage of the closeness of the female and copulates with her (Fig. 90).

Female fireflies of the genus *Photurus* attract the males of another firefly species, *Photinus*, by imitating their flashing code. The deceived males are then eaten by the females (J. E. Lloyd 1965).

The fly orchids of the genus *Ophrys* have a modified lower lip on their flowers which resemble the females of certain wasp species as well as their sexual odor. When attempting to copulate with these models, the pollen sticks to the males and is carried to the next flower (F. Schremmer 1960; B. Kullenberg 1956; see also Fig. 91).

The African devil's flower (*Idolum diabolicum*), a predatory mantid, mimics a flower that attracts insects. Because flies are at-

Figure 90. Courtship of the swordtail characin male, which employs the model of a daphnia that sits on the end of the elongated gill cover. (From K. Nelson [1964].)

a   b

Figure 91. (a) Fly orchid (*Ophrys insectifera*) with a wasp (*Gorytes mystaceus*); left: flower; center: wasp visiting the flower; right: head of the wasp with attached anthers; (b) the wasp (*Lissopimpla semipunctata*) on the orchid *Chryptostelis leptochila*; left: flower; center: wasp visiting the flower; right: abdomen of the wasp with adhering anthers. (From W. Wickler [1966d].)

tracted by other flies, a part of the flowerlike body of the mantid is dotted to resemble flies sitting there—a fact to which W. Wickler (1967) called attention (Fig. 92).

The saber-toothed blenny (*Aspidontus taeniatus*) looks so similar to the cleaner fish (*Labroides dimidiatus*) that it is mistaken for a cleaner by other fish and can approach them easily (see p. 146 and Plate V). It rushes at them and bites pieces out of their fins (I. Eibl-Eibesfeldt 1959). J. E. Randall and H. E. Randall (1960) discovered that the mimic even resembles differences in the races of various cleaner fish species. In the Tuomotus region the cleaner has an orange-red color around the middle; so does the mimic. In

Figure 92. African devil's flower (*Idolum diabolicum*) in the prey-catching position. (After a color plate by P. Flanderky in Brehm's *Tierleben,* 4th ed., vol. 2, p. 80.)

other areas cleaner and mimic have a dark stripe at the base of the pectoral fins, which is not found in members of these species in other areas.

The African cichlids of the genus *Haplochromis* are such highly developed mouthbreeders that the females take the eggs they have just laid into their mouths even before the male has an opportunity to fertilize them. However, the male has "imitations" of the eggs on his ventral fin. Once the female has taken the eggs into her mouth, the male spreads his ventral fin before the female, exposing the dummy eggs. She tries to take them into her mouth; the male milts and in this way the eggs already in her mouth become fertilized. In this example of intraspecific mimicry, discovered by W. Wickler (1962a), the conspecific is deceived (Plate V).

In *Tilapia macrochir* the female also picks up the eggs immediately after spawning. Here the male fertilizes the eggs in another way. He produces filamentlike spermatophores which the female picks up if she finds them; but many of them are lost. Again, fertilization is ensured through the existence of another deceptive signal. The males possess long, filamentlike spermatophore models that protrude from the genital region. These are even stronger releasers for the females than the actual spermatophores, just as in

PLATE V

Top right: Plectorhynchus diagrammus being cleaned by cleaners. A cleaner mimic attacks the tail of the fish in the foreground. Two places where chunks were bitten from the edge of the caudal fin are clearly visible. (H. Kacher, artist.) *Bottom right:* the cleaner Labroides dimidiatus and below its mimic Aspidontus taeniatus. (H. Kacher, artist.) *Left:* egg-spot cichlids (Haplochromis burtoni). Top: The male presents the egg spots on his ventral fin to the female during courtship. Below: The milting male exposes his egg models, which the female tries to pick up. (From W. Wickler [1964a]; H. Kacher, artist.)

PLATE VI

Top right: the Estrildid finch *(Pytilia melba),* to its left the paradise widow *(Steganura paradisea).* Below them: 30-day-old young (right: Estrildid finch; left: paradise widow); on the left the gape markings of 1-day-old young; top, paradise widow; below, Estrildid finch. Whereas the adults look very much different, the gape markings of the young are almost identical, and the young fledglings are quite similar. (After J. Nicolai [1964]; H. Kacher, artist.)

the case of the dummy eggs. The male presents these dummies to the female; she takes them into her mouth and in this way receives the spermatophores which are between the dummies (R. Apfelbach 1967b; see also Fig. 93).

W. Wickler (1965b) was able to explain some releasers that function in keeping groups together as examples of intraspecific mimicry. In hamadryas baboons signs of females in estrous became appeasing signals which are also present in the males. Males have strongly vascularized skin areas that are similar to the swollen parts of females in estrous, but in specific instances the areas in question are not homologous. Males present their posterior to other group members in the manner of females. This has definitely an appeasing effect. The swelling in these instances no longer have a sexual "meaning." W. Wickler (1965e, 1966b) discovered this same principle in other monkeys, also in carnivores and fish. Males of the spotted hyena (*Crocuta*) present their slightly erected penis during each greeting encounter with others. Females do the same and possess a pseudopenis which looks deceptively like a real penis. This makes it almost impossible even for a trained observer to distinguish males and females by external signs (see p. 112).

The concept of mimicry must be considerably broadened in line with what has been said so far. Not only protective similarities, but

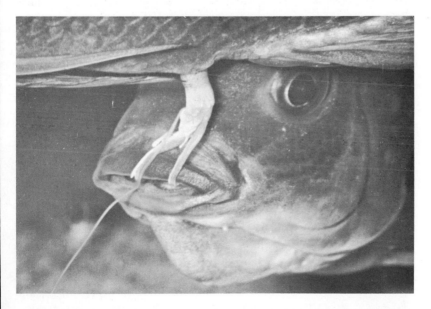

Figure 93. Spermatophore model and spermatophore of *Tilapia macrochir*. The female has just taken the spermatophore-like attachments of the male genitals, including a spermatophore (long filament), into her mouth. (Photograph: W. Wickler [1966b].)

all similarities which involve the falsification of signals, are included (W. Wickler 1964c).

An interesting ethological example of mimicry is discussed by J. Nicolai (1964). The widow birds (*Viduinae*) are breeding parasites of various species of grass finches (*Estrildidae*). Their young so closely resemble the host species with respect to plumage as well as the gape markings in the mouths that the young are raised by the host parents along with their own (Plate VI). The Viduines also imitate the courtship song of the host species in every detail, thus attracting their own females to the correct host-species pairs (p. 24). The larvae of the *Lomechusa* beetle, as guests of ants (*Lomechusa strumosa* and *Atemeles pubicollis*), mimic the begging behavior of ant larvae as well as their attractive odors. Thus they are fed by the hosts in the manner in which they feed their own larvae (B. Hölldobler 1967).

Finally, there are many examples of mimicry in the traditional sense of protective resemblance. Songbirds that had an unpleasant experience with wasps will avoid them and harmless mimics as well, for example, those from the group of Diptera and Lepidoptera (see Plate IV). Toads learn quickly to distinguish a mealworm from a bee, and henceforth will also avoid mimics of bees (L. P. Brower and J. v. Zandt-Brower 1962). Some mimics of wasps imitate not only the appearance but also the buzzing sound of the poisonous model (A. T. Gaul 1952). Some species of bad-tasting moths make themselves known to bats by warning sounds: They click in a special manner when they are hit by a burst of sounds from a bat. Some species of edible butterflies also do this, and they are probably mimics (D. C. Cunning 1968). In experiments bats avoided edible prey when it was presented in conjunction with warning sounds. It is fascinating to observe that a mimic forms several morphes, thus imitating different models (M. Tweedie 1966; see also Plate IV).

The caterpillar of the moth *Leucorampha ornata* raises its posterior end, imitates the head of a snake, and curves its body into an S when it is alarmed (Fig. 94). The snake-head model has two eye spots on its underside which are turned toward the predator. The caterpillar of the moth *Pholus labruscae* imitates a snake with the anterior and posterior ends. The anterior portion reminds one of the snake head. A small black "tongue" which protrudes from a dark-colored area at the posterior end wiggles like the tongue of a snake (E. Curio 1965a).

A remarkable instance of behavioral mimicry has been described by T. Eisert and J. Meinwald (1966). The darkling beetle (*Eleodes longicollis*) defends itself against predators by standing on its head and spraying an irritating substance from the tip of its abdomen. Another darkling beetle (*Megasida obliterata*), which lives in the

Figure 94. Mimicking a snake by the caterpillar of *Leucorampha ornata (Sphingidae)*: top: resting; bottom: when alarmed, turning the underside toward the observer. (After A. Moss [1920].)

same desert region of Arizona, imitates this behavior, although it does not have a defensive secretion.

Many insects have conspicuous, dark eye spots on their wings which they expose suddenly when in danger. These patterns are more feared by several species of songbirds than patterns that are unknown to them (A. D. Blest 1957). The same principle applies to the threat posture of the spectacled cobra (*Naja*). The argument could be made that the eye spot on the wings of butterflies which are suddenly exposed when danger appears should not be considered under the category of mimicry but are merely warning signals which present a sudden optical stimulus. This optical stimulus could then drive off the predator, independently of a similarity with another structure. The observation that songbirds fear "eye spots" more than other unknown patterns could be a result of the greater effectiveness of concentric circles as a pattern on a given plane. They would thus be the strongest stimulus for a mechanism attuned to gestalt perception. There are, however, butterflies whose eye spots also have unsymmetrical "reflections" (for example, *Caligo eurylachus*) which produce a most deceptive similarity with a vertebrate eye. This is perhaps the strongest argument for considering these spots as examples of mimicry (Plate IV). An excellent treatment of these problems, which are only outlined here, has recently appeared in W. Wickler (1968a).

8   REACTION CHAINS

When a living organism responds to a stimulus, the releasing stimulus situation is frequently changed because the animal then comes into a new position in which additional stimuli become effective. We know, for example, that a bee is visually attracted to a piece of colored paper, but rarely will she alight on it. On closer approach she knows from odor cues that no nectar will be found here. If an appropriate odor is added she will land and continue to search. Other stimuli then are needed to release the sucking movements. The bee-hunting digger wasp (*Philantus triangulum*) flies from flower to flower in search of bees and reacts first only optically to moving objects, including small flies, upon which they do not prey. If the wasp perceives a moving object it positions itself leeward from it at a distance of about 10 to 15 cm in the air and tests the wind. If the appropriate odor is present—such as a model with bee odor—the attack is made. However, the wasp only stings if a real bee is present. This reaction could not be released by the model (N. Tinbergen 1935). In all these instances the animal comes into new releasing stimulus situations by its own actions.

The same is true for the behavior sequence of hermit crabs during the selection of snail shells. E. S. Reese (1963a) distinguished eight different fixed action patterns which occur in a specific sequence. This sequence is dictated exclusively by the releasing stimulus situation, and at times a specific movement may not take place. Then the releasing stimuli are not present. Thus, if a hermit crab finds the opening of a snail shell when first encountering it, all those behavior patterns with which normally the exterior of the shell is investigated, are omitted. The animal proceeds at once with the investigation of the interior of the shell with its claws and the first pair of legs. Only then will it slip into the shell and raise it up. Whether or not the hermit crab will show additional appetitive behavior depends on the suitability of the encountered shell (E. S. Reese 1962b, 1963b).

In these cases the behavior is not normally terminated through an action-specific fatigue (p. 56) but by a stimulus situation that cuts it off. A single individual of a schooling fish species becomes calm when it swims within the group and stops searching for a school after it has joined one. In squirrels and agoutis the foodhiding behavior is terminated when the food object is actually buried. If the expected success is not achieved, then renewed efforts to bury it are made or conflict behavior occurs (p. 176).

In cases where two mutually attuned partners are present, for example, sexual partners, they reciprocally release certain reactions from one another, which in turn are releasing stimuli themselves. An especially good example is presented by N. Tinbergen (1951). If a stickleback female appears in the territory of a male, he at once begins a zigzag dance. This in turn releases a special display movement of the female. He then leads her to the nest; she follows; he shows the nest entrance and she slips into it. Then he butts her repeatedly at the base of the tail which still protrudes from the nest; in response to this she spawns. She then swims off; he enters and milts. Each of these stimuli can be imitated by a model. For example, one can remove the male after the female has entered the nest and release spawning by drumming against the base of her tail. If this stimulus is omitted, the behavior chain breaks at this point; the female does not spawn. Whenever the behavior sequence depends on releasing stimuli, the segments of the chain of behavior can be skipped. Even a behavior which has already taken place can be recapitulated by means of the appropriate stimulus configuration.

Figure 95 shows the two chains of reciprocally releasing actions. The sequence of actions of male and female partners has been simplified; in the natural situation it is not so precisely determined. There are many deviations from it, but the sequences are by no means random (G. P. Baerends and others 1955; D. Morris 1958).

The relationships are also very clear in water salamanders (*Triturus*) and anurans (*Bufo, Rana, Hyla*). In the toad (*Bufo bufo* L.) males and females migrate to the spawning sites in the spring, which they find, strangely enough, with the aid of their good memory for places. While still en route males react to all moving objects, jump at them, and try to clasp them. If it is a male toad, it will protest against this attempt with a quick succession of calls, and

Figure 95. Courtship behavior of the three-spined stickleback illustrating the mutually releasing actions of the male and female. (After N. Tinbergen [1951].)

the clasping grip is released. Females, on the other hand, remain quiet and continue to be clasped, the same as a quiet model, such as the finger of the experimenter (p. 74). If the male arrives unmated at the pond and if nothing stirs nearby, it begins to call, which attracts females. If it perceives a movement nearby, it again approaches indiscriminately everything that moves and clasps it. As on land, the further behavior of the clasped object determines the subsequent reactions of the male. He clasps the female behind her front legs; the hind legs assume a position from which the males can kick against all rivals to keep them away. The pair remains together until the female gives a signal by assuming a pronounced lordotic position. The male then slips back and forms a "basket" with its hind legs in front of the cloacal opening of the female, where the discharged spawn is collected and fertilized (Fig. 96). In this manner several spawnings take place, interrupted by intervals during which the female swims about with the male, again in the former clasping position. Finally there is a signal

Figure 96. Top: toad pair shortly before spawning; below: signal posture of the female and the "basket" position of the hind legs in the male. (After I. Eibl-Eibesfeldt [1953].)

posture that is not followed by spawning. The male again forms the "basket" posture, but since no spawn appears, the clasp is soon released and the male dismounts. If the female should then again be clasped by a male, she will behave like a male, making the appropriate rejection movements with her hind legs but remaining silent. The rhythmic movements are sufficient to repulse a clasping male (I. Eibl-Eibesfeldt 1950b, 1954, 1956a; H. Heusser 1960).

In water salamanders the males must be excited by the species-specific odor substance of the females before they will react further. One has only to put water from a container holding a female into one holding a male, and the latter will then react to a simple, moving model which until then has not been attended to (H. M. Zippelius 1949; H. F. R. Prechtl 1951). The male makes an olfactory examination and then blocks the path of a conspecific female and begins to fan odor substances toward the female with tail movements. If she approaches him, he turns around and waddles off slowly with his tail bent. The female follows and butts her snout at his tail. This is the signal to deposit the spermatophore: The male raises the tail, deposits the sperm packet, opens the cloaca, and leads the female in a straight line across the spermatophore, which then becomes attached at the female cloacal opening (J. Marquenie 1950; H. F. R. Prechtl 1951; I. Eibl-Eibesfeldt 1955c; see also Figs. 97 and 98).

Botanists know of comparable reaction chains. In the mushroom (*Achlya ambysexualis*) male mycelia begin to form antheridial cells

Figure 97. Deposition of the spermatophore and leading seen from below. After depositing the spermatophore the male opens his cloaca wide and is followed by the female. (From Scientific Film C698, I. Eibl-Eibesfeldt [1955], photograph: H. Sielmann.)

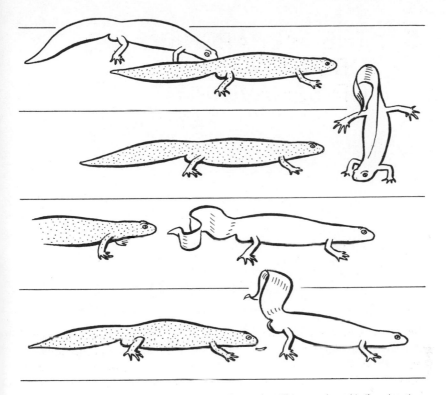

Figure 98. Courtship of the Alpine water salamander (*Triturus alpestris*) (female stippled). The sequence is to be viewed from top to bottom: Olfactory investigation of the female, blocking her way and wagging the tail, releasing the spermatophore, and leading. (H. Kacher, artist.)

when they come into contact with a substance (A) which is produced by female mycelia. The antheridial cells in turn produce a substance (B) which stimulates the female mycelium to produce oogonia. The oogonia in turn produce a substance (C) which attracts the antheridial cells chemotropically and which causes the closing off of the antheridia once the union with the oogonia has taken place. A substance (D), which is produced by the fully developed antheridia, stimulates the closing off of the oogonia, but only after direct contact with the antheridia (J. R. Raper, cited in M. Hartmann [1956]).

9 THE HIERARCHICAL ORGANIZATION OF BEHAVIOR

Behavior patterns occur in a specific order. In the reaction chains discussed in Chapter 8, the orderly sequence of the various actions was dictated by the relevant releasing stimulus situation. As was shown by the example of the squirrel hiding its nuts and the spider building its net (pp. 22–23), there are also endogenously programmed action sequences. Scratching is followed by placing of the nut, butting with the snout, covering, and stamping down, even though the squirrel has not previously dug up the ground (p. 23). Each individual movement has its fixed position in the total sequence, and each individual movement is in turn an internally programmed sequence of various muscle contractions.

The orderliness of behavior does not only consist of a temporal sequence but also of temporal parallelisms. Behavior patterns may be more or less coupled with one another, or they may exclude one another. We can also observe that behavior patterns are ordered in sets, in which each set is distinguished by a common fluctuation of the threshold for releasing stimuli. In an animal that is in a fighting mood, for example, we can see that the behavior patterns of threat, attack, biting, and so on, can be more readily released than at other times, for example, when the animal is eating. Other behavior patterns again, such as eating and nest-building behavior, are inhibited at this time. This is an indication that the behavior patterns are grouped and depend on higher, coordinating organizations, which influence one another mutually in specific ways.

162

Thus a male squirrel is not only more ready to court during the rutting season but is clearly more aggressive as well. Which behavior patterns are activated at a specific time, whether those of courting or of fighting, depends on the releasing situation. However, the male has a lowered threshold for both kinds of behaviors. We know that this is due to the influence of male hormones. In many birds the behavior patterns of nest building, courting, and fighting are similarly organized during the breeding season.

This order with respect to sequence and simultaneity reflects at the same time an hierarchical order of behavior in which several levels of integration can be recognized. An example may illustrate this: In the early summer, hatching digger wasps (*Ammophila campestris*) come into a reproductive condition and are then ready to mate and care for their brood. This mood of caring for brood consists of a number of specific drives: nest-site selection, nest building, hunting for caterpillars, egg laying, feeding of larvae, and opening and closing of the nests. Each of these specific drives in turn consists of chains of individual actions that are controlled by specific releasing stimuli. Thus the digger wasp will first search for a nest and will begin scratching and biting in order to build a nest only when she has found a suitable place. The loose sand that she dug up is carried away. When the nest chamber is completed the entrance is closed off with a clump of dirt of the proper size. Then the wasp begins a new activity. She searches for and kills a caterpillar. After that the mood to retrieve supplants the mood to hunt, and the innate actions of transporting, dropping the caterpillar before the nest, opening the nest, entering, turning within, grasping, and pulling in of the caterpillar follow. Finally, the digger wasp will deposit an egg and close the nest. Thereafter she will repeatedly visit the nest, and once the larva has hatched, she will feed it at first with small and later with larger caterpillars; when the larva pupates, the wasp closes the nest for the last time. She is able to adjust her behavior according to the demands of the situation; she will bring fewer caterpillars if some are left in the nest, or she will bring small ones when the larva itself is still small (G. P. Baerends 1941; see also p. 218).

Observations show that there are dominant and subordinate instincts. N. Tinbergen (1951) has illustrated this in his schema of the "hierarchy of instincts." He developed his ideas during his studies of the reproductive behavior of sticklebacks.

In the spring, the male stickleback comes into a reproductive mood as the length of day increases. But the change to reproductive coloration does not take place suddenly, and he shows no courting or fighting behavior. Instead, the fish migrate together peacefully in swarms from their winter habitat in deeper water to warmer and more shallow water. There each male establishes a territory, an area containing some water plants. Only when a territory has been selected does the fish acquire reproductive coloration and become receptive to new stimuli. He fights or threatens when another male appears; he courts females and builds a nest if he can find suitable material. What he will do specifically depends on the releasing stimulus situation, but he is in a state of internal readiness to perform all these activities. Fighting behavior is activated by the appearance of a red-bellied male, but which specific fighting actions take place again depends upon still more specific stimuli. If the intruder flees, he is pursued. If he beats his tail, the territory owner reacts likewise. The red male releases the readiness to fight but not the actual fight itself. One can recognize several levels of integration which lead from a more general to a more specific behavior, and certain key stimuli activate the next more specific action system or mood. For example, if one collects several migrating sticklebacks and places them into a bare tank, they remain in a group without change of color, because no territorial borders can be defined with respect to landmarks. If one adds some plants in a corner, one male will remain there, change color, and he is then in reproductive condition. He is ready to court, fight, or build a nest in response to the appropriate stimuli.

N. Tinbergen assumes that this observed order reflects an order of functional organizations within the central nervous system (Fig. 99). He speaks also of a hierarchy of center. So the term "center" is purely functionally defined. Hormones affect the highest center responsible for reproduction—the migrating center—and cause appetitive behavior in the form of migration. There seems to be no specific key stimuli for this. The appetitive behavior of migration ends when the fish perceives the key stimuli in a specific biotope. These, in turn, affect a specific innate releasing mechanism which frees the next main center for propagation that was blocked until then. Impulses can now pass to the lower centers, such as care of brood, courting, nesting, and fighting, but each of these centers is blocked until specific key stimuli release the behavior, for example, when a rival appears. The rival must then provide still more specific stimuli before the specific fighting behavior patterns are released.

More recent investigations by P. Guiton (1960) compel us to modify this model somewhat. According to Tinbergen, migration, establishment of territory, and reproduction follow in a sequence, but fighting, nesting, courting, and care of young are parallel in time. According to Guiton, establishment of territory, digging a pit, nest building, courting, and care of young follow one after the other.

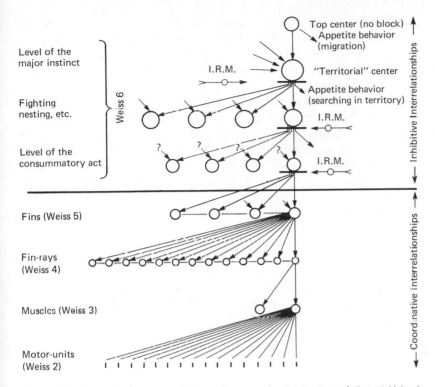

Figure 99. Hierarchical centers of the major reproductive instinct of the stickleback male. Motivational impulses are represented by straight arrows which "load" the centers (shown here as circles). These impulses may come from the external environment as well as from superordinated centers, or they may occur spontaneously within a center (this is not considered in this schema). The blocks indicate inhibiting influences, which prevent a continuous discharge of motor impulses. These blocks are removed by innate releasing mechanisms. When this has occurred the animal will show a specific appetitive behavior until more specific releasing stimuli activate the next subordinated instinct and the still more specific appetitive behavior. The concave, two-pointed arrows between centers of the same level indicate mutually inhibiting relationships and the existence of displacement activities (p. 177). Below the level of the consummatory acts a number of centers come into action simultaneously. The relation between subconsummatory centers of the same level is indicated by horizontal lines. (Additional explanation in the text.) (After N. Tinbergen [1951].)

After a territory has been established, the male digs a pit, and only this releases the carrying in of nesting material and glueing. If the pit is covered, digging is gradually reactivated, but the fish does not dig as long as before and soon begins to carry in nesting material. He is not fully ready to mate until he has dug a tunnel into the nest. These new findings do not, however, change the basic principle of Tinbergen's schema, and for this reason the schema is reproduced here unchanged, if only for historical reasons.

Based on neurophysiological considerations P. Weiss (1941a) independently developed the theory of the hierarchy of the central nervous system in which he distinguishes six levels of integration. The lowest level represents the single motor unit (1). This is followed by all the motor units of one muscle (2); next the coordinated function of muscle groups which move a joint (3); followed by the coordinated movement of one extremity (4); then the coordinated interaction of several appendages (5); and finally the movement of the entire animal (6). The sixth level includes, as Tinbergen shows, several levels of integration. In Tinbergen's schema three of Weiss' levels are represented. The horizontal line is meant to separate the fixed action patterns (consummatory actions) from the more simple and subordinated movement coordinations.

Tinbergen defines *instinct* as

a hierarchically organized nervous mechanism which is susceptible to certain priming, releasing and directing impulses of internal as well as of external origin, and which responds to these impulses by coordinated movements that contribute to the maintenance of the individual and the species.

He distinguishes between major and subordinated instincts.

In complete agreement with this statement is W. H. Thorpe (1951:3), who speaks of an instinct as

. . . an inherited and adapted system of coordination within the nervous system as a whole, which when activated finds expression in behaviour culminating in a fixed action pattern. It is organised on a hierarchical basis, both on the afferent and efferent sides.

When charged, it shows evidence of action-specific-potential and a readiness for release by an environmental releaser.

There are several ways of discerning the functional organization of the various behavior patterns. Whether or not a behavior sequence depends upon external stimuli or is programmed within the system can be determined by manipulation of the releasing situation, as was illustrated by the examples given earlier. The study of fatigue phenomena also illuminates these relationships. For example, if a behavior pattern A repeatedly occurs and a behavior B also shows fatigue but the performance of B does not influence the occurrence of behavior A, we have an indication that the actions are ordered in a sequence A-B, in which B is dependent on A. If on the other hand, an increase in the threshold of A is accompanied by an increase in the threshold for B, and if the performance of B is followed by an increase in A, then there is no strict hierarchical organization between both actions, but both are dependent upon a common higher stage (A. Kortlandt 1955). Further discussion about the hierarchical organization of behavior can be found in G. P. Baerends (1956) and R. A. Hinde (1953). Hinde emphasizes that the hierarchical order is not only manifested in simple, linear relations but in a network of relations. How this is to be understood will be discussed in the experiments by E. v. Holst and U. v. Saint-Paul (1960).

The conclusion that the organization of behavior of intact animals is a reflection of an organization within the central nervous system should be obvious. In attempts to understand this organization, experimentors at first used surgical techniques. However, the effects of lesions allow only a rough approximation of localizing certain behaviors within specific areas in the brain, and contradictory results lead to extreme interpretations. While localization theories assume the existence of strictly defined anatomical centers for certain functions, the theory of plasticity (mass action) does not contain such a strict localization of function in specific anatomical structures. For those favoring a localization hypothesis, the brain is a more or less fixed mosaic of individual centers, each responsible for specific activities. According to the mass-action theory the brain functions as a whole, whose parts have only partially overlapping functions. Both groups can base their conclusions on experimental results (see the discussion by J. Dembowski [1955]). K. S. Lashley (1929, 1931), for example, found that the learning performance was proportional to the extent of the amount of cortical lesions. Thus the location of removed cortical tissue was less important than the amount, and the learning abilities depend less on a localized anatomical structure than upon a more widespread cortical mechanism. F. A. Beach (1937, 1938, 1940) obtained similar results in the study of reproductive behavior of rats. The percentage of male rats copulating decreased with an increase in the size of the brain lesions. There were other findings, however. Small lesions in the cingulum, for example, disrupted the parental care behavior of rats quite substantially. The same is true for food hoarding (J. S. Stamm 1954, 1955). H. F. Harlow (1953) speaks against Lashley's concept of mass action. According to O. L. Zangwill (1961), there is not necessarily any contradiction. Undoubtedly there exist specific brain areas where especially instinctive behavior (emotional behavior) is localized in specific structures in the brain stem. In respect to other functions the number of ganglion cells is important. This is true with certain exceptions for the relationship of the cortex to learning and intelligent behavior.

An entirely new impetus was given to brain research by the investigations of W. R. Hess (1954, 1957), who activated many behavior patterns in the brains of largely intact cats by means of electrical stimulation with fixed electrodes. He was able to localize points in the midbrain of the cat where he could activate threat, escape, eating, and other behavior patterns. These points were concentrated in certain areas but were also widely distributed, as could

- General Locomotor
  Activity
⁄⁄ Relaxed Mood
⁄⁄⁄ Sleepiness
||| Eating
= Rage–Fear

\\ Fear
∨ Vomiting
▲ Defecation
○ Urinating

Figure 100. (a) Schematic representation of the cat brain showing the stippled area of the brain stem which was stimulated in experiments by W. R. Hess; (b) the same region enlarged. The various symbols mark the areas in which various moods were released. (After W. R. Hess, simplified by E. v. Holst [1957].)

be expected in such a complicated "wiring system" (Fig. 100). Cats raised in social isolation displayed upon electrical hypothalamic stimulation all the patterns of attack behavior that normally raised cats demonstrate in the same situation. The brain mechanisms underlying these patterns are probably innately organized (W. R. Roberts and E. H. Bergquist 1968). F. Huber (1955) was able to determine the nerve centers of some instinctive behavior patterns in crickets by brain stimulation.

By registering the activity from single neurons B. R. Komisaruk and J. Olds (1968) demonstrated a direct relationship between neural activity and observed behavior in rats. The discharge of single neurons was always coupled to definite activities, such as feeding, twitching of vibrissae, sniffing, and exploring.

An entirely new road was taken in pursuit of the question of the order within the central nervous system by E. v. Holst and U. v. Saint-Paul (1960), who used the method of electrical brain stimulation to investigate the hierarchical organization of behavior. They used electrodes that could be lowered into the brain in discrete small steps. They studied tame domestic chickens that were habituated to the experimental situation. By trial and error they placed several electrodes into the brain stem of a chicken and were able to stimulate several separate points at the same time. One group of behavior patterns, out of the many they were able to elicit, is of special interest and had the following characteristics: (1) The behavior sequence, in which various individual components occurred, remained the same even during long-lasting electrical stimulation or when higher voltages were used; (2) persons who are familiar with the species consider the movement coordinations to be "natural"; and (3) the behavior that was elicited consisted of functional patterns that have adaptive value.

It appears as if they had activated an instinct the way Tinbergen defined it, individual behavior components appearing in an orderly sequence that seems dependent upon the different thresholds required to release them. Thus the stimulation of a certain area of the brain stem releases blinking of the left eye in a rooster. If the stimulus is continued or its strength increased, the rooster will shake his head, and if stimulation is continued further, will rub his head on his shoulder. Finally he scratches his left cheek with his foot. Head shaking and scratching are then repeated as long as stimulation lasts; it appears as if the rooster is bothered by an invisible fly.

In a similar manner the authors could release the entire "disgust reaction" in chickens. Here the neck is at first stretched forward and the head is bent so that the beak points to the ground. The beak is opened slightly, and the tongue moves. During continued stimulation saliva is secreted, as if an unpalatable object had to be washed away. Finally the chicken shakes its head and scratches itself. When the stimulus is terminated the chicken makes a final wiping motion on the ground (Fig. 101a). It was possible to release quite complicated functional behavior sequences such as escape from an aerial or ground predator as well as quite simple reactions such as cackling. If only the cackling drive was activated, the animal continued to perform this behavior even when prolonged or stronger stimulation was applied. Finally a kind of fatigue set in, and a stronger stimulus was required to continually release the behavior.

During other experiments this cackling was merely a component of a more complicated behavior. The animal began to cackle when the stimulus was turned on; with an increase in strength the animal became restless, walked around, showed orienting head movements and finally flew off with calls of fear, just like a chicken that has been frightened by a ground predator (Fig. 101 b and c). In this case the escape drive has been activated whose individual parts appear in the proper order when their respective thresholds have been reached. If stimulation begins with high voltage the animal flies off at once. This escape behavior, in turn, can be a part of a drive of a still higher level of integration. E. v. Holst and U. v. Saint-Paul sometimes released only a restless wandering about. In these cases they tried to find out, through presentation of various releasing stimuli, whether they had activated a general motor restlessness or a specific appetitive behavior. In one experiment they found that

Figure 101. (a) Activation of a behavior sequence from one stimulus field. The whole complex serves to remove something unpleasant from the beak. The individual acts marked with black dots can also be obtained in isolation from other fields of stimulation. The increasingly heavier line denotes the strength of the stimulus and its duration. Several seconds after cessation of the stimulus the chicken wipes its beak. (b) Behavior sequence of fleeing from a ground enemy with slowly increasing stimulation of the field. (c) Reaction to a sudden strong stimulus. (From E. v. Holst and U. v. Saint-Paul [1960].)

the chicken was not ready to eat, drink, or court, but when presented with a fist is threatened lightly. If presented with a stuffed predator, a polecat, the chicken at once threatened and attacked. When the polecat was knocked to the ground, the chicken retained the threat posture (Fig. 102).

The conclusion that one draws is that an "attack drive" against

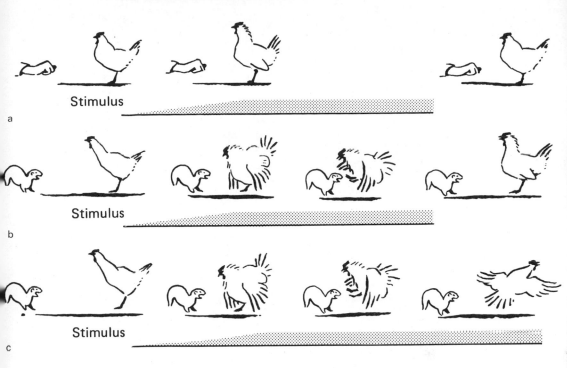

Figure 102. Centrally released ground-enemy behavior. Without a suitable object the stimulated hen shows only locomotory unrest. (a) Toward a fist she shows only slight threatening. (b) A stuffed, motionless polecat is vigorously threatened and attacked; if the stimulus ends at this moment, the hen remains standing and threatening slightly. (c) If it does not end, she checks and flees screeching, unless the polecat drops to the floor. (From E. v. Holst and U. v. Saint-Paul [1960].)

ground predators has been activated. An additional experiment, however, shows that this is not the case. If the stuffed polecat is fixed to the table so that it cannot be knocked down, the chicken will behave quite differently. After the initial attack, which has failed to dislodge the predator, the animal then calls and flies off if brain stimulation is continued. Thus "behavior directed against ground predators" has been activated, including attack as well as escape. The alternatives are ordered according to their different releasing thresholds. A very strong brain stimulus can also cause an attack to be switched to escape.

In the last three experiments cackling had been activated once by itself, another time as part of an escape drive, and finally as part of the activated ground-predator behavior complex. These levels seem to represent three stages of integration, which if arranged in a logical order, express a hierarchical organization (Fig. 103). This can be represented in the form of a diagram where additional experi-

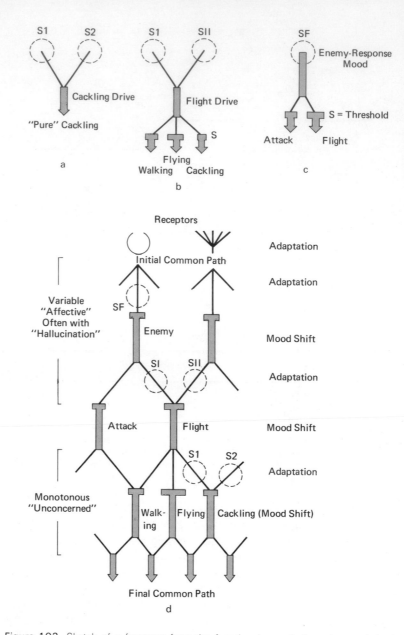

Figure 103. Sketch of a fragment from the functional organization of some behavior patterns in the fowl. On the top the behavior patterns referred to in the text, arranged side by side, which can be activated from one or more stimulus points. Below, the partial actions have been arranged in some kind of order. This order results in a diagram that illustrates how several paths lead to one and the same final action. Walking may be once a part of the activated escape drive, another time part of the feeding drive, and so on; just as sitting may be pure or part of the sleeping or incubation drive. Change of mood and adaptation can occur at various levels of the central nervous system. When the behavior patterns of a higher level of integration are activated, which are closer to the sensory areas, then the animal behaves "affectively;" for example it behaves as if it were afraid. When actions of a lower level of integration are activated, the animal behaves with indifference. (After E. v. Holst and U. v. Saint-Paul [1960].)

ments reveal details about the particular interrelationships. One and the same behavior can often be activated from two different stimulus points. By appropriate experiments one can discover in which relation to one another these points are located within the functional system. At first one tries to find out if there is a connection between the two points by stimulating both points simultaneously with stimulation of intermediate strength, which from one point alone releases only a moderately strong reaction. If the simultaneous stimulation results in a stronger response, one can conclude that the excitations from the two points flow together somewhere in the central nervous system. A behavior pattern can also be fatigued by repeated stimulation from a particular point; in order to release the behavior, stronger stimuli are required. If stimulus point 1 no longer releases a reaction and point 2 is stimulated, then the behavior can again be elicited with full intensity. In this instance only a pathway leading to the center solely from point one has been locally affected, but the stimulation did not affect the motor center.

The "fatigue" phenomenon at the point of stimulation was called a "central local adaptation" by E. v. Holst and U. v. Saint-Paul, and they distinguished it from a "change of mood," which is a different process. Repeated stimulation can result in a change of the basic mood of an animal. An animal that spontaneously cackled "angrily" could be induced to sit down after repeated activation of the drive to sit, so that it did not get up again after stimulation ceased. Mood shifts, local adaptation, and the possibility of activating behavior from several stimulus points were observed by the authors at various levels of integration. This is represented in the diagram (Fig. 103). Here the three experiments that I have discussed above (cackling, escape drive, and reactions to ground predator) have been represented in a specific order. The resulting diagram—the "functional organization of drives" (Wirkungsgefüge)—shows not only the hierarchical organization but also a network of connections. One and the same final action is the result of several initial approaches. We know from observations that sitting down, for example, can be activated once as part of the drive to sleep and another time from a drive to brood, just as running or flying can be a part of several drives. K. Lorenz called behavior patterns that are involved in several drives behavioral tools.

E. v. Holst's organization of drives emphasizes the interrelationships more than Tinbergen's hierarchical schema which is primarily linear, although even here, cross connections are shown by arrows pointing to different levels of integration. v. Holst was also concerned with the histological mapping of the various stimulus points, but his untimely death interrupted this work.

Brain stimulation studies on the opossum (*Didelphis virginiana*) achieved in principle the same results (W. W. Roberts and others 1967). The hierarchical organization, however, was much less pronounced, which might reflect the organizational difference between the avian and mammalian brain.

Hierarchical organization of behavior can be recognized in many invertebrates and in all classes of vertebrates. As one ascends the phylogenetic scale, this order becomes less and less linear. In the prey-catching behavior of the cat a linear arrangement of behavior components occurs only if an animal is hungry: lying-in-wait, stalking, catching, killing, and eating. But a satiated cat will also capture prey without eating it, and as mentioned earlier, each partial component of a sequence can become a final consummatory act because it has its own specific motivation. The other behavior components then become appetitive behavior in the service of this final action. P. Leyhausen therefore called this a relative hierarchy of drives (moods).

10   CONFLICT BEHAVIOR

Sometimes a stimulus situation will activate several drives simultaneously, for example, the drives to attack and to flee. Such opposing
behavior patterns come into conflict with one another and the resolution can take various forms. By using their method of electrical
brain stimulation E. v. Holst and U. v. Saint-Paul (1960) have
examined this problem in domestic chickens by simultaneously activating two opposing drives from two different stimulus points. In the
most simple cases the activated behavior patterns are superimposed
on one another (*superposition*), for example, pecking and head
turning. We have already discussed such a superposition in normal
animals with the example of the expressive movements and facial
expressions in dogs (p. 92). In *averaging*, two behavior patterns are
also superimposed but their intensity changes. If watching out with
a stretched-out neck and looking around with widely-sweeping head
movements are activated together, the result is watching out with a
still more stretched-out neck and looking around with less extensive
sweeping head movements. The simultaneously activated behavior
patterns may also be expressed in *alternation* according to a pattern
a-b-a-b-a-b. Such ambivalent behavior we know from the zigzag
dance of the stickleback. These instances have also been referred to
as successively ambivalent behavior, in contrast to examples of simultaneous ambivalence, where both activated systems are simultaneously
expressed. I do not think that this is an accurate choice of terminology,
because several inappropriate things are combined. "Successive"
versus "simultaneously" refers to the expression of the behavior;

"ambivalence" refers to the two simultaneously activated motivational systems. Someone might interpret the term "successive ambivalence" as referring to successive internal drives. E. v. Holst and U. v. Saint-Paul obtained alternation when they activated watching out and eating at the same time. Opposing behavior patterns such as turning right and turning left *cancel* one another.

An especially interesting aspect of behavior has been called *transformation*. If attack and escape are simultaneously activated by electrical stimulation an entirely new behavior pattern appears: The chicken runs about with fluffed feathers, calling loudly, and this is the behavior of an incubating hen if one approaches her nest. However, the term "transformation" must be used with caution. What might appear to be transformation upon superficial consideration may have other physiological causes. For instance, if in a hungry animal the escape drive is activated and at the same time the sleeping drive, which normally suppresses the drive to escape, the result is that the chicken will eat before falling asleep. Formally expressed, $a + b = c$, but in actuality c is suppressed by a, and a is suppressed by b; hence c is liberated.

We speak of masking if one behavior suppresses another without preventing it from occurring altogether. If a cackling chicken receives a brain stimulus which releases the drive to sit, it will sit down and cease cackling. If both stimuli are turned off, the chicken cackles briefly, the suppressed action thus reappearing briefly. The suppressed drive existed latently but had been blocked somewhere before the motor areas had been affected. If such an after discharge does not occur, E. v. Holst and U. v. Saint-Paul speak of *preventing* (Fig. 104).

Which behavior pattern will suppress another depends generally on the strength of the stimulus, but not exclusively so. Some dominant activities suppress others, even if they are expressed only with low intensity. This is true for several escape reactions, especially watching out and freezing.

In some cases the effect of various behavior patterns on one another was measured quantitatively. It was found that the "disgust reaction" only barely raised the threshold for head turning, but the threshold for pecking, which is part of the feeding system, is sharply raised. Thus head turning is more readily combined with the "disgust reaction" than with pecking behavior. Finally, an activated behavior can influence another behavior in such a way as to make it disappear altogether. If a spontaneously scolding hen is stimulated long enough to sit down, she will eventually calm down (Fig. 105).

Ethologists have observed very similar phenomena when observing intact animals and have interpreted the behavior correctly. One example that did not occur in the experiments discussed above will be considered. A. Kortlandt (1940) and N. Tinbergen (1940, 1952)

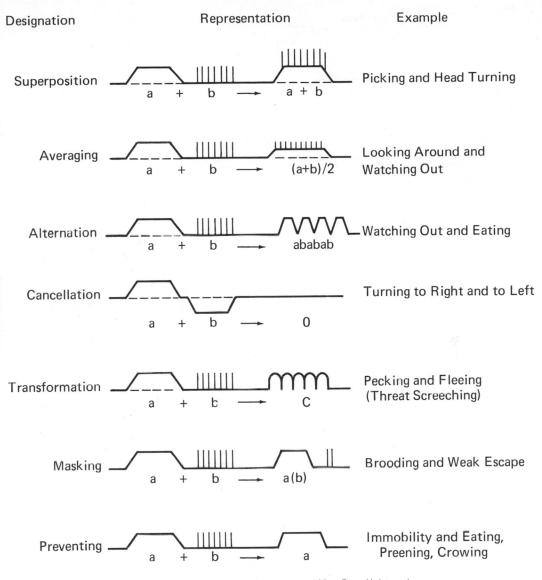

Designation	Representation	Example
Superposition	$a + b \longrightarrow a + b$	Picking and Head Turning
Averaging	$a + b \longrightarrow (a+b)/2$	Looking Around and Watching Out
Alternation	$a + b \longrightarrow ababab$	Watching Out and Eating
Cancellation	$a + b \longrightarrow 0$	Turning to Right and to Left
Transformation	$a + b \longrightarrow c$	Pecking and Fleeing (Threat Screeching)
Masking	$a + b \longrightarrow a(b)$	Brooding and Weak Escape
Preventing	$a + b \longrightarrow a$	Immobility and Eating, Preening, Crowing

Figure 104. Types of combinations of various behavior patterns. (After E. v. Holst and U. v. Saint-Paul [1960].)

discovered independently of one another that animals frequently show behavior patterns in conflict situations which cannot be attributed to competing drives. They called these behavior patterns *displacement activities* (Übersprungbewegungen). This concept is based on the hypothesis that the observed behavior patterns are not activated from their normal source, that is, autochthonously, but receive their excitatory potential allochthonously from those drives which

a

Stimulus I   Stimulus II

b

Stimulus II

Figure 105. A fowl, which in consequence of a preceding fleeing stimulus cackles spontaneously and without stopping (kut, kut, kudaw, kut, kut, kut kudaw! . . . , suggested by the upper graph). Stimulus I produces watching out; stimulus II produces sitting down. When the second stimulus is applied long enough, a change of mood takes place. (From E. v. Holst and U. v. Saint-Paul [1960].)

have been prevented in their expression by the conflicting situation in which the animal finds itself. The dammed-up excitation sparks over, so to speak, into another channel and there finds its discharge. We shall return to this hypothesis, but first the phenomenon itself will be described. Fighting roosters, which are inhibited from attacking by the simultaneously activated tendency to flee, both behaviors being released from their opponent, begin to peck at the ground. The same is done by male turkeys (H. Räber 1948). Deer engage in sham feeding when the tendency to remain conflicts with a tendency to flee (D. Müller-Using 1952). Quite often cleaning movements, shaking, bill wiping, bathing movements, and others dealing with the care of the body appear as displacement activities. Fighting starlings preen themselves vigorously between bouts of fighting. Similarly, many mammals scratch themselves in conflict situations. Sticklebacks show displacement digging when they threaten one another head down near the boundary of their territories, until deep pits result. It has been shown that the fish are in a conflict between attack and escape, and the drives are in balance. Each stickleback attacks more readily the nearer he is to the center of his territory. He is more ready to flee outside it. However, it is possible to subdue a stickleback inside his territory with a model that can resist his attacks. He then hides between the plants. If one does not move the model for some time, he will again attack as his aggressive drive becomes stronger. He will come out of his hiding place and before attacking again will show displacement digging (N. Tinbergen 1952).

Cormorants exhibit sham incubation during intervals of fighting (A. Kortlandt 1940). Avocets even assume a sleeping posture by placing their bill between the feathers in a conflict situation (G. F. Makkink 1936). Herons show prey-catching behavior when they are sexually excited (J. Verwey 1930). Since specific displacement activities occur in particular conflict situations, they have frequently become ritualized as expressive movements (displacement preening of ducks, p. 108). N. Tinbergen (1952) interprets the relieving ceremony at the nest in black-headed gulls and herring gulls as a displacement activity. The birds bring nest material to their partner. They are in an incubating mood, but the place where they could incubate is occupied. In this conflict the animal picks up nest material.

Man also shows comparable conflict behavior. R. Seiss (1915) extensively studied the behavior of speakers. A speaker exposes himself before his listeners, so he feels isolated. Escape reactions, which cannot be discharged are also activated, because the speaker is in a situation that does not permit withdrawal. This motivation to flee can in extreme cases lead to behavior reminiscent of neuroses such as sweating, trembling, and restlessness. In most cases the speaker will adjust to the situation, and several paths are open to him. He can be strictly formal and stick to his subject matter, and in this way his interaction with the environment is greatly reduced, or he may escape into his lecture by speaking to himself. He can modify his exposed position by belittling his achievements, by appearing humble and exhibiting behavior that indicates a readiness for contact, such as friendly smiles and submissive behavior. Autochthonous behavior patterns are activated, for example, those of seeking comfort (clutching the lectern, and so on). Finally, the conflict tensions lead to a large number of displacement activities. These fall into the categories of bodily care, eating, and sleeping behavior.

Into the first category belong wiping, rubbing, and scratching movements, bringing the open hand around the neck and nape of the neck, stroking the beard, even when no beard is present (!), stroking back the hair on the head, or scratching the head. In the category of eating behavior one finds biting, chewing, sucking on objects (pens), spontaneous chewing movements, licking, and swallowing. Finally, many learned behavior patterns appear in man as displacement activities, such as fondling of the tie, rhythmic pushing of the button on a ball-point pen, and many more.

Displacement activities occur not only when antagonistic drives are activated, but also when the "goal" of a behavior sequence has been reached too quickly, such as when a rival with whom the animal has been fighting leaves prematurely, or when an expected stimulus does not appear, or when a female does not follow a leading male. Tinbergen explained displacement activities with the concept of central nervous energy (p. 58); an energy surplus that cannot be discharged into its normal channel will then flow over into another channel and discharge itself in an irrelevant activity. The dammed-up energy "sparks over," so to speak, from one center to another, according to Tinbergen. P. Sevenster (1961) pointed out that in the stickleback, displacement fanning can come about by disinhibition analogous to the example described on page 176. Two drives, each independently inhibiting a third one, mutually inhibit each other in a conflict situation, loose their inhibiting capacity on the third, which is then free to be discharged.

It appears as if the choice of the term "displacement" was premature; thus for a long time all preening behavior that occurred in conflict situations was at once called "displacement preening." But when threat postures are exhibited in conflict situations feathers or hair becomes disarrayed, which presents the proper stimuli for grooming or preening behavior.

In principle all displacement activities can be explained by the disinhibition hypothesis. Tinbergen's overflow hypothesis as a basis for displacement activities has not been proved or disproved to date. Displacement activities deserve our special attention because they are interesting in themselves, independent of any interpretation of their internal mechanism.

When a releasing stimulus simultaneously activates and inhibits a behavior pattern, the result need not always be a conflict movement (B. Grzimek 1949a). An animal that is attacked or threatened by a higher-ranking one does not necessarily challenge the dominant animal but redirects his aggression against a still lower-ranking one, who in turn can pass it on. M. Bastock and others (1953) proposed the term *redirection activities*. Another example of redirected behavior is the grass pulling of herring gulls in border disputes, which Tinbergen originally interpreted as displacement activity.

11 GENETICS OF BEHAVIOR PATTERNS

What is meant by inheritance of behavior has been discussed in detail (pp. 19ff.). Developmental "blueprints" are passed on from generation to generation in the genome. They determine the prospective potentials, not all of which may be actually realized. What develops can be influenced by the environment up to a certain degree but not in all directions with equal magnitude. Instead, ranges of modifiability are inherited. The biologist can demonstrate inheritance in several ways (pp. 22 and 187). One of these possible ways is the tracing of the course of inheritance, but the genetics of fixed action patterns has not been very well studied. More is known about the genetics of movement anomalies in dancing mice, shyness and aggressivity in mice and dog races, the fighting prowess of rats, and so on. The reason for this dearth of studies of fixed action patterns may be that species and subspecies which are readily crossed are generally not qualitatively distinguished in their instinctive movements. The observed differences often are quantitative. One rat population may be slightly more ready to fight than another; another may learn a little faster (R. C. Tryon 1940); and so on. These examples are presented especially in the books by J. L. Fuller and W. R. Thompson (1960), as well as J. P. Scott and J. L. Fuller (1965), all of which deal with behavior genetics.

There is a certain tendency to trace behavioral effects to genetic effects on sensory organs and adaptive morphological characteristics. This certainly holds true in many instances. The preferred temperature ranges of various races of mice were explained by K. Herter and K. Sgonina (1938) as a result of differences in skin characteristics of these animals. Undoubtedly there exist many pleiotropic effects of sense and motor organs, but from the point of view of behavior genetics such findings are rather "trivial," in the words of E. Caspari (1964). It is more interesting to trace the inheritance of qualitatively different fixed action patterns in closely related species.

W. C. Dilger (1962) crossed the parrots *Agapornis roseicollis* and *A. fischeri*, which are well distinguished by the manner in which each species transports nesting material. *A. roseicollis* tucks strips of nesting material cut from leaves or paper under the rump feathers, which have small hooks and can hold them in place. *A. fischeri* carries the nesting material in the bill. The F_1 hybrids cut strips from leaves in the manner of their parents and try to tuck them under their rump feathers, usually failing. They show the usual tucking movements, but do not let go of the strip. After repeated attempts they finally drop the nesting material and cut a new strip. Often they perform the tucking movements at the wrong place, against their breast, say, or they do not press the feathers down tightly enough against the tucked-in strips. Finally, the movements of tucking in nesting material often change into those of preening the feathers, or the animal carries the strips in its bill, and eventually gives up all attempts to carry nesting material by tucking it in between its feathers. The hybrids show a mixture of behavior patterns; unfortunately, they could not be paired, perhaps because they are sterile.

G. Osche (1952) crossed two races of nematoda *Rhabditis inermis inermis* and *R. i. inermoides*. Only the latter show the so-called "waving" above the substratum with the raised anterior part of the body, a behavior pattern that results in contact with the carrier insects. In the F_1 generation all animals show this behavior; thus this "waving" is dominant. Backcrossing with the recessive parent resulted in some animals that "waved" and others that did not, which indicates monofactorial inheritance.

E. Clark, L. R. Aronson, and M. Gordon (1954) crossed the platy (*Xiphophorus maculatus*) with the green swordtail (*Xiphophorus helleri*)—two fish that differ in some respect in their reproductive behavior. The results indicate a polygenic inheritance. The same seems to be the case in the finch hybrids of R. A. Hinde (1956).

Of special interest are the investigations of S. v. Hörmann-Heck (1957), which can serve as a model for ethological genetics. She was able to cross two closely related species of crickets (*Gryllus campestris* and *G. bimaculatus*), which differ in several behavior patterns quantitatively as well as qualitatively, and to trace the inheritance of behavior patterns through the F_1 and F_2 generations and through backcrosses. Four behavior patterns were investigated. Antennal vibration during the postcourtship period and larval fights are only quantitatively different. *Gryllus bimaculatus* fights little or not at all during the juvenile period. *Gryllus campestris,* on the other hand, fights very intensively. This characteristic shows monofactorial inheritance, as does the antennal vibration in the postmating courtship, where both species are only quantitatively different in this respect. The pendulumlike movements of the thorax during mating are only seen in *campestris,* and in respect to this character backcrosses indicate polygenic inheritance. However, the stridulating sounds preceding courtship, which only occur in *bimaculatus,* seem to depend only on one pair of alleles.

Also of special interest are the more recent investigations of W. C. Rothenbuhler (1964). He crossed two different races of bees which differed clearly in their "hygienic" behavior. The hygienic bees opened cells that contained dead pupae and removed the dead. The nonhygienic bees left the dead pupae in the closed cells. When Rothenbuhler crossed the two races he obtained an F_1 generation which contained only nonhygienic bees. One F_1 queen produced four different kinds of drones. When the F_1 generation was backcrossed with the hygienic form Rothenbuhler obtained an F_2 generation that contained four groups of bees. One group was hygienic. Another group opened cells, but did not remove the dead pupae. Another group did not open the cells, but removed the pupae when the cells were opened. The last group was nonhygienic. These groups occurred in approximately equal proportions. The inheritance of the behavior patterns of uncapping (u) and removal of pupae (r) should then each depend on the homozygotic occurrence of a recessive gene:

	♀ (queen)		♂ (drone)
P	hygienic	×	nonhygienic
	(uncapping and		(pupae rot)
	removing		
	uu rr		U R
F_1		nonhygienic	
		Uu Rr	

The backcross of the four kinds of drones obtained (UR, ur, Ur, uR) with a queen of the hygienic race (uu   rr) resulted in the following F_2:

F_2:	hygienic	uncapping, but no removal	no uncapping, but removal	nonhygienic
	1:	1:	1:	1:
	uu rr	uu Rr	Uu rr	Uu Rr

It is not to be expected that the complicated neuronal mechanism underlying the behavior patterns of uncapping and removing is dependent for its development on deriving from a single gene. Indeed unhygienic workers in rare cases perform the hygienic activities, when the stimulus situation is very powerful. But the threshold of the uncapping pattern is virtually determined in an all-or-nothing fashion by the alleles U and u.

The investigation of F_2 male hybrids between Mallards (*Anas platyrhynchos*) and Pintails (*Anas acuta*) revealed a significant positive correlation between the inheritance of behavioral and plumage characteristics, indicating a genetic control for both groups of characteristics depending on relatively few genes (R. S. Sharpe and P. A. Johnsgard 1966).

12 PHYLOGENETIC DEVELOPMENT OF BEHAVIOR PATTERNS

General remarks

Before evolution of behavior patterns can take place, there must first exist a genetic variation in behavior on which natural selection can act. A. Manning (1961) was able to change the time between the first meeting of male and female fruit flies (*Drosophila melanogaster*) and copulation by artificial selection. He produced population-specific courtship periods of 80 minutes and 3 minutes. By rigorously destroying intermediaries between these two populations, he created a reproductive barrier whose members preferred one another. In *Drosophila obscura* positively or negatively geotactic populations can be obtained by appropriate selection (additional examples in A. Manning [1965]).

Each behavior pattern that changes the selective value of a species can initiate a phylogenetic development and can be subject to adaptive changes. It is not necessary for them to be always new mutations of behavior patterns. Many behavior patterns which are at first neutral with respect to selection, which exist as pleiotropic effects, may become subject to selection when the animal changes its habitat or when the environment changes. In retrospect this is often called "preadaptation."

B. F. Skinner (1966) writes that adaptedness is not always the irrefutable proof that a process of adaptation has in fact taken place.

Behavior patterns may be advantageous by chance without being selected for. This argument seems to be based on the premise that the process of adaptation takes place in many adaptive steps and aims toward a specific adaptation. Adaptations are present, however, whenever a selective advantage, however small, results from their presence. Measured in respect to such adaptations the result may be more or less advantageous, and when they first appear they always occur randomly. Adaptiveness is defined and measured by its selective advantage. How this came about is irrelevant.

J. Nicolai (1964) has advanced an hypothesis that might explain how the bond between widow birds and their hosts, which is based on traditions and maintained by imitations of the song of the host species by the widow species, has led to the evolution of different races in this group. It seems likely that the song dialects that can be demonstrated in various songbirds led to a certain ethological isolation of the various bird populations, which then led to an evolution of subspecies (C. W. Benson 1948; G. Thielcke 1961, 1964; P. R. Marler and M. Tamura 1964). These questions must still be experimentally tested. In a similar manner, imprinting to a certain biotope or to a specific host plant may lead to a new development. The ichneumon fly *Nemeritis canescens* normally deposits her eggs on the caterpillars of *Ephestia* moths, which it recognizes by odor. If one artificially raises ichneumon fly larvae on caterpillars of *Meliphora*, wasps that developed from these larvae will respond to *Meliphora* odor when they are ready to lay eggs, preferring this odor to others, although they still respond most strongly to *Ephestia* odor (W. H. Thorpe and F. H. W. Jones 1937; W. H. Thorpe 1938). The fruit fly can similarly be imprinted to peppermint odor during certain developmental periods (W. H. Thorpe 1939).

We already discussed how selection differentiates already existing movements into expressive movements. Behavior patterns probably often initiated a new line of development, thus functioning as "key characteristics" (G. v. Wahlert 1957). This can be seen from the fact that in many instances closely related species are more conspicuously distinguished from one another in their behavior than in their morphology, so that sometimes behavioral characteristics are used in the classification of the species. The dragonflies of the genus *Orthetrum,* for example, can easily be distinguished because they sit only on the two pairs of hind legs, while the front legs are folded against their prothorax (K. F. Buchholz 1957). Two very similar *Nereis* species can easily be distinguished by their reproductive behavior (R. I. Smith 1958). Two species of butterflies differ in the manner in which their caterpillars spin their cocoons, the time of mating, and the selection of food (C. P. Haskins and E. F. Haskins 1958). And two gallflies are distinguished by their food plants (B.

Stokes 1955). All these species can be determined morphologically only with great difficulty. Such species, which are primarily distinguished by their behavior, are called "ethospecies" (A. E. Emerson 1956). Behavior patterns frequently seem to be the "pacemakers" of certain characteristics. The rattles of the rattlesnake and the porcupine evolved on the basis of an already present movement. It is also possible that different behavior in separate populations of a species can lead to a diverging development, even if these behavior patterns are initially learned and are maintained by tradition within the group. G. v. Wahlert (1962) described the differential behavior of some Mediterranean fish of the same species which occur as cleaners in one area but not in another. In the Red Sea the white-spotted damselfish *Dascyllus trimaculatus* lives as an anemone fish between the tentacles of the giant anemones. Near the Maledive Islands and the Nicobar Islands only young fish of this species live near anemones, but they avoid contact with the tentacles. The morphological differences between populations which behave so differently are minute, but they clearly differ as ethospecies (I. Eibl-Eibesfeldt 1964c).

In this connection the formation of rites in man deserves attention. Certain clan and tribal habits separate groups of people very effectively, so that E. H. Erikson (1966) actually speaks of "pseudospecies." Thus the tribal tattoos bind the individual African to his group for the rest of his life. It is very difficult for him to emigrate, because an individual will always be recognized as a stranger in another group (P. Fuchs 1967).

Concept of homology

Behavior patterns can be compared to each other like morphological characteristics, and in this way one obtains sequences of similarity gradients which can be used to reconstruct their phylogenetic development. Earlier in this book we tried to trace the phylogenetic history of several expressive movements (p. 91ff.). But to do this one must be able to distinguish analogies from homologies, so it seems advisable here to discuss the criteria of homology. Behavior patterns in general do not leave fossils, so we are dependent on the comparison of living species when we try to reconstruct the evolution of behavior. Only in very rare instances is it possible to order the products of animals' activities into a phylogenetic sequence. R. S. Schmidt (1955, 1958) was able to do this with various termite nests.

In general those structures are called homologous which owe their similarity to a common origin. Descent in most cases implies a direct genetic relationship, where the information, which concerns the adaptiveness of the behavior pattern in question, is passed on through the genome. The homology criteria that are given below, however, only allow us to conclude that information has been passed on. They can do no more, and they are especially unsuitable for making the distinction between innate and acquired characteristics, as is shown by language homologies. Homologies which are passed on via memory have been called *homologies of tradition* (W. Wickler 1965a), as opposed to the *phyletic homologies,* which are passed on by the genome as the transmitter of the information. As no instances of homology of tradition had previously been known in the animal kingdom, this possibility had not been considered until then. Recently, however, J. Nicolai (1964) discovered song traditions (p. 24), some of which even cross species boundaries (song mimicry, p. 25). For research dealing with homologies it is only necessary that information emanating from one common source is passed on. It is not necessary for reproductive relationships to be involved. The song that is imitated by the widow birds is just as homologous to the host song as is Chinese learned by a European to that learned by Chinese themselves. For the assessment of homology it is important only that one source of information is tapped and that the animal did not individually acquire the information during its interactions with the environment. Let us assume that a predator innately possesses the neck bite with which to kill its prey; then it can be homologized with the neck bite of its mother and its siblings. This is also possible if the mother communicates this behavior in some way to her offspring. On the other hand, if each young animal acquires this information on its own, without tapping one common source of information, then we speak only of individually acquired adaptations. In the latter case, the similarities would be called analogies. The term "homologous convergent" overlaps with "innate acquired" in the manner depicted in Fig. 106. Not until a behavior pattern can be shown to be homologous and inborn (p. 22) can we infer common ancestry. In the use of the terms "analogous" and "homologous" we follow primarily G. P. Baerends (1958) and W. Wickler (1961a, 1967b), who took over the criteria of homology from morphology (A. Remane 1952) into ethology (see also K. Günther [1956]).

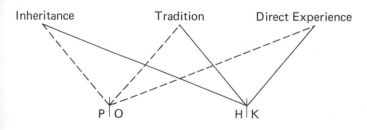

Inheritance   Tradition   Direct Experience

P O   H K

Figure 106. Schematic representation of the relationship between the concept pairs phylogeny/ontogeny (P/O) and homology/convergence (H/K), showing the source of the information. (From W. Wickler [1965].)

Remane distinguishes three main criteria. In the *criterion of position* the same relative position within a structural system indicates homology as, for example, with cranial bones. With respect to behavior patterns the position in a temporal sequence is an important criterion. If we find a regular sequence of similar movement patterns—a-b-c-d-e-f-g—in two closely related species, and if in one species one of the elements appears somewhat more modified, than the specific location within a temporal sequence indicates its homology.

The criterion of *specific* quality deals with formal similarities. Homology is indicated when more agreement is found in individual characteristics. However, we already know from morphology that adaptations to specific environmental conditions can develop independently, which then are analogies such as the fishlike form of fishes, marine reptiles, and marine mammals. This criterion will not suffice alone, but it is more likely to be valid in the study of expressive movements because their specific adaptations are mostly not a response to the nonliving environment. Here the ethologist argues similarly to the ethnologist, who does not necessarily interpret the similarity of stone axes from European, African, or Asiatic stone age peoples as evidence of a common cultural heritage. The form depends on the function; hence convergence can hardly be excluded.

However, if he finds the words *mère, mater, mother, Mutter, matka,* and *madre* with the same meaning in the languages of different people, then this indicates a common root. In a similar fashion one can consider the expressive movements of many animals as phylogenetically developed "conventions." However, many more simple expressive movements frequently come about by convergence (p. 97), for example, as was shown by the investigations of Wickler with respect to the nod swimming of cleaner fishes and their mimics (p. 110).

The *criterion of linkage by intermediate forms* is of use whenever such connecting transitions are found. In this way quite dissimilar behavior patterns can be homologized. The intermediary forms can occur during ontogenesis, and then it is possible to trace the gradual change of a behavior pattern. Were this not the case, these intermediary forms must have been derived from systematically closely related forms. R. Schenkel (1956) used this criterion in his interpretation of the courtship behavior of pheasants (p. 103).

As an *auxiliary criterion* one can use the statement that even simple behavior patterns are probably homologous if they occur in a large number of closely related species, and they are probably not homologous if they occur in an increasing number of unrelated species.

Serial organs can also be compared, such as the mandibles and walking legs of shrimp. The gradient of similarity indicates that the mandibles are actually modified legs. Such instances of serial homology are called *homonomy*. In respect to behavior there are two possibilities for homonomy, according to W. Wickler (1961a): (1) the serial homologous movements of the legs and of the mandibles derived from them, and (2) the various movements of one organ which have a common origin. The "carpentering" activity, drumming, and tapping of a woodpecker that wants to be relieved from the nest are all homonous movements.

Analogies exist when a behavior pattern is found in animals which share a specific way of life (carrion eaters, predators) or in inhabitants of a specific biotope (cliffdwellers, tree or desert inhabitants) independent of their taxonomic relationships. Analogy is also indicated when the original form of the various species, whose similarities are compared and which live similarly today, led a different way of life in earlier times but did not show the similarities in question.

A number of bottom-dwelling fish from several families developed converging adaptations in their behavior (W. Wickler 1957, 1958, 1959, 1965c). The convergent adaptations of fishes that live in rapid waters are most impressive (*Gastromyzonidae* and *Homalopteridae*). The former are descended from the *Cobitidae;* the latter from the *Cyprinidae* in the narrower sense. But the convergences in shape and behavior are so close that the members of both families at one time were combined in one category. Both groups of animals have two separate pectoral fins of large size which enable them to adhere to the substratum. When they breathe, water is pressed under the head by the strong countercurrent, which the fish pump away by very rapid fanning of the pectoral fins or the rhythmically beating posterior section of the pectoral fin, with the result that the water flows faster beneath the fish. This reduces the relative pressure, which enables them to remain in place without being swept away. Several species of *Cyprinidae, Homalopteridae, Gastromyzonidae,* and *Siluridae* have "invented" this mechanism independently, although they descend from fish which did not possess the rhythmic movement of the pectoral fin (W. Wickler 1960b). The manner of drinking by immersing the beak and sucking up the water in pigeons, sandgrouse (*Pterocles*), and pin-tailed sandgrouse (*Syrrhaptes*), which had been thought of as homologous was shown to be an analogous adaptation to life in arid regions (W. Wickler 1961c). This same "invention" was also made by other groups of birds, such as the grassfinches. It is not correct to consider sandgrouse and pin-tailed sandgrouse as relatives of pigeons because of their manner of drinking.

Homoiologies are analogies that have developed from the basis of an homologous structure. Thus the flipper of a whale is homologous with the wing of a penguin with respect to vertebrate extremities. The adaptation as a flipper, however, is an analogy, because it was independently developed.

Desert mice (*Gerbillidae*), desert kangaroo mice (*Dipodidae*), hares (*Leporidae*), several species of mice (*Muridae*), and other rodents drum on the ground with their hind legs when they are excited (aggression or escape), which is probably a ritualized intention movement to jump (I. Eibl-Eibesfeldt 1951b, 1957a). These animals evolved this behavior independently of one another, but certainly on the basis of the homologous basis of the jumping-off movement. Many other threat postures owe their similarities to convergent evolution. Tenrecs (*Echinops telfairi*), tree shrews (*Tupaia glis*), squirrels (*Sciurus vulgaris*), dormice (*Glis glis*) and the hedgehog (*Erinaceus europaeus*) threaten when they are disturbed from their sleep by hitting at the disturbing object with sudden stretching movements of their forelimbs, accompanied by shrill hissing and screeching. In tree shrews and squirrels the young already behave this way while still naked. The similarity of the movements can probably be explained from the function of frightening the attacker as well as by the fact that identical homologous origins from defensive and breathing movements served as the basis for ritualization.

Hissing threat sounds have evolved convergently in many vertebrates on the basis of their homologous breathing patterns. In an analogous manner behavior patterns of care for young (social grooming, feeding, and so on) and infantilisms repeatedly were changed in the service of group-uniting functions (see also p. 104).

If the criteria of homologies discussed above are considered, then behavior patterns are of great taxonomic value and can help to elucidate the natural relationships among animals. We refer in this connection to the studies of the *Anatidae* by K. Lorenz (1941) and the investigations of A. Faber (1953a, 1953b) and W. Jacobs (1953a, 1953b) on locusts, U. Weidmann (1951) with fruit flies (*Drosophila*), J. Crane (1949, 1952) with New World mantids and jumping spiders (*Salticidae*), W. F. Blair (1957a, 1957b, 1958) with frogs, G. K. Noble (1927, 1931) with amphibia, J. Nicolai (1959b) with serins, G. P. v. Tets (1965) with pelicans, and P. Leyhausen (1956) with cats. The detailed investigation of tree shrews (*Tupaia*) showed that they should be considered a separate order (*Tupaioidea*) rather than included among the primates, as they have been (R. Martin 1966a, 1966b). It would be difficult to assign them to any other existing order because they share many characteristics with rodents, rabbits, and marsupials. Additional examples can be found in W. Wickler (1961b, 1967c). With respect to the assessment of body form and behavior there exist opposing viewpoints. D. Starck (1959:47) writes:

To assume evolutionary relationships on the basis of behavior patterns is not justifiable when such findings clearly contradict morphological considerations. The methods of morphology will therefore remain the basis of the natural system. Its fundamental significance is based on the fact that it is the only method applicable to fossil material.

On the other hand, E. Mayr (1958:345) states:

If there is a conflict between the evidence provided by morphological characters and that of behavior the taxonomist is increasingly inclined to give greater weight to the ethological evidence.

It seems advisable to take a position somewhere between these. In principle, taxonomic relationships based on morphology should agree with those based on behavior; otherwise one or both are apt to be wrong (N. Tinbergen 1951).

Historical rudiments

Figure 107. Morphological sequence of the recent elk, showing the stepwise replacement of the canine teeth by the antlers as weapons. a, *Moschus*; b, *Muntiacus*; c, *Cervus*. (Explanation in the text.) (From G. Tembrock [1961].)

Their original function of behavior patterns can change if there is a basic change in the way of life of an animal. They may either assume a new function or may be retained in the old or little changed form as behavioral *rudiments,* as long as this is not of an immediate disadvantage for the species. The stump-tailed macaques (*Macaca speciosa, M. arctoides, M. fuscata,* and *M. maura*) perform balancing movements with their small tail stump, which of course are not actually effective. It is possible to release flying movements in ostriches (*Nandus*), although these animals have not flown in millions of years (I. Krumbiegel 1940).

Rusa and Dybowski deer and elk threaten by displaying a rudimentary organ. The oldest deer (lower Oligocene) possessed no antlers, as with *Moschus,* the most primitive living deer today. All of these primitive deer species, including the muntjac (*Muntiacus*), possess elongated upper canine teeth for slashing that are used in fighting by *Moschus* and muntjac. They display these weapons when making threats, walking to and fro in front of an opponent, nodding slowly, head raised high, gritting their teeth, and retracting their lips so that the daggerlike teeth are clearly visible. In the same manner Rusa deer, Dybowski deer, and the European elk threaten, although their canine teeth are reduced to small structures and they use their antlers when fighting (O. Antonius 1939; see also Fig. 107). In the process of ritualization of the courtship of grass finches one can trace the change of function and rudimentation (K. Immelmann 1962b; M. F. Hall 1962). Carrying nesting material for nest building evolved into the male courtship actions using grass stems. This was again secondarily reduced in some species and became rudimentary, while at the same time the song, which originally served the function of staking out territories, also underwent a change in function. These animals are gregarious and are hardly territorial. Instead of courting with grass stems, these males sing softly while sitting next to the females. In the genera *Bathilda* and *Aegintha* males are unable to court without a grass blade in their bills. They hold it continuously and perform various derived nest-building activities without actually building. Eventually they give the blade of grass to the female. The males of *Neochmia* use different material while courting and in the actual construction of the nest. This demonstrates convincingly that courtship with grass stems has acquired independent motivation. The *Lonchura* species carry a stem around for some time before commencing to court, although they actually court without it, while *Aidemosyne* still uses a stem in the initial phases of courtship. *Emblema* merely pecks at stems of grass; *Poephila* pays no attention to grass blades, although on occasion courting with grass stems occurs as a behavioral rudiment, as well as in young males. Marine iguanas make threats by opening their mouths wide, although they normally do not bite one another during fights. Many birds threaten with widely gaping mouths, even species that actually attack with closed bills. They use the phylogenetically older intention to bite as a threat display. The blue-footed boobies of the Galápagos Islands pass nest material in the form of small stones between each other during the pair formation ceremonies although they no longer build nests. In all these instances a behavior pattern survived an original function in the form of an expressive movement. Rudiments that no longer have any function also exist.

The redheaded finch (*Amadina erythrocephala*) is a nest parasite which uses the nests of other birds and no longer builds its own. When sitting on a nest, the bird still performs all the nest-building movements but in a randomized sequence. It reaches over the edge of the nest and "grasps" and pulls in nonexisting nest material as if the bird were actually building (J. Nicolai, personal communication). Ground-nesting birds roll in eggs that lie outside their nests, but some birds that nest in trees also do this. These were derived from ground-nesting species and still possess this behavior as a rudiment (H. Poulsen 1953). Wingless *Drosophila* mutants perform wing-cleaning movements, as do the wild forms (H. J. Heinz 1949). Several termites construct covered tunnels that end in culs-de-sac that are not used. In a similar position one can find passages in closely related species which still serve a function (R. S. Schmidt 1957, 1958).

We have already pointed out (p. 101) that cultural ritualization in man shows remarkable similarities to its phylogenetic parallels. Thus we observe that certain objects lose their original function and acquire new functions or are kept on as a mere rudiment. We have decorative buttons on the sleeves of suits which were originally used to button the sleeves. In the same way ribbons on hats originally were used to fasten the hat (Fig. 108). Today they are mere decorations (L. Schmidt 1952). O. Koenig (1968) collected material that documents this kind of development.

The Galápagos dove (*Nesopelia galapagoensis*) shows a distraction display near the nest, although this behavior, which is useful in deceiving predatory mammals, is no longer of use in the islands, which until very recently, were free of such predators. In the same islands E. Curio (1965b) was able to demonstrate that Darwin finches from islands that are free of predators responded appropriately when presented with them, behaving like finches from islands where predators (snakes, raptors) are present. These examples illustrate a kind of natural deprivation experiment (p. 19).

Sometimes we can gain an understanding of phylogenesis by studying the ontogenesis of behavior, although it is true even in morphology that the biogenetic law—that ontogenesis recapitulates phylogenesis—is true only in about 60 percent of the cases. There are some good examples from behavior. The young bearded titmouse at first crawls with all four limbs, which are moved in alternating diagonal pairs (O. Koenig 1951). The larks, which are derived from hopping birds, hop when they are young but run when they are older. The young peacock performs his tail display and food calling, but when older only fans out its tail (p. 104). Young marine iguanas bite one another when fighting, but when older they butt their heads.

The freshly metamorphosed glaucothoë of the coconut crab (*Birgus latro*) searches for snail shells in the manner of the hermit crab and performs the same fixed action patterns of testing shells and entering them as do the glaucothoë of the hermit crab *Pagurus longicarpus* (E. S. Reese 1968). The shell protects the glaucothoë from dessication during their migration on land. Individuals that do not find a shell die. The pattern of shell selection, although rudimentary still fulfills a function during a short period of life. Older crabs do not need the protection of a shell (E. S. Reese 1968). Additional examples of processes of behavioral rudimentation are presented by E. Curio (1960) and W. Wickler (1960a).

During artificial manipulations in bumblebee nests, A. Haas (1962, 1965) was able to elicit older behavior patterns which are no

a

b

c

Figure 108. Change of the attachment cord on the headdress of Hungarian hussars. (a) Kalpak with cap bag and cord for attachment (1700). (b) Felt cap dating back to approximately 1760. The cord has lost its original function and is purely decorative. (c) Tschako of a hussar before 1914. The cord is solely ornamental and developed to fit this new purpose. A new chinstrap is developed. (From O. Koenig [1968].)

longer performed but which in some form or other are common to all members of the genus ("generic behavior").

Research on domesticated animals and the process of domestication

Research on domesticated animals also gives us interesting leads about the phylogenetic development of fixed action patterns: The ancestor of our various races of domesticated pigeons—the rock pigeon (*Columba livia*)—possesses several characteristic courtship behavior patterns. During the display flight the male claps its wings loudly above its back. After that he glides several meters through the air with his wings elevated above the horizontal line of his body. This behavior pattern has been developed and changed further by the artificial selection of man (J. Nicolai 1965b). In several races of pouters, especially in Steller's pouter, this clapping is so extensive and vigorous—30 instead of 4 to 5 claps in the ancestral form— that the tips of the primary feathers wear off more and more from spring through the summer. Shortly before the molt only a third of the vane remains on the wing flights, so that the bird's flight is quite impeded. The position of the wings during the glide is hypertrophied: The wing tips touch and the bird loses altitude quickly. In these instances a behavior pattern has become ritualized through selection by man, because no predator is selecting against this behavior. In roller pigeons, such as the oriental roller or the Birmingham roller, the gliding display flight has been changed through selection into a continuous smooth series of backward somersaults. These somersaults can follow each other so quickly in some individuals that the bird seems to be falling from several hundred meters high like a swirling ball of feathers. The fall is broken just above the ground when the bird pulls out of its dive. This is an entirely new behavior pattern which does not occur in the rock pigeons. During courtship on the ground the cooing male turns about its vertical axis and performs a small jump when the female walks away. He may then clap his wings once or twice. In the German ringbeater this behavior pattern is hypertrophied; the cock flies up clapping noisily, flies in a tight circle about 1 meter above the hen three to four times, and returns to the ground. This flying up has become a new display flight. Turning and following has been combined into a new behavior pattern.

In the same manner new behavior patterns probably also evolved under natural conditions. In the diamond firetail finch the behavior of bringing a gift to the female is combined with the infantile begging movement (Fig. 109) into a courtship movement. The male at first courts sitting upright, flicking its tail and holding a blade of grass in its bill. When the female approaches he will bend his head down in the position characteristic of the food-begging position of this group without letting go of the blade (J. Nicolai 1965a; see also Fig. 110).

a   b

Figure 109. (a) Finches (*Carduelinae*) gape directly toward the adults in an upright body position. They are fed in small portions. Here a 21-day-old West African *Ochrospiza leucopygia* gapes toward its father, who already has regurgitated some seeds. (b) Most Estrildine finches turn their heads when begging in the manner shown and make pendulumlike movements. The adult birds pump larger quantities into their gullets at one time without removing their bills. Here a 19-day-old Estrildid finch (*Granatina granatina*) begs from its father. (After J. Nicolai [1965a]; H. Kacher, artist.)

Figure 110. Courtship of the diamond finch male (*Staganopleura guttata*). The male sits with a stem of grass on a branch and attracts a female by his song and courtship dance. The stem is then offered in the position of a begging young (see Fig. 109b). (After J. Nicolai [1965a]; H. Kacher, artist.)

Research on domesticated animals supplies still more examples of changes in behavior patterns. Fighting cocks have been selected for aggressiveness; dogs have been selected for various characteristics.

Our domesticated animals have undergone a large number of changes in behavior and appearance under the changed conditions of captivity. These domestication characteristics in animals are similar to those of man (K. Lorenz 1940, 1943, 1950a) and are the result of similarly changed selective conditions.

With the protection of pens, stalls, and other fenced-in structures, acuity of the senses and physical fitness is less important than a high rate of reproduction. One result has been a breakdown of the finely differentiated social behavior patterns. Monogamy and high selectivity in the choice of partners impede breeding and are a selective disadvantage under conditions of domestication.

Where in wild animals, for example, graylag geese, a large number of conditions must be fulfilled before their highly differentiated sexual and family life can be developed, it is sufficient in the domesticated form to lock up together any two individuals of different sexes to ensure breeding (K. Lorenz 1950a).

The selectivity of innate releasing mechanisms has been markedly reduced in domesticated animals. Domesticated zebra finches also feed nestlings which do not show the species-specific gape markings, whereas wild birds are extremely selective and they also court very simple models (K. Immelmann 1962a).

The wild form of our domesticated chicken, the jungle fowl, responds with brood-care behavior only to chicks with a very specific color pattern on the head and back. Chicks with deviations in their patterns are killed. Such selective behavior we find once in a while in Phoenix fighting cocks and dwarf breeds which are relatively close to the wild form. Our domesticated country chickens, on the other hand, accept chicks of all colors, but react selectively to the calls of chicks of their own species, so that it is relatively difficult to give them young ducks to incubate and hatch. However, the most domesticated breeds, such as the Plymouth rock, will accept duck-lings without difficulty (K. Lorenz 1950a).

Graylag geese will mate only after a prolonged courtship period and remain monogamous with the same partner thereafter. Domesticated geese pair off without showing preferences and are not mono-gamous. While the aggressive drive was often selected against because it was disruptive, hypertrophy resulted with respect to the reproductive drive and eating. The behavior becomes simpler. These behavior changes are accompanied by a large number of physical signs of domestication. K. Lorenz pointed out that there is a tendency to shorten the extremities and the muzzle, to become fat as well as a general weakening of the muscles and connective tissues (Fig. 111). All these physical characteristics of domestication can also be observed

Figure 111. Characteristics of domestication. On the left in each case the wild form, on the right the domesticated form derived from it. (a) Wild chicken–domestic chicken; (b) wild goose–domestic goose; (c) wolf–domestic dog. In the figure the relative sizes are not accurate. The domestic goose is larger than the wild form, and the pug is smaller than the wolf. (From K. Lorenz [1965].)

in many members of civilized peoples. Furthermore, in civilized man many of the "cardinal" virtues, such as loyalty to the family, courage, and moral behavior, are in danger. If someone commits unscrupled acts which are barely permissible, he will gain a certain advantage. He who produces offspring recklessly has a higher rate of reproduction, but a society can only exist as long as the number of socially responsible individuals prevails. Perhaps some of the old civilizations were the victims of such a degeneration in the final analysis. These degenerative signs, which are the result of domestication, are opposed by our innate aesthetic and ethical value judgments, according to which we disapprove the degenerative symptoms discussed above.

One must be extremely careful, however, to distinguish between the degenerative appearances which are caused by domestication and domesticated characteristics. The former have a negative selective value for the preservation of the species. The latter have a positive selective value and are adaptations to specific environmental conditions. This is true, for example, for the hypersexuality of man as a result of domestication, which has sometimes been regarded as a sign of degeneration. There is much evidence (p. 443) that this hypersexuality serves an important function in the service of preserving the bond between man and woman. By reducing instinctive behavior patterns through domestication it also became possible to clear the path for learning and education. This undoubtedly was a precondition for the development of becoming human (K. Lorenz 1943).

Commensals of man experience changes in behavior similar to those caused by domestication. The various subspecies of the house mouse (*Mus musculus*) can hardly be discriminated on morphological grounds, but more and more in their behavior. In their country of origin between the Caspian Sea and Lake Neusiedl, the subspecies *Mus musculus spicilegus* (harvest mouse) lives year-round outdoors and shows a hoarding instinct. Two to six mice collect 5 to 7 kg of seeds which they store in mounds above the ground. They cover the mounds with earth and live in nests beneath. These harvest mice are herbivorous and lack the smell of the western European house mouse (*Mus musculus domesticus*), which lives as a commensal of man year-round and lacks the hoarding instinct. Furthermore, these mice are polyphasic in their activity and, similar to domestic animals, they have their estrous cycle all year. The half-commensal central European harvest mouse (*Mus musculus musculus*) links both the other races. In summer this subspecies lives out in the fields, in winter as a commensal of man (A. Festetics 1961).

a

Behavior fossils

It is possible to draw inferences regarding the behavior of organisms by studying the fossil tracks left by those which produced them. Certain behavior patterns can be deduced from tracks left by feeding (R. Richter 1927; W. Schäfer 1965). If one studies the feeding tracks of marine organisms during geologic time, progress in the technique can be observed (Fig. 112). The tracks left by snails and trilobites

b

c

Figure 112. Feeding tracks of the sediment-feeding *Dictyodora* can be read as reflecting changes in behavior. (a) Tracks from the Cambrian age, loosely meandering paths immediately below the surface; (b) later they ate their way, in corkscrew fashion, deeper and more closely drawn around the initial spiral; (c) during the lower Carboniferous age they went deeply into the sediment. (From A. Seilacher [1967].)

of the Cambrian period as well as those of worms of the early Silurian period remind one of the scribbling of children. The animals seemed to obey an impulse to move sideways, thus avoiding a retracing of previously covered ground. A more efficient use of the grazing area is achieved when grazing occurs in a tight spiral or in a tight meandering pattern. In order that such a spiral can be followed it is necessary for the animal to maintain contact with the previously made track. If it is to meander, the direction must be changed regularly. We can now observe that the scribbling patterns disappear completely during the course of the earth's history and that they are supplanted by spiral and meandering feeding patterns. More complicated meander patterns and tight double spirals do not appear until the end of the Mesozoic age (A. Seilacher 1967).

The feeding tracks of the sediment-feeding *Dictyodora* showed a change in the feeding technique through the ages which were correlated with morphological changes. From the Cambrian age to the Devonian age (600 to 350 million years) *Dictyodora* fed immediately below the sediment surface in a loose meandering pattern. Probably as a result of numerous competitors this species escaped into deeper layers of the sediment, where the breathing tube, which can be seen in the sediment, became longer. At first they drilled into the substratum and only then did they follow their usual meandering pattern. Later they began to eat their way downward, leaving a corkscrew pattern. When they had reached the right depth they began to meander and their pattern became so changed in the course of time that the meandering paths were placed tightly around the initial corkscrew spiral. Finally they gave up meandering completely and bored their screwlike spiral deeper into the bottom (A. Seilacher 1967); (Fig. 112).

13 THE ONTOGENY OF
BEHAVIOR PATTERNS

Embryology of behavior

In general an animal has its behavior patterns ready for use when they are needed. As organs grow and mature, so does their capacity to perform their function. It is much rarer for a behavior pattern to mature before the organ that is involved in its performance. Young graylag geese will show wing-boxing behavior when fighting with wings that are yet nothing more than tiny stumps with which they are unable to hit their opponent (K. Lorenz 1943). During onto-genesis, behavior patterns develop gradually and overlap one another in time. Sense organs, coordination centers, and effectors can mature independently of each other and at different rates, as the last example demonstrated. The interlocking of fixed action patterns with the appropriate taxes may occur only after the basic patterns are estab-lished. Thus newborn mice and rats scratch at first spontaneously in the air without touching their skin (I. Eibl-Eibesfeldt 1950c). Some behavior patterns mature as organs do, whereas others owe their specific adaptiveness to an interaction of the young animal with its environment or its own body (pp. 21ff.). The development of behavior begins in the embryonic stage, so that by the time of birth or hatching a number of actions are fully functional. Little is known about the embryology of behavior. Although W. Preyer wrote a pioneer work on this subject as early as 1885, there was no followup for a long time. G. E. Coghill (1929) came to the conclusion, following numerous, careful studies, that behavior always seems to appear in well-organized patterns and that it depended on the spontaneous activity of the central nervous system. In the larva of the salamander *Ambystoma,* the undulating movement first occurs as a turning of the head to one particular side only. In fishes it is frequently the same. In bird and mammal embryos we can also observe an increasing maturation of movements from the head to the posterior parts of the body; the first behavior that can be recognized is a bending of the head. Soon thereafter follows a stage in which the entire body, legs, wings, head, and trunk are active, seemingly independently of one another. In this "mass action" repeated bouts of activity follow short intervals of rest. How such activity becomes integrated into well-coordinated movements needs to be investigated specifically. It appears certain, however, that the behavior of an animal is not, as W. F. Windle (1940, 1944), among others, thought, built up of larval reflexes, which become integrated as primary units of behavior into secondary, higher functional units, because the initial movements are always of a spontaneous nature (V. Hamburger 1963, 1966). A chick already moves when it is still a 3-day-old embryo, but not until the seventh day of incubation can a response to tactile stimulation be obtained. This interval between movement and sensitivity was noted as early as 1885 by W. Preyer (1885).

It is possible that simple learning processes play a role in the developmental physiology of innate behavior (p. 30). According to H. F. R. Prechtl and A. R. Knol (1958) later behavior in man is influenced by the position of the embryo in the uterus as a function of the relative freedom of movement of the embryo. Children who developed in a head-down position showed good flexing and extension reflexes. If they are scratched on the soles of the feet, they pull up the legs. Following a breech position the same stimulus will extend their legs, and extension movements will later predominate in the movement repertoire of the infant. There are no noticeable differences in the musculature, so these differences in behavior must depend on habituation.

Early ontogenetic adaptations (kainogeneses)

When a living organism is born or hatched from an egg it can be more or less completely developed. Sometimes the newborn animal is a miniature version of the parents and lives in much the same way as they do. This is true for most reptiles and to a certain degree for several precocial birds and mammals (Fig. 113). In many other instances the young does not resemble the parents and its mode of living is quite different. The different ways of life of many larval and imaginal insects or frogs are well known. Many of these larvae possess highly specific behavior patterns, serving functions such as a particular manner of feeding or avoidance of predators, which are completely lost after metamorphosis. We will cite a few examples here: Ant lions build pits and throw sand at their prey (H. J. Nieboer 1960). Larvae of the fly *Arachnocampa luminosa* live on the ceilings of caves in New Zealand. They are luminous and produce long, dangling silky threads which are suspended with drops of a sticky substance attached at short intervals. Insects that are attracted by the light are caught in these traps and are consumed along with them (J. B. Gatenby and S. Cotton 1960; V. B. Wigglesworth 1964; see also Fig. 114). The butterfly caterpillar of *Aethria carnicauda* builds several fences consisting of hairs arranged in whorls around

Figure 113. (a) Precocial animal: a newborn hare; (b) altricial animal: a newborn rabbit. (From F. Bourliere [1955].)

Figure 114. Larva of *Arachnocampa* (*Bolithophila*) *luminosa* rests on a horizontal web from which are suspended catching filaments with attached drops of glue. From the abdomen of the larva light is emitted which attracts prey (insects) which become entangled in the filaments and are eaten. When in danger the larva retreats into a hiding place in the wall of the cave. (After J. B. Gatenby [1960] from W. Wickler [1967a]; H. Kacher, artist.)

the branch, on either side of itself, which protect the pupae from predators (Fig. 115). Larvae of the long-horned woodborer beetles (*Cerambycidae*) position themselves close to the surface of the wood within their passages just prior to the pupal stage. The adult beetle is unable to drill through the wood and would perish if the larvae pupated deep within the wood. The pea weevil (*Bruchus pisi*) metamorphoses within the pea. The developed beetle cannot eat its way to freedom; this is done by the larva, which eats a passage up to the thin peeling of the pea. Here it gnaws a round furrow so that the beetle merely has to push the cover open from the inside (J. H. Fabré 1879–1910).

Figure 115. Caterpillar and below it the pupae of *Aethria carnicauda* Butler with the protective fences that were built by the larva before pupation. (After W. Beebe [1953].)

Even the larvae of primitive invertebrates can perform surprising actions. The nematode (*Dictyocaulus viviparus*) lives in the lungs of cattle. The larvae, which hatch in the bronchial tubes, emerge via the intestinal tract onto the pasture. Because cows avoid grazing in the proximity of feces and the larvae are unable to migrate, further transfer takes place in the following manner. The larvae climb, positively phototactically oriented, to the tops of pilobolus fungi, which grow in great abundance on the feces. These fungi are known to project a spore packet into the air when exposed to light. The larvae wait on top of these spore packets on which they are then shot out to land some distance away. Over 50 larvae have been found on one spore packet (J. Robinson 1962).

The cercarian larvae of the leech (*Dicrocoelium dendriticum*) control the behavior of their last intermediary host, the ants. Whereas most of the cercaria are in the abdominal cavity of the ant, one of them, with a different appearance than the others, is always found in the subesophageal ganglion near the nerves that innervate the mouth parts. This "brain worm" seems to change the behavior of the ants so that they climb upward on grass stems and there hold on with their mandibles. This ensures that grazing sheep will eat them so that the parasite is safely delivered to its host (W. Hohorst and G. Graefe 1960).

Larval adaptations in respect to behavior are of various types and are lost after metamorphosis. Whether some of these abilities are retained in some changed form we do not know, but it is known that what is learned in the larval stage can be retained through metamorphosis. Mealworm larvae (*Tenebrio*) which were trained by W. v. Borell Du Vernay (1942) in a T maze retained the learned habit when they were tested as beetles. Fruit flies (*Drosophila*) prefer odors that were added to their larval food (W. H. Thorpe 1939); J. E. Cushing 1941).

Another intriguing observation that should be followed up comes from E. Fischer (cited by R. Fletcher [1948]). The pupae of *Hoplitis mühlhauseri* are enclosed in very hard cocoons that are cut open with a thorn located at the headpart of the pupae. Two pupae that were removed by Fischer from the cocoon did not perform the necessary movements, but the butterflies performed them after hatching. They stopped, with their head against the wall of their container, performed butting and sawing movements as if they were going to cut a hole. Normally, these butterflies unfold their wings right after hatching. In vertebrates we often find early ontogenetic adaptations in the behavior of fish and amphibian larvae whose way of life is quite different from that of the adult animals. An early ontogenetic adaptation seems to be the specific movement coordinations involved in hatching of birds (V. Hamburger and R. Oppenheim 1967), as are the gaping reactions and begging movements of young birds, which sometimes reappear in the behavior repertoire of the adults as so-called infantilisms (p. 122). Specific adaptations prevent soiling of the nest. In the black woodpecker defecation is released when the adult bird stimulates the anal region of the young, after which the adult simply eats it (H. Sielmann 1956). Young bee-eaters walk backward to the nest wall and defecate there, thus keeping the center of the nest clean (L. Koenig 1951). Other young birds defecate over the edge of the nest. The young cucoo pushes its nestmates out of the nest. This instinct is evident from 10 hours after hatching and is no longer present 4 days later.

Young mammals also possess a number of early ontogenetic adaptations; some of these animals are often born in very early stages of their development. The young of the great gray kangaroo is born as a 2-cm-long embryo but is able to climb into the mother's pouch unaided by means of its powerful front legs. Young mammals often possess a searching mechanism, a rhythmic head movement, which leads to the teat. When they suck, they show treading movements with the forepaws or butting with the head into the udder (H. F. R. Prechtl and W. M. Schleidt 1950, 1951; H. F. R. Prechtl 1958; J. Adler, G. Linn, and A. V. Moore 1958). Young mammals often have special alarm calls, and can hold on when in danger. Prematurely born human babies are able to hold on to a rope with all fours as well as hands alone without any aid. This ability is later lost and may again develop secondarily (A. Peiper 1951, 1961).

Frequently young animals can produce sounds which enable them to maintain contact with their mother in a kind of dialogue. A young graylag gosling will give a two-syllable wi-wi call from time to time even when asleep under its mother, which she will answer. If a gosling is kept alone the wi-wi calls become more and more urgent. Such "lost calls" are widespread, and human infants also have them. Thus nightly unrest of the child comes from a phylogenetically very old need to be reassured by the presence of the mother. To be left alone meant the greatest danger for the breast-fed infant, and this crying was a contact call which alarmed the mother and aided her in finding her child. Today we place our children into beds, but the old mechanism is still operating, and the child seeks to bring about the calming contact with the mother with its crying. The child can be calmed down by rocking or by providing it with a pacifier, which is a model of the mother's breast (A. Peiper 1951).

Maturation of behavior patterns and "instinct-learning intercalation"

New behavior patterns are built up as the result of maturation and learning processes (p. 21), while infantile and larval behaviors are completely or partially fragmented. They can disappear completely or they can reappear as expressive behavior in the repertoire of the adult (see food begging during courtship, p. 122). Sometimes infantile behavior patterns appear as *reggressions* in adult animals and man (M. Holzapfel 1949; J. Adler, G. Linn, and A. V. Moore 1958; D. W. Ploog 1964a). This proves that they existed in latent form, that is, that the mechanisms on which they were based were retained. In older people who suffer from degenerative processes of the central nervous system a recurrence of the infantile searching mechanism, oral orientation, and sucking movements were observed

(H. F. R. Prechtl and W. M. Schleidt 1950; S. Wieser and T. Itil 1954; S. Wieser 1955; G. Pilleri 1960a, 1960b, 1961; D. W. Ploog 1964a, 1964b).

The maturation of new actions and the breaking down of old fixed action patterns can overlap. Young nestling sparrows peck when they are still gaping. They do this especially when they are satiated, after feeding. If they are hungry, they gape again. Gaping initially inhibits pecking. K. Lorenz (1935) reported that young hand-raised starlings, which had been picking up food on their own when their caretaker was away for several days, again gaped continuously after he returned. They had eaten on their own until then, so he did not think it necessary to feed them until he noticed that they were becoming weak. The drive to gape, which was released again by the presence of the caretaker, blocked the pecking reaction (see also M. Holzapfel [1949]).

Study of the behavior patterns of food intake in cormorants reveals a gradual transformation from infantile behavior patterns to an increasing integration of the individual acts with the appropriate motivations. Until the third week of life the young bird begs and gapes. Between the third and fifth week they begin to cudgel fish, and from the sixth week on they begin to catch their own. Begging drops out when they are 6 months old. During the ontogenesis of the nest-building behavior a similar picture emerges. The final action of "trembling" occurs first; it is followed by the fastening of the twig, bringing of the twig, and so on. This integration is not the result of individual learning. In this connection it is remarkable that a disintegration in reverse order occurs within the annual cycle toward the end of the breeding season (A. Kortlandt 1940, 1955).

In species that metamorphose there is at the same time a complete reorganization of the behavior, and behavior patterns appear for which there were previously no indications. As A. J. Rösel v. Rosenhof (1746–1761) pointed out long ago, butterflies do not have to learn to fly after hatching. In species with incomplete changes the differences are less startling and behavior patterns of the imago appear during the larval stage (W. Jacobs 1953). The larvae grasshoppers perform the leg movements necessary to produce songs, although their "song" is at first silent, because the morphological structure that is necessary to produce the song develops much later (A. S. Weih 1951; see also p. 201). The fully developed cricket, on the other hand, does not begin to sing immediately after the hardening of the elytra but several days later. In this instance the organ has matured before the behavior (S. v. Hörmann-Heck 1957). We have already seen a converse in the wing boxing of the graylag gosling. Behavior patterns can be activated prematurely by the injection of hormones. Young male dogs raise their hind leg when they urinate in the manner of adults following an injection of testosterone (S. Freud and J. E. Uylert 1948). Fourteen-day-old male rats begin to mount females after testosterone injections, and in 21-day-old female rats the copulatory posture (lordosis) can be released following injection of follicle hormones (F. A. Beach 1947). According to these results the neuromotor mechanism underlying these behavior patterns is readily available long before the behavior normally occurs. Immelmann's observation that zebra finch females sing the songs which they have heard when they were young, following treatment with male sex hormones, is also very interesting in this respect (p. 231).

In the white butterfly (*Pieris napi*) the ability to fly improves as the wings harden. Until this occurs, spontaneous flight seems to be suppressed (B. Petersen, L. Lundgren, and L. Wilson 1957). The locust *Schistocerca gregaria* is unable to fly properly until several days after reaching the adult stage. Only then are the cuticle of the wings and thorax sufficiently hardened (J. S. Kennedy 1951).

In many instances the developing behavior patterns appear very clumsy. Only gradually does the initial lack of coordination become lost, which may be the result of maturational processes, learning processes, or a combination. Whether or not improvement depends upon learning or maturation can only be decided on the basis of experiments. No distinction can be made by mere observation of the phenomena. Newly hatched chicks peck at small objects, but their aim is not very accurate. When allowed to peck at the head of a nail embedded in soft clay, the impressions of the beak initially are more scattered around the nail than they are later on. On the fourth day they are clustered very closely around the head of the nail. E. H. Hess (1956) was able to show that learning was not involved in this improvement. He fitted chicks with hoods containing prisms that displaced the object to the right. The initial peck marks were scattered on day 1 to the left of the target. On day 4 they were tightly clustered but still displaced from the target so that it was outside the peck marks (Figs. 116 and 117). In this case the improvement is not due to learning but to a maturation of the aiming mechanism. The animals never learned to hit the nail on the head. In some cichlids (*Cichlidae*) the selectivity of the innate releasing mechanism for the following reaction is also a function of maturation (E. Kuenzer, and P. Kuenzer 1962).

K. Lorenz (1935) spoke of "instinct-training intercalation" when innate and acquired components become integrated into one behavior sequence. We have discussed on page 25 how these components can be distinguished experimentally.

Objections that are based on the assertion that a behavior can be generally modified down to its smallest units (T. C. Schneirla

Figure 116. Goggle experiments of E. H. Hess: a chick with prism goggles. (Photograph: E. H. Hess.)

a

b

c

d

Figure 117. Space-perception studies of E. H. Hess. (a) In the 1-day-old chick the impressions of the pecks are more scattered around a target (nailhead) embedded in soft clay, as is the case (b) in the 4-day-old animal. A similar scattering, although displaced some distance away from the target due to the displacement by the prisms, in a 1-day-old (c) and 4-day-old (d) chick. (Additional explanation in the text.) (Photographs: E. H. Hess.)

1956), can not be upheld. Behavior cannot be indiscriminately shaped, and the numerous examples which show the great resistance to all attempts at modification disprove the hypothesis about the general modifiability of even phylogenetically adapted behavior mechanisms. P. Weiss (1941a) severed the leg muscles of newts from their tendons and grafted them onto their antagonists while keeping their original innervation intact. It was found that each muscle continued to react according to its original function, which led to the inappropriate leg movements. The newts were unable to learn the new coordination. Corresponding muscle transpositions and artificial crossing of nerve fibers in rats and monkeys also did not result in corrections. In man partial reorganization has been observed following muscle transposition (R. W. Sperry 1958). When the right and left limb buds were exchanged in newts at the time the anterior-posterior axis was already determined, this resulted in the front legs, which were now facing to the rear, walking toward the hind legs so that the animal was unable to move. No change occurred throughout the year following the operation. The newts were not able to learn to reverse the movement of the front legs and thus become able to move forward. Afferent impulses from the periphery were unable to change the central coordination and the legs moved in the manner they would have in their original position. Many of the functions of the central nervous system therefore are predetermined as phylogenetic adaptations (p. 21) and are quite resistant to change. P. Weiss (1939:558) expresses this in the following comparison:

The nervous system reminds us of an industrial complex, not only because of the multiplicity of cooperating agencies and the harmonic coordinations of their activities, but also because the organization is already present in its basic outline without prior experience and indeed before the beginning of its functions. The condition of the central nervous system, at the time it is first functionally activated, can be compared to a ship at the time of its launching. While it still remains to be completely outfitted it is already capable of floating, moving and being steered. In the same way the central nervous system, when activated, can transmit impulses, coordinate them and can control the musculature in a general way.

Just how rigid the various movement patterns are determined depends on the specific species and is different even within the same species in different functional cycles. Usually additional learning takes place in the sense of Lorenz's instinct-training intercalation.

Squirrels (*Sciurus vulgaris*) possess the movements of gnawing and prying, but they must learn how to employ these behavior patterns effectively when opening a nut. Experienced squirrels can do this with a minimum of wasted effort. They gnaw a furrow on the broad side of a nut from base to tip, possibly a second one, wedge their lower incisors into the crack, and break the nut open into two halves (Fig. 118). Inexperienced squirrels, on the other hand, gnaw

Figure 118. Hazelnut opened by an experienced squirrel using the wedging technique. (Photograph: I. Eibl-Eibesfeldt.)

without purpose, cutting random furrows until the nut breaks at one place or the other (Fig. 119a and b). They already try to wedge their teeth into the opening attempting to pry them open only leading to success if the furrows have been gnawed in the proper way. The first improvements in the technique can be seen when the furrows run parallel to the grain of the nut and are concentrated on the broad side of the nut (Fig. 119c and d). The squirrel follows the path of least resistance, and in this way the activity of the squirrel is guided in a specific direction by the very structure of the nut. The squirrel continues with its attempts to pry, and it keeps repeating those actions which have led to success. In this way most squirrels acquire the most efficient prying technique (I. Eibl-Eibesfeldt 1963). There are, however, individual deviations. Some squirrels learned to open nuts by gnawing a hole by a few closely spaced furrows (Fig. 120). One squirrel achieved almost instant success by gnawing a hole into the base of the nut and continued to use this technique.

Figure 119. A view of (a) the base, (b) the tip of a hazelnut opened by an inexperienced 66-day-old squirrel, the fourth nut tried. Numerous gnawing marks have produced many furrows on the nut in a random pattern. (c) The thirteenth and (d) the fourteenth nuts opened by the same animal, showing parallel gnawing tracks in line with the grain of the wood. (Photographs: I. Eibl-Eibesfeldt.)

Eventually it learned to gnaw the hole with a few closely spaced furrows (Fig. 121). Finally it changed to treating the thin-walled tip of the nut in the same fashion. Similarly other rodents develop techniques to open hard-shelled fruits (E. Petersen 1965).

I refer in this connection to the food-hiding attempts of squirrels (p. 22), which showed that a fairly large chain of innate behavior

Figure 121. (a) Hole-
gnawing technique; (b)
the hole-cracking tech-
nique derived from it.
(Photographs: I. Eibl-
Eibesfeldt.)

patterns matures predominantly in the absence of experience. The animal adds little to this technique by learning. In this functional cycle the animal possesses a primarily phylogenetically acquired behavior program, and it is quite obvious that in this case selection pressures work against adaptive modifiability. This is unnecessary because one behavior blueprint is sufficient for the performance of the task. Any modification of the tried and successful contains a certain risk. This is quite different in the case of nut opening. The variability of the available nuts requires various techniques and hence individual adaptation.

In England titmice learned to open milk bottles. By a recombination of innate behavior patterns of food getting they developed various methods of solving the problem. The acquired habit of bottle opening spread geographically, which supports the hypothesis that the skill is passed on by tradition (J. Fisher and R. A. Hinde 1949). Such learned behavior sequences can be called *acquired coordinations,* although they contain fixed action patterns as elements. They have been integrated into new functional units. Additional examples for acquired coordinations were discussed in connection with bird songs (p. 24) and begging movements (p. 112). In higher mammals we can generally observe that fixed action patterns consist of very short movement sequences, which are combined by learning into acquired coordinations.

The rat possesses all nest-building behavior patterns innately but learns the appropriate sequence of the individual components (I. Eibl-Eibesfeldt 1963). In a similar manner the inexperienced canary shows all the behavior patterns of nest building, but it must learn to integrate them into one functional whole (R. A. Hinde 1958).

An animal may possess an innate behavior pattern, but its application must be learned. The woodpecker finch of the Galápagos Islands (*Cactospiza pallida*) uses a tool to probe insects out of wood. Once it has opened the tunnel of an insect, the bird picks up a cactus spine or a straight, thin piece of wood, breaks it into the proper length, and proceeds to impale the insect in attempts to remove it (I. Eibl-Eibesfeldt and H. Sielmann 1962, 1965; see also p. 279). A male bird which I received when it was quite young did not fully master the technique. It searched for sticks, probed in crevices and holes, but only in a playful manner after feeding. When it saw an insect in a crack it dropped the stick and attempted to catch it with its bill. The bird learned only gradually to use the tool for this purpose.

D. Morris told me that chimpanzees which were born in captivity in the London Zoo poke sticks into cracks and holes in a playful manner. In the wild they use thin twigs to fish termites out of holes in the ground (p. 267). It seems probable that this behavior is based upon an innate disposition to use a tool, but that the specific application must be learned.

The inexperienced raven possesses a specific nest-building behavior, but it must learn what to use as nesting material. The bird tries everything—broken glass, cans, twigs, pieces of ice, and so on. These objects are pushed into place with sideways, trembling movements over the substrate at the prospective nest site. The frequency of trembling increases when the object meets with some resistance. If the object is wedged or caught, the raven stops. Glass and cans do not work as well as twigs, so the raven quickly learns which to choose. The night heron, on the other hand, innately knows the quality of the nesting material, but it must learn the best place for building a nest (K. Lorenz 1954c).

Frequently an orienting component of behavior is learned. Polecats (*Putorius putorius*) and other mustelidae kill rodents (such as rats), which are quite able to defend themselves by grasping them at their necks and killing them (Fig. 122). This orientation of the killing bite toward the neck of the prey they learn by trial and error. Isolated polecats, which have never killed a prey, attack a rat when it runs away, but they bite it in any part of the body they can get hold of. If the rat defends itself the polecat lets go and tries to grab again. It learns very quickly how the prey must be grasped in the neck so that it cannot bite back. Polecats that had an opportunity to play with litter mates can learn this much faster (I. Eibl-Eibesfeldt 1963). The polecat also learns to recognize its prey. At first it pursues any fleeing animal. A quietly sitting rat is approached and sniffed inquisitively. If the rat runs toward the polecat, it runs away. Only after having killed a rat is it recognized as prey, whether it sits still or approaches. When one raises polecats from early life

Figure 122. Female polecat killing a rat (neck bite). (Photograph: H. Sielmann from Scientific Film C697, Göttingen.)

with rats, they are clearly inhibited to bite them; they accept the member of the other species as a social companion (p. 343). In all other respects, however, the repertoire of prey-catching behavior remains unchanged; only different objects are able to activate them. Z. Y. Kuo (1967) observed the same in cats that he raised with rats. However, he interprets this observation by saying that undoubtedly no prey-catching instinct exists, or it would not be possible to change the "nature" of the cat in such an obvious way. He concludes that the body structure alone explains why a cat behaves like a cat and that it is not necessary to postulate additional instincts in the form of structures in the central nervous system: "The behavior of an organism is a passive affair. How an animal or man will behave in a given situation depends on how it has been brought up and how it is stimulated" (Z. Y. Kuo, 1932:37). However, this is only partially true if one observes carefully. Even the cats that were friendly to rats could still capture prey according to Kuo's account. Their behavioral repertoire was changed apparently as little as that of my polecats. The only thing that was changed was the object that released the behavior.

Polecat males grasp the females in the neck region during courtship, which immobilizes them. Animals raised in isolation grasp the female anywhere and have to learn as a result of the female's defensive behavior to grasp her at the right place. Polecats that were raised with siblings until they were 2 months old and were isolated after that time grasp the females correctly. They learned during play fighting that a conspecific will remain still when grasped at the neck (Fig. 123). The behavior patterns of mating itself are identical in experienced as well as inexperienced animals (I. Eibl-Eibesfeldt

a   b

Figure 123. (a) Polecat mating. The male holds the female in the neck. (b) Playfully wrestling polecats; one holds the other in the neck. During play the animals learn the neck bite, which later becomes a normal part of the mating behavior. (Photographs: (a) H. Sielmann, Scientific Film C697, Göttingen; (b) I. Eibl-Eibesfeldt.)

1963). F. A. Beach (1958) and K. Larsson (1959) found that inexperienced male rats copulated as well as experienced ones. This is also true for guinea pigs (*Cavia*), hamsters (*Cricetus*), golden hamsters (*Mesocricetus*), and rats (*Rattus*). The behavior patterns of mating are innate, but inexperienced animals often mount incorrectly oriented and clasp the head of the female, for example (E. S. Valenstein, W. Riss, and W. C. Young 1955; F. Dieterlen 1959; I. Eibl-Eibesfeldt 1953a, 1963; F. A. Beach 1942).

Rhesus monkeys raised in isolation are excited by females in estrous and try to copulate with them but they do not mount them correctly (Fig. 124). They are unable to learn this later in their lives (W. A. Mason 1965; H. F. Harlow and M. K. Harlow 1962), this in contrast to hand-raised chimpanzee males which can achieve intromission with the aid of experienced females who help them to overcome initial difficulties. Here social isolation has less permanent

Figure 124. The sexual behavior of male rhesus monkeys who (a) grew up normally with others; (b) who had no such social experience. These isolated animals become excited in the presence of a female in estrous and try to mount her, but intromission fails because they do not hold themselves properly on the hind legs of the female. (After W. A. Mason [1965] from I. De Vore [1965].)

effects, perhaps because normal mating behavior is more variable to begin with than in the lower monkeys (R. M. Yerkes and J. H. Elder 1936).

During normal ontogenesis more than learning and maturation with respect to motor performance is observed. Innate releasing mechanisms mature as well and they become increasingly more selective through learning, as we have discussed on page 67 with the example of the prey-catching behavior of toads and clawed frogs. New releasing mechanisms can be acquired by learning, as is discussed by W. M. Schleidt (1962).

From all the examples presented so far, it is clear that various species of animals are equipped with various innate dispositions to learn. These innate capacities to learn will be discussed in more detail in Chapter 14. Learning is not the result of a passive reception of stimuli by an organism. All observations support the view that there are often quite specific learning dispositions and internal motivating mechanisms, the latter being expressed in curiosity and play behavior (p. 238).

Innate disposition to learn

Species-specific learning capacities

Different species of animals possess different learning capacities. In many species the behavior is largely determined by phylogenetic adaptations in the form of fixed action patterns and innate releasing mechanisms, and very little is left to learning. Animals that are so equipped have the advantage that they do not have to acquire adaptations through time-consuming, risky learning experiences. The Yucca moth (*Pronuba yuccasella*) "knows" as a result of its phylogenetic construction that it must collect pollen prior to laying its eggs into the seeds of the yucca plant and rub it onto the stigma of the flower. Only in this way will the seeds develop on which the larva will feed. Adaptations of this nature are of advantage when the environmental conditions to which these adaptations are adjusted do not change appreciably. The more variable the environment is, the less precisely can the behavior be adjusted in advance. Changing environmental conditions require individual capacities for adjustment. Stenotopic forms can afford to run on tracks like "trains," but eurytopic forms are specialized for adaptive modifiability of behavior; they are the "specialists of nonspecialization" as K. Lorenz (1959) observed appropriately.

When adaptations of behavior are precisely determined, any change may be detrimental to adaptation. J. H. Fabré has shown this long ago in many experiments with insects. The digger wasp (*Ammophila*) opens and inspects the cavity which it has dug before it deposits caterpillars in it to serve as food for its larvae. It arrives with the caterpillar, drops it near the entrance, enters the cavity, inspects it, reappears head first, and pulls the caterpillar inside. If one removes the caterpillar to a place some distance from the nest while the wasp inspects the cavity, it will search for it until it has found the caterpillar, bring it back to the entrance, and the entire sequence of dropping, inspecting, and so on, is repeated. This can be repeated 30 to 40 times, at which time the wasp will finally carry the caterpillar directly into the cavity without prior dropping and inspection (G. P. Baerends 1941). The animal can adapt to the new situation only with great difficulty; its behavior follows a quite rigid program. Normally no disruptions occur, so the wasp achieves its goal quite readily. However, this rigidity of behavior is not necessarily found in all functional cycles. Just as squirrels learn very little when hiding food and much when they open nuts (p. 210), there are behavior systems in which the digger wasp can perform astounding learning tasks. For example, it learns the way home with its prey on its flight away from the nest. These digger wasps take care of several nests at the same time during the phases of caring for the brood (p. 163); they supply each of their larvae, which differ developmentally from each other, with the appropriate amount of food. What the digger wasp will do at a particular nest throughout the day is decided during the first inspection visit in the morning. Before beginning with hunting activities, the wasp visits all nests not yet permanently sealed and checks the contents. G. P. Baerends (1941) was able to induce wasps to bring more caterpillars than they normally would have by removing caterpillars from a particular nest. The wasp would bring less food when he added caterpillars. These manipulations influenced the behavior only when they were performed before the morning inspection by the wasp. Later manipulations were unsuccessful. This means that the wasp's behavior is determined for the rest of the day at the time of the morning inspection, and that the wasp is able to remember the condition of up to 15 nests for the entire day.

This is a memory task that could not be detected with the usual methods of delayed-response experiments. In this method, which was originally developed by W. S. Hunter (1913) and is still widely used today, an animal is trained in a multiple-choice apparatus to select one of several doors which are marked by specific stimuli (for example, a light that has been turned on briefly). Once the animal has mastered the basic task, it is prevented from responding as long as the positive stimulus is presented. The animal is allowed to respond only after various intervals of time, and the longest possible delay is considered a measure of the memory capacity of an animal. N. R. F. Maier and T. C. Schneirla (1935) questioned the value of this method according to which a gorilla capable of 48-hour delays would have a memory 576 times as good as an orangutan who mastered only a 5-minute delay. N. Tinbergen (1951), who cited this example, agrees with this criticism.

These observations teach us that standardized learning methods are not always appropriate for the comparative study of learning capacities in many species (see p. 257). If one wants to test the learning capacities of a heron, a rat, and a frog, one would be ill advised to run each of these animals through a maze. The rat, which normally lives in burrows, will do better than the heron or the frog. If we test the frog instead on bad-tasting prey models, we will discover a very rapid learning capacity.

There exist species-specific learning capacities for various functions. Predators are very intelligent in the functional cycles dealing with capture of prey, and animals that live in a specific area of their habitat are very good at learning paths. Some social lemurs show great social intelligence (A. Jolly 1966), which is in marked contrast to their otherwise lower level of intelligence. If one wants to obtain knowledge about the degree of adaptive modifiability of a species, one should first observe the animals under natural conditions. As stated earlier, learning capacities are adapted to the demands of the ecological niche of the species and its other inhabitants.

Most species whose members live in a specific place or which repeatedly visit a landmark acquire knowledge of their surroundings. The limpet *Patella* just follows its own mucous track back to its original resting place like a tracked vehicle (W. Funke 1965). The bee-hunting digger wasp (*Philanthus triangulum*) remembers the landmarks surrounding its nest and is thus enabled to find its way back. If one surrounds the nest with a circle of pine cones or stones while the wasp is in the nest and then transposes this configuration by 1 meter after the wasp has flown away, it will search upon return within the circle and not find the nest entrance (N. Tinbergen and W. Kruyt 1938; I. A. Chmurzynski 1967; see also Fig. 125).

When bumblebees find the small flowers of *Cynoglossum* they undertake a short orientation flight, apparently to learn the exact location. If, on the other hand, they find the large flower of *Digitalis,* which can be seen at great distances, they dispense with the exploration of the surroundings (A. Manning 1956).

Bees remember the location of good foraging grounds for a long time. Continuously dancing honeybees reproduce from memory the distance and angle to the sun making allowance for changes in the

Figure 125. (a) After leaving the nest the digger wasp makes an orienting flight and learns the position of the nest entrance; (b) after the circle of pine cones has been moved from around the entrance the returning digger wasp searches wrongly in the center of the circle now in a new place. (From N. Tinbergen [1951].)

sun's position, although they have not visited the place whose position they indicate with their dances for days. M. Lindauer (1963) cited the case of a continuously dancing bee, which had visited a food source for the last time on November 1, 1959, and which indicated the position correctly in the closed hive on December 8. When such continuously dancing bees had visited two food sources at different times of the day, they indicated within the closed hive the position of the feeding place whose visiting time was closest to the time of dancing (M. Lindauer 1957).

A young mammal which leaves its den for the first time begins by learning the return path to the nest. If the marine gobi (*Bathygobius soporator*), which lives in the tidal zone, is surprised in a tidal pool, it will jump out of the pool in the direction of the sea, from one pool to the next, until it reaches the sea. They cover distances of 10 meters, using up to 11 pools. If placed into an aquarium the fish will not jump, but they will jump if returned to the tidal pool, even after a stay in captivity of 14 days. If changes have been made in the environment, such as emptying a pool, they jump into the dry pool. They learn the path during high tide, as was demonstrated in elegant experiments by L. R. Aronson (1951). The path can be learned in darkness and may be remembered for 40 days. Equally extraordinary learning performances are shown by animals that return to their spawning grounds (for example, salmon, p. 381).

Many species learn while obtaining food, both with respect to the knowledge of the object and the manner of handling it. We will add some examples to those already cited (nut opening of squirrels, p. 210; prey catching in the polecat, p. 214, in the frog, p. 67).

Snapping turtles (*Chelydra serpentina*) which had been fed for 12 days with worms or fish, after hatching, subsequently showed clear preferences for the specific food, fish or worms, which they had been fed previously (G. M. Burghardt and E. H. Hess 1966). Many mammals and some birds learn from the mother what to eat and in this way traditions become established that are restricted to groups of animals in a specific locality. Mother squirrels, rats, and other rodents tolerate the seizing of food from their mouths by their young (I. Eibl-Eibesfeldt 1958; see also Fig. 126) and so do the parents of squirrel monkeys (*Callithrix*). Wood pigeons must learn that acorns can be eaten. The picking-off movement is innate, but they do not attempt to eat large fruit without the presence of adult pigeons which they have observed eating acorns (O. Heinroth and M. Heinroth 1928). Rats quickly learn as a result of bad experiences to avoid poisonous bait. If new bait is offered at first only a few animals of the group will try sublethal amounts of it. Usually they are not poisoned but merely become ill, and the bait is avoided thereafter. It is interesting that other members of the group also avoid the bait, taking their cues from those who have had the bad experiences. In this way local traditions come about, so that specific types of bait are rejected in certain districts of a city for several generations. On the North Sea Island of Norderoog, Norway rats specialize in catching birds, which mainland rats do not normally do. They overcame their innate fear of fluttering birds and learned to creep up to them and ambush them (F. Steiniger 1950).

Figure 126. Young Norway rat (albino), pulling a piece of food from the mother's mouth. (Photograph: I. Eibl-Eibesfeldt.)

In the Manyara game preserve in Tanzania lions have the otherwise rare habit of sleeping in trees, which leads one to suspect a local tradition. Female meerkats (*Suricata*) actually present food to their young, which snap it out of their mouths. In this way they learn to eat the same foods eaten by the mother and thus develop food preferences. The young of one litter did not accept bananas which the caretaker offered them but snapped at them when presented by their mother in her mouth. From this moment on, they accepted bananas readily (R. F. Ewer 1963). Actually, teaching seems to occur in the behavior of the mountain gorilla. G. B. Schaller (1963) observed a female removing an inedible hagenia leaf from the mouth of her young. Another animal helped a juvenile to dig up a root. The gorilla mother Achilla in the Basel Zoo also taught her son to move by retreating gradually. When he was older she placed his hand on the cage bars so that he could hold on. She also enticed him actively to climb the cage bars and watched over him. By pulling him in the small of his back, she stopped him from doing certain things. She was definitely interested in various activities of her child and either encouraged or limited the direction in which they could develop (R. Schenkel 1964). The origin and spread of new feeding habits through groups of Japanese macaques (*Macaca fuscata*) was well documented. A troop of these monkeys on Kôshima Island was regularly fed with sweet potatoes beginning in 1952. In 1953 Imo, a 1½-year-old female, was first seen washing the potatoes at the edge of a freshwater brook. She held the potato in one hand and cleaned off the sand in the water with the other. This new habit spread in the group during the course of the next several years, at first within the closer families and within the groups of playmates. Later it was passed on from mother to her children. In 1962 three quarters of all monkeys 2 years and older were washing potatoes (Fig. 127). At first the monkeys washed their potatoes only in fresh water. Later they also used seawater, which some monkeys seemed to prefer because of its salty taste. They gradually began to salt their potatoes by repeatedly dipping them into the water before the next bite.

a   b

Figure 127. (a) and (b): Stump-tailed macaques of Koshima Island (Japan) washing sweet potatoes. In (b) the young is observing the mother. (Photograph: Japanese Monkey Center, Aichi.)

At the same time these animals were fed wheat in the same area. It was spread along the shoreline. At first the monkeys carefully picked up each individual kernel until the same female, Imo, who had invented the potato washing and who was now 4 years old, began to gather the wheat with the sand and throw it all into the water, where the sand quickly separated from the lighter wheat. Until the time of the last report 19 of 49 monkeys are using this invention (M. Kawai 1965; S. Kawamura 1963; J. Itani 1958). The habit of swimming in the sea and begging with specific postures developed as patterns specific for the group and are passed on by tradition. Near Kyoto the macaques learned to keep warm at an open fire after the fashion of their keepers. In 1958 a female began joining the keepers at a fireplace and now all have the same habit. Group specific habits are known from free living macaque populations. The monkeys of Mt. Takasaki in Kyushu eat the fruits of the Aphananthe trees but spit out the kernels. The monkeys of Mt. Arashi, however, crush them to get the seedling. The monkeys of Mt. Minoo feed on eggs; those of Shodoshima do not eat eggs. Even social behavior patterns are influenced by traditions. High-ranking males of the Takasaki group carry young monkeys around, which are already weaned, but still need protection. This is not observed in other groups (D. Miyadi 1965, 1967). It is only necessary to show a chimpanzee once how to operate a push-button water faucet (R. M. Yerkes 1948). This behavior is then passed on by social tradition to others.

Knowledge of possible danger is also passed on by tradition. K. R. L. Hall (1965) frightened a monkey (*Erythrocebus*) mother by popping open a box containing a snake. Her young were present but could not see what was in the box. The mother was startled when she saw the snake and her response was observed by the young; from then on they avoided this box, although they had previously opened it frequently and without hesitation.

P. H. Klopfer (1957) trained ducks with electric punishing stimuli to avoid a bowl filled with water. Other ducks that were in the same cage and observed the procedure avoided the bowl without being punished themselves. Other examples about learning through observation can be found in P. H. Klopfer (1962).

Many animals learn about matters related to social behavior. Birds learn their songs, for example. Gregarious animals often learn to know their group members personally and form emotional attachments or bonds with other individuals. They also learn their rank within the rank order of the group (p. 343) and often the attachment to a particular locality which they defend as their territory. J. P. Scott (1963) in this context spoke of a process of "primary socialization," which serves to organize and control agonistic behavior.

Such habits are often tenaciously adhered to once they have been formed, and this is advantageous up to a certain degree, because in this way "tried-and-tested" behavior persists. Retention of such a habit is probably facilitated by the fact that each deviation from it is accompanied by feelings of displeasure and fear. K. Lorenz (1963) cites some interesting examples in this respect. One of his graylag geese, which had lived in his room, had become used to a certain detour. At first she always walked past the bottom of the staircase toward a window in the hallway before returning to the steps, which she then ascended to get into the room on the upper floor. Gradually she shortened this detour, but persisted in initially orienting toward the window, without, however, going all the way to it. Instead she turned at a 90° angle once she was parallel with the stairs. Once Lorenz forgot to let the goose into the house at the usual time. It was beginning to get dark and the goose ran, against her usual habit, directly toward the staircase as soon as the door was opened and began to climb up.

Upon this something shattering happened: Arrived at the fifth step, she suddenly stopped, made a long neck, in geese a sign of fear, and spread her wings as for flight. Then she uttered a warning cry and very nearly took off. Now she hesitated a moment, turned around, ran hurriedly down the five steps and set forth resolutely, like someone on a very important mission, on her original path to the window and back. This time she mounted the steps according to her former custom from the left side. On the fifth step she stopped again, looked around, shook herself, and performed a greeting display behavior regularly seen in graylags when anxious tension has given place to relief. I hardly believed my eyes. To me there is no doubt about the interpretation of this occurrence: The habit had become a custom which the goose could not break without being stricken by fear (p. 112).

Margaret Altmann (cited by K. Lorenz) was obliged to unload and reload her old horse symbolically at a location where they had previously camped on several occasions. Unless this was done the horse would not continue. A dog belonging to an acquaintance once scared up a mouse when he had accompanied his master to the shed to fetch coals. Since that time the dog performs a "mouse jump" each time he comes to the shed. At another time he had seen a rabbit running ahead in the car's headlight. Since then he always chases ahead wildly in front of the car when it returns at night. The force of habit is no less strong in man.

These examples should suffice to illustrate the learning dispositions of different animal species. The basic assumption underlying all classical learning theories that all responses are about equally conditionable to all stimuli is no longer tenable. Further support for this view is presented in K. Breland and M. Breland (1966) and J. Garcia and others (1969). One of these innate learning dispositions is illustrated by the imprinting phenomenon which occurs under special conditions and is most likely the result of specific learning mechanisms. It will be discussed in the following sections.

Imprinting and imprinting-like learning processes

The course of learning may be determined in various ways by phylogenetic adaptations. We learned earlier that chaffinches in a choice situation prefer the song of their own species, the preference being based on innate knowledge as to the type of song to be learned. Chimpanzees know innately that threats are made by producing noise, but they must learn the method of producing it.

It is somewhat different in the zebra finch, which learns its song from those who feed it. If a society finch feeds zebra finch young, they will learn the society finch song, although zebra finches are singing in the adjacent cage. If they are fed by both society finches and zebra finches, they will learn the song of the zebra finch. Thus a preference for the song of the species as the model becomes evident even here (K. Immelmann 1967).

In many cases learning is genetically programmed so that the animals are able to learn in specific sensitive periods and possess specific learning capacities only at this time. If this period is restricted to a developmental period after which the animal can no longer learn, we speak of a critical period. Such periods come to an end, even when no learning has taken place, as the result of internal changes. Zebra finches that have been isolated from species members before they are 35 days old are unable to distinguish males and females of their own species. They are chased by their own father when they are between 35 and 38 days old, and from then on males become the object of their aggressions. If they have not had this particular learning experience, they are unable to learn it at a later date (K. Immelmann, personal communication).

If an animal has learned something during a sensitive or critical period of its life, the readiness to learn ceases, and in general the animal firmly adheres to that which it has learned. Many examples are known of animals learning particular details of an object of a fixed action pattern during a specific developmental period. After this they seem to be fixed on the particular object in respect to the particular drive (object imprinting). In later research on specific sensitive periods for learning, motor patterns were discovered. These phenomena will be discussed in the following sections. For reviews, see P. P. B. Bateson (1966) and W. Sluckin (1965).

Object imprinting

Many innate behavior patterns can be released by unspecific key stimuli. Rhythmic calls and the most diverse moving objects release the following reaction in a young graylag gosling shortly after hatching. It follows a man as readily as it would a goose or a moving box. If it follows such an object even for a short time, it will remain with it. Once the gosling followed a person it could not later be induced to follow its own mother (K. Lorenz 1935). In respect to the following reaction it has become imprinted to man. The same is true for chicks and ducklings. The reaction to auditory stimuli, while modifiable, tends to be more selective. A number of species clearly prefer the call notes of their own species (G. Gottlieb 1965a, 1965b; see also p. 68). This innate preference reduces the chance of imprinting to the wrong object.

The statement that an animal is imprinted on something always refers to a specific reaction whose releasing stimulus situation becomes determined. In the example just cited this was the following reaction. Cichlids frequently distinguish their own young from those of other species and eat the latter. *Hemichromis bimaculatus* pairs preferred young of another species if they were given eggs of the other species instead of their own during their first breeding cycle (A. A. Myrberg 1964). In this case the reactions involved in parental care become imprinted to an object. This is also true for behavior patterns of other functional cycles, such as those of reproduction and food getting. A jackdaw that has been raised by human hand from the nestling stage will join a flock of jackdaws when becoming fledged if an opportunity is available. In the following year, however, it will court humans during the reproductive season, even if other jackdaws are available. In respect to its sexual reactions the bird is imprinted to man and prefers him over his own species (K. Lorenz 1931). This is all the more remarkable when you consider that this kind of imprinting seems to take place at a time when the animal does not yet show sexual behavior (Fig. 128).

The fixed action patterns themselves do not undergo any noticeable changes in imprinted animals. A human-imprinted ring dove (*Streptopelia risoria*) courts the human hand with the same behavior patterns with which he would normally have courted a female of its kind, and human-imprinted female doves invite courtship feeding from the human hand and squat before it in the copulation posture (E. Klinghammer 1967).

K. Lorenz (1935) discovered the phenomenon of imprinting and emphasized several criteria which distinguish this learning process from normal association learning. E. H. Hess (1959) has further clarified these differences.

a   b

Figure 128. Examples of sexually imprinted animals: (a) The rooster, which is imprinted on mallard ducks, waded regularly into the water to join them. He did this most frequently in this duck cage because it was possible for him to approach the ducks very closely. Otherwise they would usually swim away when he approached. (b) Even when 7 years old these three male wood ducks still behaved homosexually, although numerous females were present. They inspect a nesting box in search of a nest site. (Photographs: F. Schutz [1965a, 1965b].)

1. Imprinting takes place only during a specific *sensitive* period. If this time passes the animal can no longer be imprinted. This sensitive period does not have to be restricted absolutely to the first few days or weeks of life. The time and duration of this period can vary depending on the reaction even within the same animal. For the following reaction the sensitive period is between 13 and 16 hours after hatching. E. H. Hess (1959) determined this by imprinting ducklings in a special apparatus after keeping them in darkness for varying periods of time from 1 to 35 hours after hatching (Fig. 129). For 1 hour they were allowed to follow the model of a mallard male duck, which called by means of a built-in loud speaker. Then they were returned to darkness. For the test they were once again placed into the apparatus, where they could choose between a male and a female model. Both models were at first silent; after 1 minute both began calling, the male model with an artificial gock, gock, gock, gock call, the female model with the recorded natural call of a mother duck. During the third phase of the test the female model alone called, and in the fourth test the female model also moved while the male model remained stationary. If the ducklings that were imprinted on the male model approached the male model, in all four cases the imprinting was scored as 100 percent. The results can be seen in the graphs (Fig. 130). The strength of imprinting increases proportionally to the amount of distance followed. This could account for the increase and decrease of imprintability with age. Initially the young animal is too weak to follow. The older it becomes, the better it is able to do so. The decrease of imprintability could

Figure 129. Imprinting experiments of E. H. Hess. The experimental situation: Chicks or ducklings that were maintained in a dark box from 1 to 35 hours after hatching until they were to be used in the experiment were introduced onto the circular runway. Here a male mallard model with a loudspeaker inside was moved in a circle around the runway. Each animal remained for 1 hour with the model. After a certain time interval, during which they were kept in the dark, they were tested as described in the text. (From E. H. Hess [1959].)

Figure 130. Sensitive period is indicated by the peaks of the curves around 13 to 16 hours after hatching. Left: Average of positive responses; right: percentage of animals that had errorless scores in all four test situations. (From E. H. Hess [1959].)

be due to maturing avoidance responses (E. Fabricius 1951; E. H. Hess 1959).

2. The acquired knowledge of the releasing object is retained for life, while normally, forgetting is common after learning. What is learned is not only retained but the imprinting object is also preferred during the rest of the animal's life; imprinting is then *irreversible*. It is possible to induce two hand-raised parakeets to mate and incubate eggs. But if they are again exposed to man both will court the human, the pair bond is broken and the brood is neglected (G. Hellmann, cited by K. Lorenz [1954c]). F. Schutz (1965) raised male ducks of various species with chickens, ducks, and males of other duck species and confined them all together for several weeks. Then they were released onto a lake on which conspecifics which they had not seen until then were also present. Now they had a choice. In the following spring most of the males courted members of that species with which they had been raised, but normally not those individuals with which they were raised. In numerous instances the effects of imprinting were shown after the animals had mated with a female of their own species and had lost her through accidental death or by separation by the experimenter. The imprinting effects of early experience in such cases showed up even in males that were normally paired to females for a period of two years (F. Schutz 1968). This shows quite clearly the difference between imprinting as opposed to association learning via the usual reinforcement contingencies. In additional experiments F. Schutz (1965b, 1968) imprinted male ducks on other males of the same species and then induced them to mate with females of their own species. The imprinted homosexual preference remained, however. W. M. Schein (1963) imprinted three male turkeys on people and three to conspecifics. All were equally tame. In the absence of other turkeys they all courted people, and they courted turkeys when no humans were present and mated with them. However, when humans and turkeys were present at the same time the human-imprinted males courted humans, and turkey-imprinted turkeys courted their own kind. The turkeys still showed these preferences when they were 5 years old. K. Immelmann (1966) had male zebra finches (*Taeniopygia guttata castanotis*) raised by female society finches (*Lonchura striata f. domestica*). All males courted only society finches. Then each male was kept isolated with zebra finch females. They courted, bred, and raised young. When their preference was tested later in the free-choice situation, they preferred society finches to conspecific females.

3. During imprinting the animal learns only supraindividual species-

specific characteristics. A male mallard duck that was imprinted to a male sheld duck courted all male sheld ducks. A human-imprinted graylag goose follows all humans.

4. Only specific reactions become imprinted to a particular object. A jackdaw raised by K. Lorenz (1935) regarded humans as parent and sexual companions.[1] The bird flew about with hooded crows as flight companions and accepted young jackdaws as child companions.

5. The determination of the object for an instinctive activity can take place, as in the case of sexual imprinting cited above, at a time when the appropriate behavior pattern is not yet matured and thus has not yet been performed by the animal.

6. During imprinting of the following reaction of ducks, there was a greater effect when the imprinting experience of following a model took place in massed trials than in distributed trials, which is usually more effective for association learning. Painful stimuli strengthen imprinting, whereas punishment usually results in the avoidance of the associated stimulus (E. H. Hess 1959).

Of the criteria for imprinting cited above some also pertain to other forms of learning, such as the sensitive period and the learning of supraindividual characteristics, which is by no means present in all imprinting-like learning processes. Imprinting is principally recognized by the criteria listed under the second point, especially those pointed out by E. H. Hess, that is, that primacy of the imprinting experience is more important than recency, while the reverse is true for normal associative learning. From this fact derives the observation that imprinting is irreversible.

The assertion made in point 5, according to which imprinting of an object can take place for an action that is not manifested until it has matured later, cannot be regarded as proved. In the example of sexual imprinting it is possible that the initial orientation of a young animal to an object already constitutes the first link in a chain of sexual behavior patterns. Such an association could be rewarding in itself while being quite independent of the actual mating behavior. The motivation of such an initial orientation to another must be studied before we can make statements about the presence or absence of sexually motivated behavior in young animals.

As is the case with all "injunctive" concepts (B. Hassenstein 1955), it is not always possible to draw a sharp distinction between imprinting and other learning processes. Transitions to normal associa-tion learning can be expected (E. Klinghammer 1967).

Object imprinting takes place especially when the object of an instinctive action shows a lack of releasing signals. Female ducks of

---

[1] Companion (Kumpan) is a partner in a specific functional cycle.

different species are often very similar to one another, while the males have very conspicuous releasers (green head and white neck ring in the mallard). Thus the imprintability of the various sexes is also different. The male mallard must learn to recognize the plain-looking conspecific female, and this knowledge is acquired during the imprinting experience. If raised with other species it will accept them as sexual companions. The female, on the other hand, cannot be imprinted with respect to her sexual behavior. She knows innately the releasing, species-specific signals of the male. Even if she is raised with other species she will only court males of her own species when adult (K. Lorenz 1935; F. Schutz 1964; see also p. 87).

Imprinting of motor patterns

Imprinting-like learning processes also exist for the acquisition of motor patterns. Chaffinches learn their song only during the first 13 months of their lives. Toward the end of their sensitive period their learning capacity increases for several weeks. It is sufficient, however, that they are exposed to the song at a time when they do not as yet sing themselves. Chaffinches that were isolated from their parents during the month of September, long before they had begun to sing, produced a normal song during the following spring. What they have heard is not forgotten (W. H. Thorpe 1958a). O. Heinroth (cited by K. Lorenz [1954b]) once recorded the song of blackcaps (*Sylvia atricapilla*). In the same room he kept 12-day-old nightingales (*Luscinia megarhynchos*) which only uttered the begging call at that time. The total time these birds were exposed to the blackcaps was about 1 week. When they began to sing the following spring they surprised Heinroth with the complete blackcap song, which was identical with the recorded song. The white-crowned sparrow (*Zonotrichia leucophrys*) possesses differing local song dialects which are learned during a sensitive learning period. Several months separate this critical period and the time when the birds begin to sing. The birds therefore must have retained what they have heard (P. R. Marler and M. Tamura 1964). Zebra finch males which have been raised by society finches learn the song of their foster father. This acquisition is completed by the time they begin to sing on their own. When a zebra finch has heard the song of a society finch for the first 35 days of its life, and from then on only hears the song of zebra finches, it will always sing like a society finch (Fig. 131). The females, too, remember what they have heard when they were young, although they do not normally sing. If, on the other hand, they are given male hormones they will sing the song of the species on which they have become imprinted (K. Immelmann 1965). According to M. Konishi (1965b) the same is

Figure 131. Sound spectrograms (a) of a society finch; (b) of a zebra finch raised by (a); (c) of a normally raised zebra finch [natural father of (b)]. Abscissa: time, in seconds; ordinate, frequency range, in cycles per second. (From K. Immelmann [1965].)

true for the white-crowned sparrow. In motor "imprinting" apparently the relevant information is acquired at a time when the behavior which is learned, the song, does not even exist in rudimentary form.

Imprinting-like processes in mammals

Only a few observations have been made about imprinting in mammals (U. Grabowski 1941; G. Grzimek 1949b). Hoofed animals such as horses and sheep can be imprinted to man in respect to their following reaction in the same way as graylag geese, if one assumes the parent role shortly after birth. Dogs pass through a critical period with respect to the development of social relationships during weeks 4 and 6. During this time they form a close social bond to conspecifics or to man as a substitute, regardless of whether they are punished, fed, or treated indifferently. J. P. Scott and J. L. Fuller (1965) have emphasized that the internal process on which this readiness for contact is based seems to be more important than the external factors. It is possible that imprinting exists also in man

(P. H. Gray 1958; W. H. Thorpe 1961a), but there are as yet no experimental investigations.

Psychoanalysts were perhaps the first[2] to recognize the existence of sensitive periods during the course of individual human development. During the early childhood years certain environmental influences can affect the behavior of a child in decisive ways and can cause possibly irreversible disruptions. The well-being of an infant is not only dependent on hygiene, but personal contact provides an important stimulus for further development. During the first year of life a short separation from the mother can result in a serious retardation. A separation of several months can often lead to irreparable damage, and infant mortality is increased in such instances (R. A. Spitz 1945, 1946; R. A. Spitz and K. M. Wolf 1956; W. Goldfarb 1943; J. Bowlby 1969, 1952; A. Dührssen 1960). It is especially during the second half of the first year of its life that the child develops a personal bond with its own or a foster mother. This bond is the basis for the development of "original trust" (E. H. Erikson 1953), the basic attitude toward one's self and the world. The child learns that it can depend on another person and this positive attitude is a cornerstone of a healthy personality. If this relationship is disrupted, a basic attitude of distrust develops. Such a development can also be caused by a prolonged stay in a hospital in the second half of the first year of life. The child will attempt even there to achieve close contact with a substitute mother, but no nurse can bestow such intensive care on an infant so that such a bond could be established. The nurses change so frequently that any growing relationships are continually broken off. A child that has been repeatedly disappointed in its efforts to make lasting contacts becomes apathetic after a short rebellion. During the first month in the hospital it will cling to a caretaker and whine often. During the second month there is much crying and a loss of weight. During the third month of separation such children only whimper a little and at last they become completely apathetic. When they return home after 3 to 4 months, they recover, but after longer periods of separation, the damage is irreversible. These children remain retarded in their development despite the best of hygienic care and nutrition; no one plays with them and no one carries them around. Of 91 children in one orphanage who were studied by R. A. Spitz (1946) and R. A. Spitz and K. M. Wolf (1965) and who were separated from their mothers from the third month of life, 34 had died by the end of their second year. The

[2] M. Montessori, in her description of the human development, also employed the concept of a sensitive period: "One is here concerned with a special receptiveness which occurs during development, that is, during childhood. These periods are transitory and serve to acquire certain abilities" (M. Montessori 1952:61).

developmental index of the surviving ones was 45 percent of normal. The children practically functioned at the level of idiots. Even after 4 years many could not stand, walk, or speak. In this orphanage one nurse took care of 10 children. In another home where the mothers were able to care for their children much of the time, none died and the children developed normally. When these socially deprived children survive, they are frequently lacking in affect and avoid contact. They are only capable of superficial contacts and may completely shun all close relationships. Certainly these observations of R. A. Spitz need further examination, especially concerning the question if the retardation is primarily due to emotional factors and whether the damage is indeed irreversible. Contradictory evidence has been published by W. Dennis (1960).

J. Bowlby (1969) describes the effects of separation from the mother upon children aged 15 to 30 months. In an initial phase of protest the child appears acutely distressed at having lost his mother and tries to establish contact by crying. He behaves as if he expects the return of the mother any time and normally rejects other persons. This period may last for over a week. It is followed by the phase of despair during which the child is quieting down. He is withdrawn and inactive. The final phase of detachment is characterized by an increased interest in the environment. The child accepts nurses and is sociable, however with no strong emotional tie to anyone. When the mother visits him, the child seems hardly to know her. "He will cease to show feelings when his parents come and go on visiting day; and it may cause them pain when they realize that, although he has an avid interest in the presents they bring, he has little interest in them as special people. He will appear cheerful and adapted to his unusual situation and apparently easy and unafraid of anyone. But this sociability is superficial; he appears no longer to care for anyone" (J. Bowlby 1968:28). It seems that the human child is innately biased to attaching himself to one figure and if this attachment is prevented, serious disturbances of social behavior are the consequence.

In this connection the investigations of H. F. and M. K. Harlow (1962a, 1962b) deserve attention. They raised rhesus monkeys without their mothers. The animals had only surrogate mothers, which were either covered with terrycloth or consisted of bare wire mesh. Attached to these models were bottles with nipples from which the baby monkeys could suck their milk. The monkeys that were raised in this way later proved to be poor mothers. They allowed their young to be removed without protest, did not nurse their young or only after some time, and even mistreated them (Fig. 132). Here, too, an early childhood experience led to substantial disruptions of later social behavior.

Figure 132. (a) and (b): Normal rhesus monkey mothers care for their young even in captivity; (c) and (d) rhesus monkeys that grew up in isolation remain indifferent toward their own young; sometimes they are even rejecting or aggressive. (Photographs: Sponholz, Primate Laboratory, University of Wisconsin; H. F. Harlow and M. K. Harlow [1962a, 1962b].)

In human beings the case of hospitalism that was cited above is undoubtedly the crassest case of loss of love, but during the normal course of development there are various degrees of early childhood experiences of this nature which only show their effects much later. Infants are certainly adapted to a close contact with the mother, and children experience such contact among all peoples in which children are carried about. The child that is placed into a carriage or into a bed by itself misses this feeling of security in our type of culture. We do not know what the consequences are for the later development, but it is entirely feasible that our characteristically sober, detached, and critical attitudes, as well as many neuroses, have their root in these practices. M. Mead (1965) has attempted to interpret some idiosyncracies of various cultures in line with their differential early childhood experiences. However, with the current incomplete state of our knowledge, one must be careful not to equate a plausible interpretation with a causal explanation. M. Mead writes, for example, that the Mundugumur of New Guinea nurse their children only reluctantly, remove them abruptly from their breasts, allow them to cry, and in general treat them roughly. This, she feels, is the root of their later aggressiveness. The Arapesh of New Guinea, on the other hand, have a very close, warm, and permissive relationship with their children, and they are very docile as adults. Seen in this light the explanation seems to fit, but whether or not this different treatment of infants is actually the cause of later behavioral differences has not been proved. There exist a number of warlike tribes who nevertheless treat their children with utmost love and kindness. Included here are the Nilotic tribes, where the Masai, are especially aggressive. They in no way mistreat their children; on the contrary, children are carried on the mother's body, spoiled, and cared for, including by the fathers.

In the development of the human child further sensitive periods can be found. Between the age of 2 and 3 years European children usually begin to explore actively and investigatively interact with their environment. If this activity is excessively suppressed and the child punished for its often destructive experimentation, and if not enough guidance is given, then the impulses for independent actions soon become surpressed. Creative activity is impeded, and such children may lose the capacity to live according to their own perceptions and ideas, and they may be disrupted in their initiative and work attitudes (A. Dührssen 1960).

Around the fifth year of life the child experiences a critical period (the oedipal stage, S. Freud), which is important for later sexual life. The sexual drives begin to mature, and sex-specific behavior is determined. Members of the opposite sex become the object on which appropriate behavior directed to the other sex is practiced. A boy may seek physical contact with his mother during this period, crawl into bed with her, and be very tender. At the same time he identifies with his father, who is his model. If the mother is too rejecting, homosexual preferences may result. On the other hand, permissive or even seductive behavior of the mother may occasionally lead to behavior disorders. A boy may become strongly sexually attached to the mother, and this may lead to feelings of guilt toward the father. In this way at least the demonstrable sexual aberrations have been interpreted by psychoanalysts.

Girls seek contact with the father, who then becomes the object for practicing behavior toward male partners. At the same time they begin to assume the role of the mother. If the mother does not readily accept her female role or acts in a pronounced masculine way, identification with her is more difficult and female homosexuality can develop. A rejecting father can influence the behavior of a girl in the same homosexual direction.

The experiences of the oedipal stage affect the process of self-identification. Object fixations for the sexual motivations occur probably much later. Sexual pathologies such as handkerchief or shoe fetishisms can often be traced to earlier sexual experiences (R. v. Krafft-Ebing 1924). Homosexuality may also be traceable to an object imprinting in the final analysis; a predisposition for a specific sex role may have been acquired during the oedipal phase. The parallels to imprinted homosexual animals are striking. This seems to be supported by the fact that such preferences are highly resistant to therapy. To what degree a true cure is possible should be tested with the pupillometric tests of E. H. Hess (1965).[3]

During puberty there is probably another sensitive period in which the young person is especially receptive to new values. Young people are apt to identify with a social group, and often the final political and religious attitudes are determined for the rest of life.

Knowing about the strong fixations caused by imprinting, one is confronted with the question whether it is right to expose a child during the respective sensitive period with ethical concepts. Any fixations of that sort hamper the development of free will. This is certainly dangerous in modern times.

In discussing this point H. Hass (1968) wrote:

To be fair towards our children one should expose them up to the age of 16 only with those ethical concepts that are agreed upon by all men. One should furthermore point out to them all the dangers of early fixations and emphasize that they have a right and even the duty to judge on their own, even if they should come into opposition with the parents of the community. This is, of course, a utopic concept for the moment. But such development is perhaps already indicated in our young people (p. 206).

---

[3] E. H. Hess found that the pupils of a test subject are dilated when some interesting material is viewed but contract when something unpleasant is seen. If a normal man is shown pictures of girls, men, and other objects, the pupil response is positive, especially to girls. Homosexual men, on the other hand, react positively to pictures of men. One should test how cured homosexuals respond.

Curiosity behavior and play

A conspicuous disposition of animals for learning is expressed in curiosity and play. If we place a new object into a room inhabited by rats, all rats will explore it very soon. They will cautiously approach the new object repeatedly at first. Finally, they will sniff, chew, climb, and urinate on it—hesitatingly at first, later with more confidence. Once they have explored the object in this way they lose all further interest. In a similar manner many higher mammals investigate a new object by sniffing, looking, chewing, biting, scratching, or manipulating it in some other way. What are the typical characteristics of this exploratory behavior?

Close observation reveals that the exploring animal alternately approaches and withdraws from the object of its interest. It makes contact with the object by means of its sense and effector organs, withdraws, and approaches again, with a new view of the situation, so to speak in order to become acquainted with all aspects of the object. The animal is attracted by the object, but it does not get interlocked with it rigidly by a particular behavior, retaining the ability to withdraw again. This ability to withdraw to a distance is the basis for any dyadic interaction. It is typical for curiosity behavior and play. During the ontogenesis of the human child development of this ability can be readily observed. When the child first reaches for an object, the first actions are fairly stereotyped. The child grasps the object, brings it to its mouth, and begins to suck it; at first this is all that is done. Soon afterward the child is able to remove the object from its mouth again, to look at it, suck it some more, perhaps drop it, and reach for it with the other hand. By now the rigid action sequence has become more flexible and the child is able to explore. We shall discuss later that this ability to withdraw to a distance is one of the roots of human freedom of action.

At least during their youth, most mammals are pronounced "creatures of curiosity," actively seeking out new situations for exploration under the compulsion of some internal drive. In respect to curiosity various groups of animals behave differently both qualitatively and quantitatively. Primates and carnivores are more curious than rodents, and within the rodents porcupines are more curious than mice. Squirrels are somewhere in between. Rodents gnaw at new objects and occasionally hoard them. Monkeys look at the objects and try various things with their hands, the *Cercopithecinae* more than the *Colobinae*. Some fish and birds are also curious. Once a new object is explored, all interest gradually wanes (S. E. Glickmann and R. W. Sroges 1966; A. Wünschmann 1963; see Fig. 133).

There seems to be a drive to learn that can be called a "curiosity

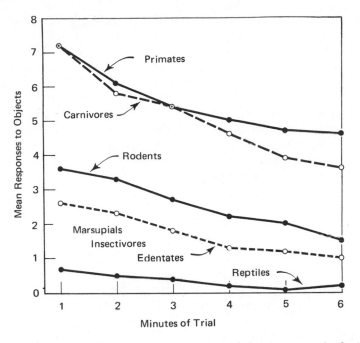

Figure 133. Average readiness to react to test objects during the course of a 6-minute period. To measure the approach to an object the authors divided the 6 minutes into 78 5-second intervals. When an animal reacted to an object within such a 5-second period this was noted. In this way a value scale was obtained which represents the ordinate. The curves show the average values of the length of contact. They were computed from four presentations each with different objects. Two conclusions can be drawn: (1) Higher mammals more intensively interact with new objects than do lower mammals and reptiles; (2) within one 6-minute period the interest in a new object already begins to wane. (From S. E. Glickmann and R. W. Sroges [1966].)

drive." It can be demonstrated that there is a drive to learn new motor skills as well as to receive new perceptual impressions and thus to acquire knowledge. Some experiments have supported this view. Rhesus monkeys learned a puzzle game without any other reward than the performance of the task itself (H. F. Harlow, M. K. Harlow, and D. R. Meyer 1950). They also learned a task when as a reward they were allowed to look out a window of their cage (R. A. Butler 1953; additional examples in E. R. Hilgard [1956]).

In exploring rats, neurons in the lateral hypothalamus and in the preoptic region are active. Electrical stimulation at these locations is rewarding and rats learn readily to self-stimulate by pressing levers (B. R. Komisaruk and J. Olds 1968).

This drive to learn also seems to underlie the play behavior of animals. It is usually easy to recognize when an animal plays and when it is more seriously occupied; nevertheless it is not easy to give a definition of play. "The animal works," writes F. Schiller, "when the behavior is motivated by some need, and it plays when an excess of energy provides the motivation." This statement aptly characterizes one of the roots of play. An animal only plays when it is satiated and when it is not occupied with other tasks. Play is not dictated by any immediate need, but it is nevertheless of enormous importance for the normal development of an animal. The nature and value of play becomes clear when we study how widespread it is and when we look more closely at the various forms it can take. Only the most highly developed "learning animals"—those which seek new situations out of their own initiative—play in the true sense. They are curious and try out new behavior patterns, learning in the process. Insects do not play; neither do fish and amphibians. However, many mammals do, especially when young, and so do some birds. The taxonomic distribution of play seems to indicate that play is related to learning, and we shall see that an animal interacts with its environment during play. It experiments with things in its environment and acquires knowledge about their characteristics. The animal collects experiences during play with conspecifics and learns the possible range of its own movements. Play always implies a dialogue with the environment, and this dialogue is always the result of an internal drive. One could even assume a separate drive to play, but I am inclined to believe that the drive to learn, which is the basis of all curiosity behavior coupled with an excess of motoric motivation, will suffice to account for the phenomenon of play. Learning takes place during playful experimentation in the same way as during endlessly repeated movement games. In this connection it is interesting to observe that animals develop special modes of action during play; that is, they play at particular times and only until they fully master what they have practiced; then they lose interest and try something new (I. Eibl-Eibesfeldt 1950a).

An animal may fight with a companion during play or flee before a predator, to mention only two examples. But in most instances the play version is clearly distinguishable from the actual behavior: The serious aspects of the behavior are lacking. An animal that plays at escaping does not actually flee. A rat that actually flees underground will not reappear as quickly and then only with hesitation. The rat playing escape reappears at once. When two polecats pursue one another in play they frequently change roles of the follower and the pursued, which does not happen when the chase is in earnest. When engaging in rough and tumble play they have an inhibition to bite, and the threat behavior, which accompanies a real fight, is missing. This is true for play-fighting rats, as it is for

PLATE VII

Right: chimpanzee paintings (Tierpark Hella-brunn, Munich). Top: painting of a high-ranking female. The sequence in which the colors were offered was red, blue, and green. The animal filled the available space. Bottom: painting of a low-ranking female. Color sequence: red, black, blue, and yellow. The animal painted one color over the next. (Photograph: I. Eibl-Eibesfeldt.) *Bottom:* protective coloration and behavior. The somatolytically colored perch *(Cephalopholis miniatus)* remains still when the diver approaches. The colorful *Paracanthurus theutis* tries to hide and offers the observer only a view of the narrow back. (Photograph: I. Eibl-Eibesfeldt.)

PLATE VIII

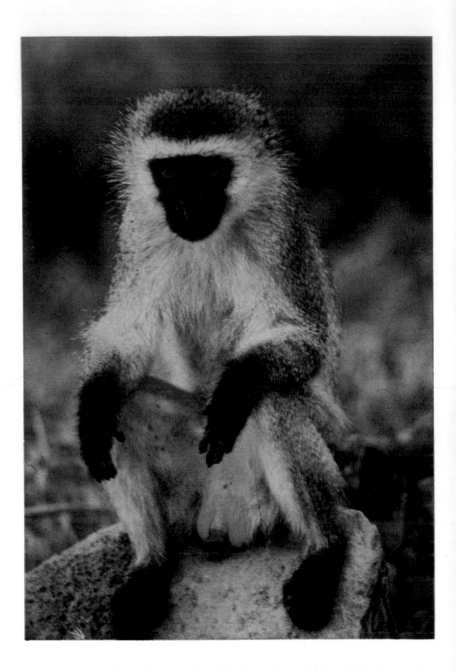

Vervet monkey *(Cercopithecus aethiops)* displaying the colorful genitalia when sitting (Tsavo National Park, Kenya). (Photograph: I. Eibl-Eibesfeldt.)

squirrels or any other play-fighting predator. Squirrel monkeys squeak continuously when playing. If the squeaking stops, a serious fight has started (P. Winter, D. W. Ploog, and J. Latta 1966).

Play behavior may also be repeated many times over, which is especially noteworthy because usually instinctive actions are quickly fatigued. A dog will retrieve a stick many times. It may play fight with another dog until sheer exhaustion or a new exciting stimulus brings it to an end.

How can this be interpreted? At first, one might think that in play we are dealing with instinctive behavior which has not as yet matured. Frequently this is not the case. Animals may still play fight with all the accompanying social inhibitions at an age when they are already able to fight seriously; this is readily observed in dogs. During play a number of fixed action patterns appear which are in fact part of serious fighting behavior, but these behavior patterns occur independently of their usual function. They do not appear in the sequence in which they appear in the real fight and sometimes behavior patterns from different functional cycles are combined which are mutually exclusive in an actual fight, such as when sexual behavior is mingled with fighting and prey-catching behavior. It appears as if these behavior patterns are activated in something other than their normal fashion. The movements of lower levels of integration are more or less independently activated and not by the arousal of an entire instinct (in Tinbergen's sense) (I. Eibl-Eibesfeldt 1951c, 1953a; M. Meyer-Holzapfel 1956). This could explain why subjective correlates that accompany the real fight are lacking during play. These subjective correlates are perhaps only aroused when the behavior is activated from a higher level of integration, as was implied by the brain-stimulation experiments of E. v. Holst and U. v. Saint-Paul (p. 72). Only when behavior patterns are thus independent of the normally involved superior centers can an animal freely manipulate its various movements, recombine them in new ways, and thus experiment with its movement repertoire. It has been shown that the animal actually learns many things which are indispensable for the acquisition of behavior needed in later life.

Polecats, for example, learn during play the orientation of the neck bite which is necessary for successful mating (p. 214), and the woodpecker finch learns the use of the tool (p. 213). Chimpanzees who have not handled sticks do not discover that they can be used to pull in a banana into their cage. After playing with sticks for three days they can solve this problem within 20 minutes (H. G. Birch 1945). A prerequisite for this freedom to play is the condition that an animal is free of stimulation to fight, flee, or hunt. All this takes place, as G. Bally (1945) so aptly says, in a "relaxed field." True, innate behavior patterns are used, but the "instinctive need" is to be relaxed (G. Bally). The relaxed field is at first provided by the parental care and protection which frees the animal from the need to find its own food, and shields it from predators. Adult animals only play when they feel safe, and especially when they are satiated. My own woodpecker finches regularly played after feeding. They took leftover mealworms and placed them into holes and crevices and poked them out again. They often placed a mealworm between a split branch and pushed it to and fro with their sticks. Outside the reproductive season many songbirds sing their most variable and beautiful songs. They "compose," as bird lovers say. My young polecats after feeding regularly engaged in rough-and-tumble play, as did my badger (I. Eibl-Eibesfeldt 1950a, 1956c).

The playful actions are performed as a result of an inner drive, but this drive does not seem to be identical with that which motivates the behavior patterns when they are performed seriously. The playful actions seem to have become "unhooked." This "independence of actions from the underlying motivation" is, according to A. Gehlen (1940), especially characteristic of human behavior. In this way a hiatus between needs and their fulfilment is created in which planned human thought can function rationally without the disruptive influence of drives. The root of this specifically human freedom of action exists in the play of animals.

Play versions and their counterparts differ then in the basic foundations of their drive motivation. In the pure case play is undirected at first. The animal may engage in play fighting, hunting games, or it may experiment in play, each according to the nature of the environmental condition prevailing at the time. There are exceptions to this rule, however. Frequently play is directed toward certain functional cycles, so that there are transitions between play and nonplay behavior. My hand-raised badger often changed from play fighting to actual fighting, including threatening calls, especially during the time when he was beginning to become independent. When no play partner was available he often sought out a substitute object. I observed several times how he threatened a bush, butting and pulling at its branches. He invited the bush to play fight in typical badger manner. In young rabbits an initial escape game may begin with several cross-jumps and end in "serious escape," where the animal presses itself to the ground in a dark corner.

The expressions of play are quite variable and we found many examples in the summaries of M. Meyer-Holzapfel (1956), K. Groos (1933), F. J. J. Buytendijk (1933), and E. Inhelder (1955). The various theories of play are treated in detail by M. Meyer-Holzapfel (1956) and C. Allemann (1951).

We may state the following criteria for play behavior:
1. During play we see instinctive behavior patterns along with ac-

quired behavior; neither, however, are performed in earnest, that is with their specific function although there are transitions between play and nonplay activities.

2. Behavior patterns that appear during play seem to be devoid of the motivating mechanisms to which the behavior is normally subject. I interpret this as implying that the entire instinct, in Tinbergen's sense, has not been activated. Instead, the behavior patterns have been individually activated at a lower level of integration.

3. This results in random combinations of behavior patterns which can include actions from different functional cycles and which allows a random change of activities or roles. This seems to be the root of the ability to *withdraw,* which is a prerequisite for any kind of dialogue. This ability is lost at once when actual appetitive behavior becomes aroused.

4. The animal seems to learn things that are of use to it in its later life, and the development of some behavior patterns seems to include "preprogrammed" play activities with siblings (for example, prey catching in the polecat, p. 214). This aspect of learning will be further discussed when we discuss fads (movement games, p. 245).

5. There exists a specific motivation for play which is based on a curiosity drive, that is, a mechanism that moves the animal to seek new situations and to experiment with new objects. Added to this is a strong motor motivation. The urges to play and to learn have a common source. Play is an active form of learning.

If one applies these criteria, then real play can only be demonstrated in some higher mammals and in some birds, such as the raven and the kestrel (E. Gwinner 1966; O. Koehler 1966). Whether or not fish play remains to be checked. The fish *Gnathonemus* balances objects on the tip of its snout which was interpreted as play activity, an interpretation I hesitate to accept (M. Meyer-Holzapfel 1956). Not all mammals play, however. In most species of mice I have never observed play behavior (I. Eibl-Eibesfeldt 1958).

Animals that fight with members of their own species as adults practice the actions during play fighting (Figs. 134 and 135). These are clearly distinguished from serious fighting by the inhibitions to bite and by the lack of threat behavior, as well as the repeated interchanges of roles. If one polecat or rat unintentionally bites another too hard, then the bitten animal will utter a cry and thus inhibit the partner. Between bouts of play there are periods of chasing, where the roles of pursuer and pursued change often. Polecats and badgers jump to and fro with humped backs facing the partner, which may be interpreted as an invitation to play. Dogs play passing and cutting-off games (J. Ludwig 1965). Here it is difficult to distin-

Figure 134. Play-fighting young polecats. (Photographs: I. Eibl-Eibesfeldt.)

Figure 135. Young lions playing with one another and with adults. (Amboseli, East Africa.) (Photographs: I. Eibl-Eibesfeldt.)

guish between play fighting and play prey catching. In predatory species the pursuer seems to be more in earnest when trying to catch up with the partner during the chase, which is in contrast to many vegetarian animals, where the pursued animal tries to escape. Squirrels run up trees during escape games in which each tries to get the tree trunk between itself and its pursuer. Neither animal makes an effort to actually catch the other. Instead it runs into hiding as soon as the other appears around a corner. Games of hide and seek are widespread and so is playful defense of a particular place. My tame polecats liked to hide under blankets and peek out with their heads. If another animal came close it was attacked. In this way they defended wastebaskets and other objects they occupied. Young deer and goats play king of the castle (F. F. Darling 1937).

During hunting games movements of prey catching are practiced, such as closing in, throwing down prey, shaking, and creeping. Often the animals use substitute objects which are treated like prey: the cat with a ball of yarn, a lion with its siblings, the dog with its ball. Sometimes impressive feats are performed. I knew a poodle that would carry a ball to the top of an embankment, push it down, and chase after it. Sea lions of the Galápagos Islands dive for stones, throw them into the air, and catch them again. These "hunting" games can be readily distinguished by their overall appearance from the usual rough-and-tumble games and chases. Usually there are no invitation gestures, although transitions exist.

Of special interest are the movement games and playful experimentation with objects. In the former the animal experiments with its own abilities. It jumps, frequently changes the direction, rolls on the ground, and invents new movement coordinations. Wild sea lions ride the waves. My young badger discovered forward somersaults accidentally, and he practiced them until he was able to roll down a long incline in one continuous series. On another occasion he discovered that he could slide down an icy road, after which he practiced this tirelessly. Under these circumstances animals behave like children who also practice all manner of locomotory actions: walking on heels, balancing, and so on, learning many new things thereby (Fig. 136). It is also possible that animals acquire knowledge about their own body during play. They play with their limbs as well as with their own shadow.

New inventions may be retained as fads in both animals and people. W. Köhler (1921) reported that his chimpanzees retained newly invented games for some time. At one time they fished for objects with long sticks; after they had discovered that chickens could be attracted and chased away, the habits changed. After my hand-raised badger had invented the forward somersault he concentrated exclusively on this game for awhile. This animal always sought contact with its caretaker and attempted with great persistence to get to him when he had been locked up. Once he accidentally discovered that he could enter the room by climbing to the low windowledge. After this discovery he lost interest in his caretaker for a while. Instead he practiced playfully leaving through the open door

Figure 136. Playing
young gorillas. (From
G. B. Schaller [1963].)

and climbing in again through the window. Similar behavior was
observed in a Beech marten. When we ate dinner on an elevated
platform the marten used to visit us, to beg for food and to play.
When he had accidentally discovered a shorter route he forgot us
and continued to practice climbing his new path. During play an
animal manipulates various objects. It chews and throws them about
until it has acquired mastery over them. In most animals, objects
are more commonly taken apart than put together in a constructive
way, but among the apes games involving construction exist. As I
already mentioned, new objects are especially likely to arouse curiosity, but gradually interest wanes and the animal plays less (E.
Inhelder 1955; A. Wünschmann 1963).

During play animals make inventions that later prove useful.
The classical example is provided by W. Köhler (1921), whose
chimpanzee Sultan had the task of reaching for a banana outside
its reach by means of two sticks that could be joined. After several
unsuccessful attempts at this, he withdrew from the task, began to

play with the sticks, and succeeded in putting one end into the other. Then he returned to the former task and reached for the banana. E. Inhelder (1955) described how a monkey (*Macaca silenus*) discovered that a ball could be put into a pail, how the animal repeated the game, and how others imitated it.

In man, too, play is an experimentation with one's own abilities, by interaction with the environment as well as with other people. A new element is that play is often constructive and based on models (Fig. 137). One should study in more detail to what degree these construction games of humans are based on specific innate predispositions. For example, children of a certain age especially like to build tree houses and other structures. City children who never had an opportunity to observe adults performing the behavior begin to build tree nests or leaf shelters when they go to the country during vacation. Is there a bias in favor of this behavior?

Figure 137. Construction game of El-Molo children. Top: scenery near Lake Rudolf (Kenya) with one of the typical huts; bottom: children playing "house." (Photographs: I. Eibl-Eibesfeldt.)

In humans imagination plays an important role in all play activities. Children attach different meanings to objects and assume changing roles during play. Such reality is superimposed by phantasy (A. Gehlen 1940). We also know from introspection that we play with our imaginations—in our daydreams. We will discuss the biological significance of this capacity in Chapter 18.

Although innate dispositions of various kinds provide a framework in which play fighting, hunting games, and so on, occur, play is primarily liberated action (G. Bally 1945; J. Huizinga 1956). The independence of the basic motivation permits an exploratory change from approach to withdrawal, as was mentioned earlier. It is interesting to note that during play new traits are revealed which normally cannot be observed. Chimpanzees produce paintings during play that are quite pleasing to the eye, illustrating a basic aesthetic appreciation for symmetry and balance (D. Morris 1963). In addition to this, the pictures reflect individual expressions, and in three chimpanzees of different social rank within a group, this ranking was also expressed in the individual styles. The high-ranking male covered the entire canvass and began turning the brush after each stroke, including curved lines in the picture. The high-ranking female also filled all available space with bold strokes from the center to the periphery. When given several colors she painted something reminiscent of a rainbow. The low-ranking female, on the other hand, limited herself to painting a small spot at the lower edge of the canvass. She pressed down so hard that the paper became roughed up (Plate VII). When given additional colors she did not fill the still vacant space but instead filled in the center of the spots she had painted earlier. This reminds one of the tree test that psychologists use: Healthy children draw trees that spread out their branches in all directions, whereas children with psychological problems produce unharmonic or crippled pictures.

14 MECHANISMS OF LEARNING

In everyday usage learning refers to the acquisition of new skills and knowledge. Scientifically speaking this term can be used whenever the probability of the occurrence of certain behavior patterns in specific stimulus situations has been changed as a direct consequence of encounters with this or similar stimulus situations and not as a result of maturational or fatigue processes (E. R. Hilgard 1956; P. R. Hofstätter 1959).

This modifiability of behavior presupposes special adaptations of the organisms. First, an organism must have the ability to memorize. Furthermore, it must be programmed in such a way as to be able to distinguish experiences that are positive from ones that are negative with respect to the preservation of the species. This means, among other things, that an animal does not indiscriminately associate each environmental stimulus with specific perceptions. This is actually not the case. A rat that has tasted poisoned bait will henceforth avoid the bait but not the place where it was found (F. Steiniger 1950). That taste and olfactory impressions are associated with visceral conditions, on the one hand, and pain stimuli are associated with auditory and visual stimuli, was demonstrated in the experiments of J. Garcia and F. R. Ervin (1968) and J. Garcia, B. K. McGowan, and others (1968). When people have become seasick they tend to associate the nausea with specific odors or foods but not with the ship itself. Differences in this kind of programming from species to species have been discussed in Chapter 13. In accordance with these differential dispositions many species behave differently even in artificially uniform experimental situations (K. E. Grossmann 1967). What is positively or negatively reinforcing during training an animal must "know" before all experience. As we have seen in Chapter 14, an organism can be so constructed that it not only learns passively from events taking place, but also actively searches out the unknown. It is *curious*. This requires the existence of appropriate motivating mechanisms. In the following sections we will discuss the motivations underlying learning processes and the nature of memory.

The experimental analysis of learning and its motivation

The simplest learning process is called *habituation*. The animal learns passively to refrain from responding further to repeated stimuli which are not accompanied by positive or negative reinforcement, that is, those events which biologically are unmeaningful to the animal. A clawed frog will startle when one taps the side of its container. If this stimulus is repeated several times, the animal will no longer show escape behavior. An additional example is the habituation of graylag geese to silhouettes of birds flying overhead (p. 78).

In higher learning processes two basic types are generally distinguished:

1. The formation of conditioned reactions or reflexes by a process of *classical conditioning*.
2. Reinforcement learning as a consequence of the organism's own activity, called *instrumental conditioning, trial-and-error learning, or conditioned reflex type II*.

Complex innate behavior patterns can be released by conditioned stimuli after appropriate training. Japanese quail (*Coturnix coturnix japonica*) will show species-typical courtship behavior in response to a buzzer when it was previously paired with the appearance of a female. The various components of the behavior become linked in a specific sequence, which approximately follows their appearance during ontogeny, with the conditioned stimulus, and they are extinguished in the reverse order (H. E. Farris 1967).

The formation of conditioned reactions has been studied in great detail by I. P. Pavlov and his school. We cited the case of a dog that salivates in response to the sound of a bell as an example of such a conditioned reaction (p. 66). A previously neutral stimulus which precedes the presentation of food repeatedly will eventually elicit the unconditioned response to the food, even when the food itself is not presented. The initially neutral stimulus is coupled with the unconditioned stimulus for the response. If one illuminates the eye of an animal, the pupil will contract. If this light is accompanied by the sound of a bell, then the stimulus becomes associated with the unconditioned stimulus, and the pupil constricts eventually in response to the sound alone. From time to time the unconditioned stimulus must again be paired with the conditioned stimulus, or the conditioned reaction will gradually extinguish, after which the animal will no longer react to the conditioned stimulus. Negative associations are retained longer without reinforcement than positive ones.

In circus training an unconditioned reaction is frequently linked up with new conditioned stimuli. In this way circus horses can be brought to perform innate behavior patterns in a reliable fashion by presenting conditioned stimuli. Rearing up on the hind legs, for example, occurs naturally during rival fights in horses (K. Zeeb 1964).

Behaviorists in the United States, in contrast to Pavlov, base their investigations of the learning processes upon the spontaneous actions of the animal. Typically, an animal is placed into a closed cage from which it can escape by pressing a certain lever (K. S. Lashley 1935) or where it can obtain food or water by pressing a bar (B. F. Skinner 1938; see also Fig. 138). However, it is possible to reward any other movement of the animal, for example, when the animal turns to the right. Thus a pigeon can be taught various increments of a sequence of steps by reinforcing each increment with food, so that within a short time the pigeon will walk in a

Figure 138. One form of Skinner box. When the rat presses one of the two levers a food pellet will drop into the food cup. The rat learns, for example, to activate the levers in response to various signals—here light signals. (Photographs: I. Eibl-Eibesfeldt, laboratory of T. I. Thompson.)

Figure 139. Trained blackfish whale (*Globicephala*) during a performance in Marineland (California). (Photograph: I. Eibl-Eibesfeldt.)

circle, as was demonstrated by Skinner. If a pigeon is rewarded each time it raises its head above a certain line, it will soon run about with its head raised. If a pigeon is to be trained to peck at a marble, one initially rewards the mere turning toward the marble, then looking at the marble, approaching the marble, and finally the pigeon is rewarded only when it pecks at it. It has been possible to train whales (*Globicephala*) at Marineland, California in the same way. For example, they will slap their fluke on the water surface in response to a command and continue until given a signal to stop. Initially the animals were rewarded whenever they accidentally hit the water surface with their tail. They very quickly learned on what the reward depended. Other toothed whales (*Delphinidae*) learned in a similar way to jump together out of the water, to dive down head first, propel themselves backward while standing on their flukes, and many other tricks (H. Hediger 1963; see also Fig. 139).

Looking at a learning curve of, say, bar pressing in a rat, one can generally see that the first few successful responses seem to have had no effect. However, after a few successful performances the rate of bar pressing increases sharply (Fig. 140). B. F. Skinner (1953) has proposed to lead school children through learning programs that have been broken into small steps, where the reward consists of the student being allowed to progress to the next step in the program after successfully completing the preceding ones. The immediate reward of success strengthens the learning process. The results of such programmed learning depends, of course, on the abilities of the programmer. Furthermore, one has to realize that in this way of teaching the forming of individual opinions has been precluded.

A favorite method for the study of learning processes is the maze experiment, which was first used by W. S. Small (1900) to train rats.

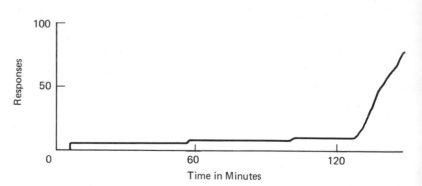

Figure 140. Learning lever presses in a Skinner box setup. Each correct response is indicated by a rise in the level of the curve. It can be seen that the first three correct responses apparently were without effect. After the fourth success, however, the number of correct responses (bar presses) increases sharply. Ordinate, number of responses; abscissa, time, in minutes. (After B. F. Skinner from N. L. Munn [1950].)

In such an experiment the animal must learn the path to a goal which cannot be seen from the starting point. Complex mazes have many dead ends, and in time the animal learns the shortest route to the reward in the goal box. Very simple mazes are of Y or T form. The animal may run in open, closed, or on elevated runways. Figure 141 shows several of the types of mazes that are frequently used (see also N. L. Munn [1950]). In the evaluation of maze experiments, the biology of the animals used should be considered. Rats, which normally live in tunnel systems, bring a specific learning disposition to the experimental situation which many other species do not have.

a

b

c

Figure 141. (a) Hampton Court maze of Small (depth maze); (b) elevated maze made of narrow boards; (c) elevated maze made up of T-maze components. (From N. L. Munn [1950].)

Maze experiments have shown that such reinforcement learning is not limited to the specific situation in which the learning has taken place. If mice have mastered one maze, then they can find their way, without additional learning, if all angles of paths leading away from the choice points have been changed from 90° originally to 45 or 135°, or if the lengths of the runways have been doubled. Even the mirror image of an originally learned maze can be mastered. At the first choice points they make errors but then seem to understand the principle and are able to transpose. In the house mouse this ability to transpose is maintained even after the mice are blinded, but if the visual cortex is destroyed, the ability is lost (W. Dinger and N. Heimburger, cited by O. Koehler [1951, 1953]). Rats that have learned to run through a maze can still find their way when they must swim it, and no additional learning seems required (D. A. MacFarlane 1930). E. C. Tolman (1932) has pointed out in this connection that the animal does not learn movements but meanings. An animal that runs through a maze, he said, does not learn movement patterns but a concept of the path to the goal that it seeks. That higher animals are guided by conceptions of the goal can be demonstrated by the fact that rhesus monkeys, who have seen a banana hidden under a cover, will continue to search for it after they have discovered a leaf of lettuce that has been surreptitiously substituted for the banana. The animals are quite agitated during the search, undoubtedly filled with some expectation. When learning a new task animals often try out what brought success last time. Rats in a new maze will, for example, repeatedly turn right or left, depending on what has most recently led to success. Or they may alternate. Thus at the beginning of training there is no random trial; instead, the animals behave as if they were trying out "hypotheses" (I. Krechevsky 1932). Improvements are made only after the rats abandon their hypotheses.

Based on the various experiments a number of learning theories were developed. Much discussed was the question of what were the preconditions necessary for an animal to learn. C. L. Hull (1943) emphasized his theory of the strengthening of reaction tendencies along the lines of E. L. Thorndike (1911), asserting that learning will take place only when the correct response is rewarded in some way, where the reward leads to a reduction of a specific drive, for example, hunger. Such a reward leads to reinforcement of the response that has brought it about.

In contrast to this, the theory of contiguity of E. R. Guthrie (1952) stated that such a reinforcement is not a prerequisite for learning. Associations are formed by the close contiguity of stimulus and response. The behavior is not strengthened by the reward but is prevented from decaying. For example, if an animal escapes from a puzzle box, it will not forget the behavior that led to the escape because it is removed from that environment and has no opportunity to form new associations. F. D. Sheffield and T. B. Roby (1950) claim to have demonstrated that a reduction of a physiological need is not necessary for learning. They rewarded rats with saccharine. Furthermore, it was found that rats will learn a maze not only because they receive a special reward. If they are allowed to run about in a maze without any reward, they will later learn the maze much quicker when the goal contains food than will control animals which have not had this prior experience. During exploration, learning is said to be "latent" (M. H. Elliot 1930; H. C. Blodgett 1929; E. C. Tolman and C. H. Honzik 1930). This latent learning has been cited as evidence against Hull's reinforcement theory. However, it is possible to say that in this case the "curiosity" of an animal becomes satisfied. It is certain, however, that the opinions expressed by C. L. Hull (1943) and B. F. Skinner (1938) that hunger, sex, and avoidance of pain and a few other drives are the only motivations for learning is surely incorrect. My own experiments with squirrels have shown that inexperienced animals will learn to open nuts, although the kernels have been removed and the shells glued together again (I. Eibl-Eibesfeldt 1967). If gnawing is activated in rats as a result of brain stimulation, they learn a task if their only reward is to be able to gnaw at wood or cardboard (W. W. Roberts and R. J. Carey 1965). If the rats happened to be eating during the experiments, they stopped in response to the brain stimulation and sought out objects to gnaw, which clearly demonstrated that the drive to gnaw was activated rather than the feeding drive. It seems then that the mere performance of fixed action patterns is already rewarding, a point that had been made by ethologists (p. 54). In a Skinner box the rats learn to press a lever if they can administer an electrical brain stimulus to themselves by means of implanted electrodes. When the electrodes are in a certain location the frequency of bar presses increases rapidly. The self-applied stimulus seems to be rewarding. The animals will even cross an electrified grid to gain access to the lever by which they will obtain the brain stimulus. In one specific location the frequency of self-stimulation decreased until testosterone was injected, after which it increased. In other areas the frequency of self-stimulation increased when the animal was hungry. Undoubtedly there is an activation of mechanisms resulting in pleasurable sensations which are normally associated with mating or feeding (J. Olds 1958). Finally there are acquired secondary motivations. One can become used to a certain kind of food and have a very specific desire for it. One becomes attached to one specific environment and is homesick when one is away from it. If all these possibilities are included under the heading of motivation, then the theory of reinforcement by rewards is undoubtedly correct in respect to learning. Simple conditioned reactions and some other learning can also be explained according to Guthrie's principles: A temporal or spatial contiguity of objects or events is remembered without any demonstrable reward or punishment. Sometimes animals are taught unusual postures or movements by passively forcing them to perform these forms of kinesthetic learning. Circus animals are taught in this way to stand on their heads. Usually progress in learning is facilitated by additional rewards (H. Hediger 1954). Balinese girls learn their complex dances by guidance from a teacher by means of kinesthetic training (G. Bateson and M. Mead 1942; see also Fig. 142).

Figure 142. On Bali girls learn the complicated dances (a) by watching examples and by direct leading by the instructor (kinesthetic learning, b–d). (Photographs: I. Eibl-Eibesfeldt.)

Up to a certain point one can use visual paired comparison experiments to compare the learning achievements of various animal groups. The animal is presented with two stimuli side by side, for example, a circle and a cross, and allowed to choose. Choice of one stimulus leads to food; the other leads to no reward or to electric shock. The stimuli are randomly switched in respect to position. In this situation the animals learn to choose the rewarded stimulus. An octopus mastered three tasks, and was able to discriminate six different stimuli (J. Z. Young 1961). Trout (*Trutta iridea*) mastered up to 6 tasks, iguana (*Iguana iguana*) 5, large chickens up to 7, and an Indian elephant and a horse up to 20 (B. Rensch 1962). Memory has been tested in many species; the octopus made 83 percent correct choices after an interval of 27 days. A carp was able to distinguish a circle from a cross after 1 year and $8\frac{1}{2}$ months and selected the positive stimulus significantly more. A trout still retained a task after 150 days, a rat after 1 year and 3 months; an elephant retained 12 of 13 visual discriminations after 1 year, and a horse remembered 19 of 20. The quantitative maximum performances of these animal groups are quite similar and do not reflect the great differences commensurate with their phylogenetic level of organization (B. Rensch 1962). M. E. Bittermann (1965) believes that he has demonstrated qualitative differences in learning abilities in fish, reptiles, birds, and mammals. Fish do not learn a habit reversal, a task in which after learning that one of two stimuli is positive, the previously negative one now becomes positive; then when the new task is learned it is again reversed, and so on. Monkeys, rats, and pigeons quickly learn reversals. Turtles did not learn a visual reversal but were able to make spatial reversals, where they had to chose between two identical signals on two different sides. Bittermann has only tested a few species, so that generalizations are not well founded. Gobies and blennies (*Gobiidae* and *Blenniidae*) probably would surpass many a reptile in their learning abilities.

Nature of the engram

A basis for all higher accomplishments is memory, and this has been demonstrated in all animals with a central nervous system, including planaria. Opinions about the learning capacity of protista diverge (W. H. Thorpe 1963). According to B. Gelber (1965) paramecia gather about a platinum wire which had been baited repeatedly. H. Machemer's (1966) experiments to train hypotrichous ciliates, on the other hand, yielded negative results. However, he agrees in principle that protozoans can learn.

In vertebrates the learning performance is clearly correlated with the size of the brain, whereby the actual size of the brain and the number of ganglion cells seems to be more important than the systematic position (B. Rensch 1962). It seems certain that several levels of the central nervous system are capable of learning (R. Hernandez-Peon and H. Brust-Carmona 1961). Frogs are able to learn with only their spinal cord (L. Franzisket 1955). In mammals, most, but not all, experience is stored in the neocortex. By electrical stimulation of various regions of the temporal neocortex, W. Penfield (1952) was able to elicit acoustical and optical hallucinatory images in epileptic patients. The patients also remembered these sensations after excision of the stimulated area. One can conclude, therefore, that the memory trace is also present in the temporal region of the other half of the brain. That memory traces of one brain hemisphere are also projected to the other has been demonstrated by the experiments of R. W. Sperry (1964) and R. E. Myers (1956).

If the optic chiasm of a cat or monkey is cut in the sagittal plane, then the stimuli impinging on one eye are only transmitted to the homolateral half of the brain. After this operation it is possible to teach the animal a simple discrimination with one eye such as distinguishing a circle and a square. If it has learned with the right eye, this eye is covered and the animal is then tested for discrimination with the left eye. Such tests of transfer are successful, which proves that during learning a projection of the information from one hemisphere to the other has taken place. However, if before the training one also cuts the corpus callosum (Fig. 143), then the animal can perform the task only with the trained eye and completely fails with the other; in fact, the animal behaves as if it had two brains.[1] Each eye learns independently, so it is possible to teach each eye something different. If, on the other hand, one transects the corpus callosum after training with one eye, the animal remembers what

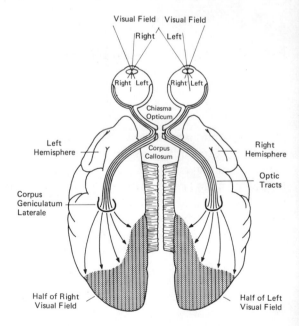

Figure 143. After cutting the optic chiasm and the corpus callosum the information obtained through one eye is transmitted only to one cerebral hemisphere. (See explanation in the text.) (After R. W. Sperry [1964].)

Visual Field Visual Field
Right   Left
Right Left   Right Left
Chiasma Opticum
Left Hemisphere   Corpus Callosum   Right Hemisphere
Optic Tracts
Corpus Geniculatum Laterale
Half of Right Visual Field   Half of Left Visual Field

[1] This is also true for humans in which the *corpus callosum* has been cut (M. S. Gazzaniga 1967).

it has learned also when the other eye is tested. This memory copy does not seem to be as sharp, however. For difficult problems this projection of the memory trace is not sufficient; direct sensory information is more effective. In humans one hemisphere is more specialized for the storage of memories than the other. This is especially true for the memory of words.

The nature of a memory trace or engram is not really known today. According to J. C. Eccles (1953) memory consists of electrical reverberating circuits which, once activated by a specific excitation, continue. This reverberating-circuit hypothesis is opposed by theories that assume structural changes at and in the ganglion cells. Morphological changes have frequently been demonstrated at the cellular level. Changes occur in the number of microsomes, apical dendrites of pyramidal neurons swell, and changes take place at the synapses themselves (R. W. Gerard 1961). In recent years biochemical hypotheses have also been advanced. Thus experiments with planaria (*Dugesia*) suggest a basis of memory in some substance. Planaria were cut in half behind the pharyngeal region after they had learned a task. When the brainless posterior part had regenerated a head, this animal was said to have retained the task (J. V. McConnell, A. L. Jacobson, and D. P. Kimble [1959]; J. B. Best 1963). The changes that accompany learning must have been distributed throughout the tissues of the entire animal and not restricted to the central nervous system. The biochemical learning hypothesis for the nature of the engram seems to be supported by experiments which showed that if trained flatworms were fed to untrained ones, the latter learned faster (J. V. McConnell 1962). A. L. Hartry, P. Keith-Lee, and W. D. Morton (1964) found, however, that planaria in general learn better after cannibalism, even when they ate untrained planaria.

There were indications that ribonucleic acids (RNA) are the carriers of information. If animals were trained, then cut in half, and their tails were allowed to regenerate in a ribonuclease solution, then no memory was found after regeneration; this in contrast to the tails of control animals, which regenerated in normal pond water (W. C. Corning and E. R. John 1961).

A. L. Jacobson and others (1965) and F. R. Babich and others (1965) trained rats to search for food when a clicking sound was presented. When they had learned the task they ran toward the food dish upon hearing the click sound, even when it was empty. The trained rats were killed and RNA was extracted from their brains and injected into the peritoneal cavity of seven untrained rats; 8 control animals received the extract of untrained rats. In 25 tests per animal, conducted by experimenters who did not know to which group an animal belonged, those injected with the brain extract of trained rats approached the food dish on the average of 6.86 times when the click stimulus was presented, while the control animals did so only once on the average. The effect took place 5 hours after the injection and lasted over 24 hours. In a second experiment the conclusion is probable that specific information is transferred. Inexperienced rats, which had been injected with extracts from rats that were trained to respond to the clicks, responded in a critical test (during a random series of light and click stimuli) significantly more to click stimulation with approach to the food magazine. In contrast, animals that had been injected with RNA from rats trained to respond to light stimuli approached food significantly more when a light stimulus was presented. F. R. Babich and others (1965) were able to transfer memory contents from hamster brains into rats. All these experiments have been criticized, however, and are not yet conclusive.

There seems to be a correlation between an inhibition of RNA metabolism and a decrease in learning ability. An increase of this metabolism by drugs strongly enhance the learning capacity of rats. It has also been shown that the RNA content of nerve cells in rats increases during the first year of life as a result of stimulation. In animals that grew up with a lack of this stimulation, the RNA content remained low (H. Hydén 1961; H. Hydén and E. Egyhazi 1962).

A mechanism for these changes is proposed by Hydén as follows: Impulse series from motor and sensory cells change the ion balance of the cytoplasm of the affected cells. A given impulse could produce a permanent change in an RNA molecule. This new RNA molecule, although only slightly changed, would direct the synthesis of a protein molecule differing slightly but significantly from that previously produced. Hydén assumes that the new protein has the property of responding to the same electrical pattern that created the change in the RNA. When the same electrical pattern does occur again, the new protein dissociates rapidly, causing an explosive release of the transmitter substance at the synapse. This allows the electrical pattern to bridge the synapse and be passed along by the second cell, then by a third cell, and so on. The nerve cells then respond differentially, depending on whether or not the arriving electrical pattern is new or has occurred previously.

More recently doubts have arisen as to whether RNA is the transmitter of information. The synthesis of RNA depends on the code which is present in deoxyribonucleic acid. It is possible that DNA changes during learning and that this causes the changes in RNA synthesis (J. Bonner 1964; J. Gaito 1964). RNA would then only be a link in a chain of events and not the storage substance for the engram. This role would be assigned to DNA, which is also the carrier of the "inherited memory." As early as the 1890s E. Hering

(1896) suspected fundamental similarities between memory and inheritance, which he referred to as organic memory. Whether or not the ability, acquired during phylogenesis, to store information on a molecularly programmed basis is used during learning has to be shown by additional experiments. There is much in favor of this view, but some caution is advisable in the interpretation of the work cited so far, especially because a number of investigators who repeated the rat experiments of Jacobson and Babich could not confirm their results (W. L. Byrne and others, 1966).

The structural hypothesis and the reverberating circuit hypothesis earlier discussed supplement one another. It has been found, in the meantime, that a distinction can be made between short-term and long-term memory, which, as was shown in new experiments with the octopus, are located in different parts of the brain (B. Boycott 1965).

Short-term memory could depend on reverberating circuits. This hypothesis is supported by the observation that supercooling to the point of cessation of all electrical brain activity and electroshock eradicates it. This short-term memory precedes long-term memory; the more time that has passed after training, the more difficult it is to eradicate what has been learned.

Rats and hamsters show a completely normal learning curve if they are subjected to an electroshock or supercooling 4 hours after a set of trials. If the same is done 1 hour after training, learning is slower. At intervals of 15 minutes between training and shock, learning is significantly disrupted, and if the animals are shocked 5 minutes after each training session, no learning improvement occurs. If a goldfish brain is injected with a drug that inhibits protein synthesis following a learning trial, then the fish forgets what it has learned; no long-term memory develops (B. W. Agranoff 1967).

It is quite possible that excitation initially activates reverberating circuits, whose activity over a longer period of time leads to structural and biochemical changes (R. W. Gerard 1961; D. O. Hebb 1949). In line with this theory a dynamic principle is postulated for short-term memory as contrasted with structural changes as a basis for long-term memory.

Abstraction, nonverbal concepts, and insight behavior

A notable result of training experiments is the realization that animals are capable of achievements usually placed in the category of human "higher brain functions."

This is true first of all for the ability to abstract and to generalize. In the previously discussed visual discrimination experiments (p. 256), an animal learns to distinguish two simultaneously presented figures, patterns, or colors. The approach to one pattern is rewarded (positive stimulus) and to the other is punished (negative stimulus). During the course of this training it was found that the animals can recognize similarities between figures. They abstract and draw, so to speak, the averbal conclusion "this is similar to that" (B. Rensch 1965). Fish are already capable of forming such averbal concepts. Minnows that were trained to distinguish a triangle as the positive stimulus from a square as the negative stimulus reacted positively to an acute angle versus a straight line, in line with the previous training. They apparently had "abstracted" the cue "pointedness" as the positive stimulus. An elephant that had learned to differentiate an x as a positive stimulus over a circle, later responded positively to any stimulus that contained crossing lines. In these cases the animals seem to learn only the most characteristic cues contained by the positive stimulus and ignore the rest. This also seems to take place in the natural situation. Toads that have been taught to avoid inedible prey models "abstract" various different cues. Initially they react quite specifically. Only later do they begin to generalize. Toads that I trained at first recognized negative models only in a specific location and only when they were moved in a certain way. They had to learn to recognize the models in other situations. Some learned to distinguish the models according to color, and they recognized this color on other models. Other animals avoided everything presented after only a few negative experiences, everything, that is, except mealworms. They even rejected grasshoppers. One female that responded in this way later learned to distinguish a moving model that was pulled past the animal from prey that moved on its own. It eventually ate grasshoppers, but it avoided mealworms that were pulled on a string. Mealworms that moved by themselves were eaten (I. Eibl-Eibesfeldt 1951a). Rhesus and capuchin monkeys are able to recognize pictures of insects and flowers, implying a capacity to generalize (E. Lehr 1967).

Animals do not necessarily learn all the characteristics of the positive or negative stimulus. Sometimes they learn the relation of the patterns to each other. If wider stripes are presented versus narrower ones, the animal does not learn the specific width of a stripe but instead recognizes the wider or the narrower of the two as the training stimulus. If the narrower stripe has been learned as the positive stimulus, and a new stimulus pair is now presented which contains a still-narrower stripe, then this new stimulus and not the original positive one (which now is negative) is responded to. In a similar manner, animals learn to choose the lighter of two shades of gray, where the positive stimulus is always lighter than the negative stimulus.

Abstractions are at the same time generalizations. B. Rensch and G. Dücker (1959) have shown how far this "averbal" concept formation can be carried. They trained a civet cat to develop the concept "bent straight" by training the animal to respond to two parallel half-circles as the positive stimulus and to two parallel vertical stripes as the negative stimulus. When the animal had learned this, the patterns were rotated 90°, then in a different arrangement of the stimulus components, so that the animal was forced to pay attention to the one constant feature "bent in a circle." Finally, the animal preferred any bent lines over straight ones even when they appeared against an entirely different background. In a similar manner, Rensch and Dücker were able to teach the civet cat the concept "equal–unequal" (Fig. 144).

Figure 144. Civet cat choosing between two patterns. (Photograph: B. Rensch.)

In this connection investigations dealing with "value concepts" in primates are important. Chimpanzees learned the differential symbolic value of chips of various colors and sizes (J. B. Wolfe 1936; J. T. Cowles 1937). The animals were able to insert blue, white, and brass discs into a food dispenser and receive one or two grapes or none at all. Soon they preferred the blue chips, which brought the biggest reward. The animals could be trained to perform certain tasks by rewarding them with chips which they then exchanged for food. These experiments were further varied by teaching the animals that some chips were for food, others for getting out of the cage, and others for playing with the keeper. Some animals utilized the chips according to their needs. One female used light-blue chips which she could drop into a slit in the door so it would open whenever the cameraman, of whom she was afraid, arrived to take her pictures. The chimpanzees undoubtedly had learned the value of the various chips. This same ability has been demonstrated in some other lower monkeys by T. Kapune (1966, see also for additional references). A female rhesus monkey finally mastered six different value concepts.

The highest achievements of generalization seem to have been discovered during the investigations of counting abilities of several animal species by O. Koehler (1943, 1949, 1952, 1954b, 1955). Pigeons, parrots, ravens, and squirrels learned to take only a certain number of kernels or pieces of food from a larger number. Ravens were taught to take only as many as were indicated on a sign. If the sign contained two dots, they chose the food dish whose cover also contained two dots. After further training it was no longer necessary for the size and arrangement of the dots on the sign to coincide with those on the cover. The birds selected solely on the basis of number. One gray parrot learned to open as many food dishes out of seven and take out as many kernels as was indicated by the number of times a bell rang. The bird also learned to take food kernels in response to two or three light signals, and then transpose, without further training, to one acoustic signal.

The spontaneous development of nonverbal concepts without trial-and-error learning is already a hint that *insightful learning* is taking place. This is also the case when an animal masters a detour and combines several, independently acquired, experiences (Fig. 145). When I once prevented my badger from entering my living quarters, it soon gave up its useless scratching at the door, and ran around the building and entered through a window on the other side. Such spontaneous insightful behavior was also shown by a dog who entered a garden that had been avoided until then by opening the door leading to it, after discovering that its rival had been tied up there (W. Gnadenberg 1962).

H. H. Hsiao (1929) built a maze in which a rat could reach the food box by three different routes. Two shorter paths led to a common entrance that could be blocked by a gate; the third path was much longer. All three paths were well known to the animals and they learned to use the shortest route. If the common entrance was now blocked, the rats at once used the third, longer, path without trying the second, shorter, route. They behaved as if they possessed a spatial picture of the maze and knew that with the closing of the gate the second route was also blocked. A similar maze was used by E. C. Tolman and C. H. Honzik (1930a, see also Fig. 146).

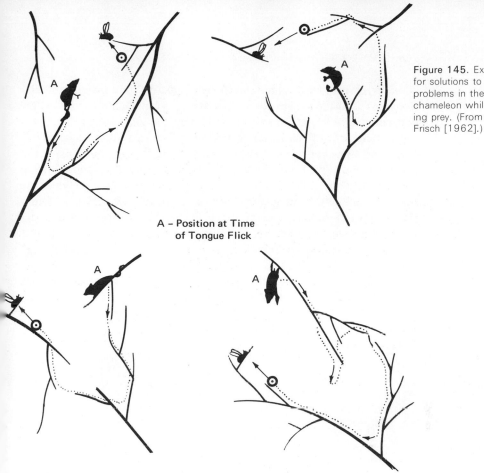

A – Position at Time of Tongue Flick

Figure 145. Examples for solutions to detour problems in the chameleon while stalking prey. (From O. v. Frisch [1962].)

Figure 146. Maze used by Tolman and Honzik to demonstrate insightful behavior. Rats learned each of the three paths that led to the goal. When the straight path 1, which led directly to the goal, was blocked at A, the animals could reach it via paths 2 or 3. When path 1 was blocked by gate B, the rats at once selected the only correct path, 3, which they normally did not prefer, and this was considered evidence for insight. Only a few rats were able to solve this problem in the elevated maze shown here. (After E. C. Tolman and C. H. Honzik [1930a].)

Correct insightful behavior in the sense of understanding relationships is often observed in tool using. Here W. Köhler's (1921) experiments with anthropoid apes showed the way. His chimpanzees used sticks to pull in bananas that were outside their cages. They could put two short sticks together to make a longer one. They could put boxes on top of one another to reach a banana suspended from the ceiling (Fig. 147). From the descriptions one can see that the behavior sequences were not learned by trial and error. A chimpanzee would sit and look around, at the box, at the place under the banana, and at the banana, until the solution had been found. In these examples the behavior sequence is thought out as though the trial is internalized. The ability to plan mentally was demonstrated by chimpanzees which had even learned to visualize complex mazes covered by Plexiglas. The animals discovered the shortest path to the goal solely by looking at the maze and were later able to maneuver a piece of iron to the goal with a magnet (B. Rensch and J. Döhl 1968).

Figure 147. (a) Chimpanzee putting sticks together; (b) chimpanzees in an experiment to try to reach a suspended piece of food by placing boxes on top of each other and using a stick. (After W. Köhler [1921].)

a

b

Planning with foresight was also described by M. P. Crawford (1937). Two chimpanzees had to pull in a food box by ropes. The animals had to cooperate, because the box was too heavy for one alone. At first one animal tried to pull the box alone. When this failed, it directed the other's attention to the rope, gesturing at the rope. Additional examples about insightful behavior and tool using in captive chimpanzees can be found in N. Kohts (1935) and R. M. Yerkes (1948).

More recent observations from the wild are now available (J. van Lawick-Goodall (1968). The chimpanzees she observed in the wild pulled termites from their tunnels with thin twigs or grass stems. They opened a tunnel which the termites use when swarming, pushed the tool into the tunnel, and pulled it out, with the termites clinging to it (Fig. 148). This was repeated over and over. They were very

a   b

Figure 148. Chimpanzee fishing for termites: (a) Selection of the tool; (b) insertion of the tool into an opened passage of a termite hill. (Photographs: Baron and Baroness van Lawick-Goodall, with permission of *National Geographic*.)

careful in their choice of tools. In that same area chimpanzees use leaves to soak up water from holes in trees, which they cannot otherwise reach with their lips. They use the leaves to clean themselves. In one of the films made by Mr. and Mrs. van Lawick-Goodall, a chimpanzee suffering from diarrhea picks leaves and cleans itself. In captivity one female chimpanzee cleaned her young with a cloth after each elimination (K. Heinroth-Berger 1965). Tool using alone is no criterion for intelligence, but a varied, individually modifiable utilization of tools, as found in the chimpanzees, certainly is.

Interesting observations about tool using in chimpanzees were made by A. Kortlandt (1962, 1965, 1967a, b). Chimpanzees that were captured in the savannah and kept in a large enclosure in Guinea directed well-aimed blows from above toward a stuffed leopard while they were standing upright (Fig. 149). Chimpanzees of the forests in the Congo were much clumsier in the same situation. They also beat about with sticks, even threw them in the direction of the predator, but their aim was bad and they never directed a blow from above at the predator, and they never hit the stuffed leopard. Kortlandt is of the opinion that the use of sticks as weapons originated in the open savannah. When the ancestors of the forest chimpanzees were pushed into the forest through competition with man's ancestors, their skill in using weapons atrophied.

Figure 149. Savannah chimpanzee beating a stuffed leopard with a stick. (Photograph: A. Kortlandt.)

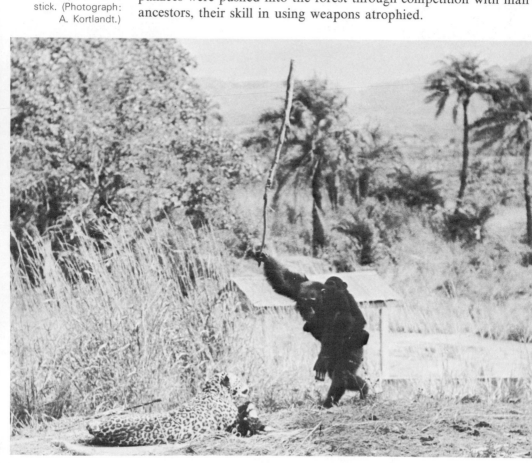

15 ECOLOGY AND BEHAVIOR

In previous chapters we have talked about the functions of various behavior patterns, but primarily the causal and historical aspects were considered. In this section I want to discuss the factors with which an organism has to deal in its environment and how its behavior must adapt to the various contingencies. Each organism, first of all, must maintain its internal milieu constant and defend this homeostatic balance against various disruptive influences. The organism must grow and reproduce itself. We will discuss the dependence of these functions on diverse environmental factors. The function of behavior in preserving the species will be illustrated with some selected examples.[1] Much will inevitably be omitted. The presentation of the various locomotor behavior patterns alone, such as running, swimming, jumping, climbing, and flying, would go beyond the scope of this book.

For the purpose of clarity we will group the relationships to the nonspecies environment separately from the relations to species members. We must remember, however, that many behavior patterns are not really restricted to one or the other grouping. Many serve in several functional cycles, such as the behavior patterns of locomotion. An animal may run to catch up with a rival or to escape from a predator (see p. 173).

(Additional literature on this topic is the unsurpassed work of R. Hesse and F. Doflein (1943) and the summaries by W. Kühnelt (1965), A. Kaestner (1965), and G. Tembrock (1954).

Relationships to the environment other than the species

Adaptations to nonbiological factors

That each animal species has its own preferred temperature was demonstrated by K. Herter (1943, 1952, 1953) in numerous investigations. Animals in a temperature-choice apparatus were allowed to choose a room with a specific temperature. They gathered in a room whose temperature matched the temperature optimum of the species. In the same way they selected a certain humidity, they avoided dryness or sought it, and they sought exposure to the sun or avoided it. In short, a number of behavior mechanisms that enable animals to select their appropriate biotope (cliffs, loess walls, meadows, brush, and so on) are inborn. The New World mouse *Peromyscus maniculatus bairdi* selects grassland; *Peromyscus maniculatus gracilis,* on the other hand, selects woodland. Both subspecies occur in adjacent areas, but they are so strictly separated with respect to their biotopes that they do not hybridize, although in captivity they do so readily. T. v. Harris (1950) raised mice of both subspecies in captivity and allowed them to choose between terraria that were planted differently. In one he planted small trees; in the other he imitated grassland with thin paper strips. *Peromyscus maniculatus bairdi* innately preferred this "grassland" while *P. m. gracilis* selected the woodland habitat. In a room half of which contained pine branches and the other half oak branches, sparrows (*Spizella passerina*) selected the side with pine branches, although on both sides there were an equal number of perches. The choice matched the natural preference of their biotope (P. H. Klopfer 1963; P. H. Klopfer and J. P. Hailman 1965).

Environmental adaptations also include those behavior patterns with which an animal creates a shelter against bad weather—structures and nests that protect it against heat as well as cold. It is known that many desert mice can only survive the midday heat by retreating to their dens (F. Bourliere 1955). The earth mounds around the dens of prairie dogs protect them against flooding. Beavers dig dens into the riverbanks, but if the bank is so low that no cave can be dug, then the beaver will put its den into a large mound of twigs. The animal will gather branches up to 4 meters long, heap them on top of one another, plug the holes with mud, earth, and weeds, and finally will make a den inside. The entrance to the den is underwater. The beaver ensures itself adequate protection and a reliable means of transporting his food by damming

up running streams. Dam-building activity is at once activated by
a drop in the water level (G. Hinze 1950). When building a dam
the beaver cleverly utilizes natural elevations as supports. He piles
twigs and branches on top of each other with the thick end facing
upstream. The side branches catch on each other and mud collects
at the steeper and upstream side of the dam. The dam may be
anchored with trees that have been felled across the stream and by
poles that have been pushed vertically into the bottom (P. B.
Richard 1955, 1964; see also Fig. 150). The dams are rarely higher
than 1.50 meters. They can be several hundred meters long and are
then the work of many generations of beavers. In this way the
beaver goes a considerable way to creating his own appropriate
environment.

Figure 150. (a) Beaver den in the bank of a river with the entrance lying below the
water surface. One dam dams up the water. (b)–(d) Various types of dam with trees
along the banks used as supports and with tree trunks rammed into the bottom for
support. (After P. B. Richard [1955].)

Termites provide yet another example. Some regulate the humidity and temperature within their mound so that it remains
around the optimum of 30° centigrade and 98 to 99 percent humidity. To accomplish this, water bearers carry water from tunnels
that lead down to the groundwater level. The termites protect
themselves against extreme temperature changes by building hard,
thick walls that shield them against the environment. The high
mounds, which have ridges, are made from a clay–saliva mixture.
The ridges are used for ventilation. Air ducts lead up and down

within them. The used-up air rises from the center of the nest to a central roof chamber and from there via the air ducts returns to the bottom. Carbon dioxide is given off through the pores of the outer wall and oxygen is taken up. The fresh air collects in a chamber below the nest and from there rises upward. The termites regulate the ventilation by widening or narrowing the air ducts in accordance with the need for oxygen and for warmth or cooling temperature (Fig. 151).

Figure 151. Termite hill of *Macrotermes natalensis*. One sector of the nest was cut out to allow a view of the ventilation system.

Special behavior patterns for fur and feather care maintain their insulating and water-repellent qualities. Waterbirds whose feathers are caked together quickly perish. Behavior patterns for bodily care are therefore adaptations to climatic factors (H. Dathe 1964; M. Bürger 1959). Emperor penguins (*Aptenodytes forsteri*) court and incubate during the continuous night of the antarctic winter. Storms rage with average velocities of 80 km per hour and reach as high as 140 km per hour, coupled with temperatures of minus 60°C. The penguins survive by huddling together in V formations which effectively break the wind. In this way they maintain their body temperature at about 35.7°C, whereas in individuals separated from the group the temperature dropped to 27.9°C (J. Prévost 1961). Here extreme environmental conditions are met by adaptations of social behavior (additional examples on p. 334).

Reptiles assume special sun-bathing postures so that they warm up faster. Grasshoppers (*Chorthippus dorsatus*), which in central Europe court in the fall, expose their lateral side to the sun and drop their thigh, thus exposing the dark-pigmented abdominal area to the sun rays.

Many animals of the intertidal region seek out areas that are suitable for surviving dessication at low tide. The limpet (*Patella*) and the false limpet (*Siphonaria*) have independently evolved an attachment to a home base. Both return to the same resting place after each foraging excursion. The outlines of their shells fit exactly into a seat (Fig. 152). There are also behavior patterns which are typical for survival in a particular ecological niche. Whenever an animal consistently seeks out a specific habitat, a number of changes in body structure and behavior evolve. We mention here only the parallel adaptations of fish in rapidly flowing waters, of bottom-dwelling fish, and of birds that breed on cliffs (W. Wickler 1958, 1959, 1965c; E. Cullen 1957; additional examples in Wickler 1961a). Fish that live in the high seas, where they normally never meet with an obstacle, bump against the walls of the aquarium without ever learning not to, just as some birds of the plains bump into cage walls (K. Lorenz 1959). Animals that live in structured biotopes such as woods or coral reefs behave much more intelligently by comparison.

Figure 152. Place loyalty of *Siphonaria gigas*. The young fit exactly into the seats, which were made into the shells of the adults. Left: an animal in its resting place; right: the empty seat of another snail, showing the shape of the, seat and the impression of the foot. (Photograph: I. Eibl-Eibesfeldt.)

The Norway rat and the house rat (*Rattus norvegicus* and *R. rattus*), both followers of man, come from different biotopes, which is clearly reflected in their behavior. The house rat, which originated from tree-dwelling forms, and which in southern countries today still nests in trees, prefers to settle in the upper floors of buildings, hence the name roof rat. It climbs well, and when panicked tries to climb upward even if there is no hiding place there. This animal is primarily a vegetarian. The Norway rat, which in the wild lives near water and on river banks, prefers to live in the lower floors of human dwellings—the cellar rat. It also lives in sewage systems, where it fishes for food objects in the water with special straining behavior patterns. On occasion the animal leads a predatory way of life (I. Eibl-Eibesfeldt 1953d).

Finally, some nonbiotic adaptations to environmental conditions are concerned with periodic environmental changes of the tides, diurnal rhythms, moon phases, and annual cycles. These will be discussed in Chapter 17 (p. 389). We will also discuss several questions that deal with the orientation of animals (p. 363).

Procurement of food

Just as numerous as the adaptations to the nonliving environments are the adaptations to the procurement of food, which we will illustrate by a few examples. In general, the specialists predominate. The struggle for existence of various species often leads to surprisingly specific adaptations. The small leather jacket, the filefish (*Oxymonacanthus longirostris*), specializes in picking off individual coral polyps. Many coral fish developed long snouts which enable them to pick off small animals that hide between crevices in the coral (Fig. 153). They not only have the appropriate morphological adaptations, but in their food-getting behavior show a strong preference for seeking food in crevices and clefts, even when kept in an aquarium where such crevices do not contain food. Substrate feeders are often so rigidly specialized for foraging on rocks that it is very difficult to get them to accept food that lies loose on the bottom. This is true for the Moorish idol (*Zanclus cornutus*), which Lorenz was successful in keeping after many unsuccessful attempts to feed it. Finally, he placed chopped clam meat on stones, let it dry enough so it would stick, and placed the stones into the water. The fish at once took to this food and began to forage.

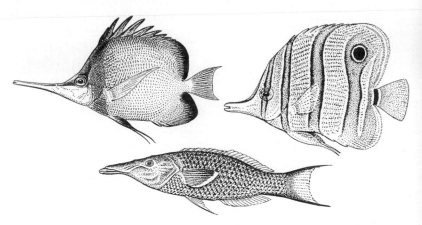

Figure 153. Pipefish, which pick their food out of the branches of corals. Top: The two butterfly fish *Forcipiger longirostris* and *Chelmon rostratus;* bottom: the lipfish *Gomphosus.*

The saber-toothed blennies *Runula* and *Aspidontus* tear chunks of flesh from the fins of other fish (p. 151). The flutemouth fish (*Aulostomus maculatus*) approach their prey by hovering above peaceful species such as parrot fish. When the latter feed, small fish approach in search of food particles that are broken off by them. At the opportune moment the flutemouth will glide from its position above the parrot fish and catch the smaller ones (H. Hass 1951; I. Eibl-Eibesfeldt 1955 see also Fig. 154). The archer fish (*Toxotes jaculatrix*) ejects a stream of water at insects that rest on leaves

Figure 154. "Mounted" trumpet fish (*Aulostomus maculatus*).

above the water surface. Standing just below the surface the fish aims, and toward the end of the shot it jerks up its head, scattering the stream of water, which increases the probability of hitting the prey (H. Lüling 1958; see also Fig. 155). The various methods used by spiders to catch their prey are as diverse as they are surprising (A. Kaestner 1965); the lasso spider *Mastophora* throws a thread with a sticky ball at its end at passing insects. Other lasso spiders, such as *Cladomelea* and *Dicrostichus,* sit on a horizontal thread from which they suspend another thread with a sticky ball on the end.

Figure 155. Spraying archer fish (*Toxotes jaculatrix*). The fish is able to spit water for a distance of more than 10 times the length of its own body. (From J. Eibl-Eibesfeldt [1964c].)

This pendulum is swung about with the claws of one leg. Insects that bump into this contraption stick to it and are caught (Fig. 156). The complex nets of the spoke-wheel net spiders (Fig. 157), which were studied in detail by H. M. Peters (1939, 1953) and G. Mayer (1952), are well known. The spider begins by stretching one thread from one object to another. This is accomplished by projecting the hindquarter into the air and ejecting a long thread that is caught by the wind. If it does not stick anywhere, the spider pulls it in again and tries again. If it should stick, for example on another branch, it will also be fastened at the other end. Then the spider moves out onto this bridge, biting the thread in half, but holding the two ends with the

Figure 156. Australian bola spider (*Dicrostichus*) during its nightly insect hunt. The spider is suspended from a horizontal thread and swings a suspended thread, which hangs from one leg, in a circle. At the lower end a drop of sticky substance is attached. Insects, which are attracted in some still unknown manner, become trapped on the bola. (From D. Bergamini, Life Nature Library, Australia, 1964.)

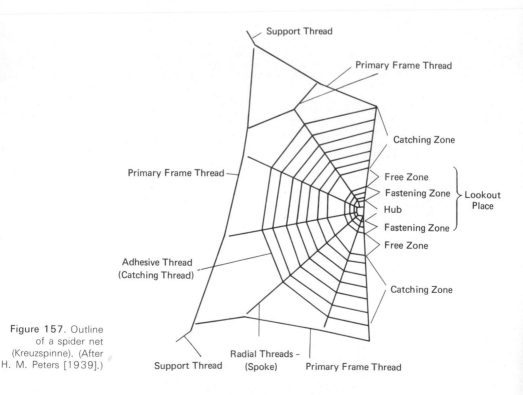

Figure 157. Outline of a spider net (Kreuzspinne). (After H. M. Peters [1939].)

legs so that the body is the link holding the two pieces together (Fig. 158a). The spider moves along collecting the thread ahead while adding additional substance to the thread behind. Once it has arrived in the center of the thread it glues the two ends together and lowers itself to the ground, where the end is fastened. Thus the first three radii of the net have been formed. Then the spider spins new spokes beginning in the center of the net and makes the primary frame. In this way a spoke-type net is made to which is added a widely spaced auxiliary spiral from the center to the periphery, which connects the radii at large intervals. This auxiliary spiral provides a hold for the spider when it spins the closer-spaced catch spiral, which is spun from the periphery toward the center and which contains sticky drops. During this work process the auxiliary spiral is again taken down (Fig. 158d).

Hunters that capture quick-moving prey frequently stalk it. My tame iguanas (*Tropidurus*) stalked flies which they took from my hand, and they never learned that stalking was not necessary. When they saw the fly in my hand they quickly approached to with 30 cm; then they stalked closer, body pressed to the ground, until they had come very close, after which they made a final lightning-quick dash to grasp the prey. We have already discussed the technique of capturing dangerous prey by polecats (p. 214). Other predatory mammals behave in a similar way (see P. Leyhausen [1956]).

The California sea otter (*Enhydra*) opens *Mytilus* clams by beating them against a stone which is balanced on the stomach while the animal is floating on its back (K. R. L. Hall and G. B. Schaller 1964). The Egyptian vulture (*Neophron percnopterus*) opens ostrich eggs with stones that are collected nearby. The bird positions itself next to the egg, raises its head as high as possible, and flings a stone down on the egg. Some stones weigh as much as 300 grams. This is repeated until the egg is cracked (J. van Lawick-Goodall and H. van Lawick 1966). The North American nuthatch (*Sitta pusilla*) uses bits of scotch pine as a lever to loosen pieces of bark that hide insects (D. W. Morse 1968).

From these descriptions it should be clear that animals which have different kinds of food needs have developed the most varied adaptations. The behavioral adaptations are more varied than those needed by fruit- and plant-eating animals, partially because prey animals evolve various forms of defense. This is correlated with the general observation that grazing and browsing animals are somewhat less intelligent than predators, which show more varied appetitive behavior, learn more, and perform spatially insightful behavior (p. 265).

These varied specializations are the result of competition. The adaptive radiation of bony fishes shows this as clearly as that of the Darwin finches, which, originating from one basic form, eventually

Figure 158. Construction of a spider net (Kreuzspinne). (Explanation in the text.)

filled the most varied ecological niches. Their differential way of feeding is expressed in the shape of their bills (Fig. 159) as well as in their behavior. The cactus ground finch (*Cactospiza scandens*) has a pointed bill which it pokes into the flowers and seeds of cacti.

Figure 159. Three Darwin finches as examples of adaptive radiation (Indefatigable, Galápagos). Notice the different shapes of the bill. Top left: Small ground finch (*Geospiza fuliginosa*); top right: medium ground finch (*Geospiza fortis*); bottom: cactus ground finch (*Cactospiza scandens*). The first three species are primarily seed eaters, which are specialized to feed on various seeds. The cactus finch with its sharp bill feeds on cactus fruits and drills into the juicy cacti. The species live side by side in the same biotope. (Photographs: I. Eibl-Eibesfeldt.)

The small insectivorous tree finch (*Camarhynchus parvulus*) searches for insects, and the woodpecker finch (*Cactospiza pallida*), lacking a long, sticky tongue, uses a long pointed twig or cactus spine for probing insects out of holes (Fig. 160). In Hawaii, where a similar adaptive radiation took place, the Akiapolaau (*Heterorhynchus*) fills the ecological niche of a small woodpecker. This bird gets at insects in a still different way. It chips at wood with the straight lower part of the bill and retrieves the prey with its curved upper bill. In a bird on New Zealand, *Heteralocha acutirostris,* the male has a short, straight bill for chipping, the female a longer, curved probelike bill. Here both sexes collaborate when in search for food (D. Lack 1943; see also Fig. 161). Many animals lay in stores; we gave the example of the squirrel. The various feeding patterns of insects are fascinating. The fungi-growing ants (*Atta*) make fertilizer from leaves that they have carried in. They cultivate fungi and live

Figure 160. Tool using in the woodpecker finch (*Cactospiza pallida*). Top left: insertion of the tool (cactus spine); top right and bottom: probing and lifting of a larva. (Photographs: I. Eibl-Eibesfeldt.)

Figure 161. Adaptations of birds that search for insects hidden in wood. The woodpecker (1) probes with its long tongue. *Heterorhynchus* (2) chips wood with the lower bill and probes with the upper bill. The woodpecker finch (3) (*Cactospiza*) uses a cactus spine or a piece of wood as a tool for probing. In (4a) (*Heteralocha*) the male does the chipping and the female does the probing (4b). (From D. Lack [1947].)

off their bulbous swellings. One could fill an impressive volume with only the descriptions of ways of feeding in hymenoptera (H. Bischoff 1927). Special adaptations with respect to body structure and behavior are found in those animals which live as parasites of other animals. We will discuss this separately (p. 302).

Defense against predators and interspecific competition

In competition with the pursuer, the pursued continued to evolve new adaptations in body structure and behavior that serve their defense and escape. The commonest response, of course, is to flee from the predator. There is commonly a species-specific flight distance which can be varied through individual experience (H. Hediger 1934). In general small animals have a shorter flight distance than larger ones. The less protection a species has through other means, the longer is the flight distance. A protectively colored grouper which is hard to see may be approached very closely. The same is true of a hare which hugs a depression in the ground. Conspicuously colored fish, on the other hand, will flee into a hiding place sooner when they are approached (Plate VII), but the poisonous dragonfish and other well-armored species permit a person to approach quite closely. In these fishes conspicuous coloration serves as a warning. Fish that are thus protected flee less readily and are more easily kept in captivity because of their tameness. The three-spined stickleback (*Gasterosteus*) is better protected by its long spines than the ten-spined stickleback (*Pygosteus*) and is correspondingly less shy (R. Hoogland, D. Morris, and N. Tinbergen 1957).

The direction and the goal of flight are frequently fixed as a phylogenetic adaptation in the same way as responsiveness to flight-releasing stimuli. A squirrel flees to the tops of trees, a mouse into its hole, a beaver dives, and a pheasant flies up. In two closely related geckos on New Britain H. Hediger (1934) observed that one species always fled up the tree trunks while the other fled downward and hid in a crevice.

In a coral reef many fish have specific hiding places. Above the large reefs in the Indian Ocean large swarms of blue triggerfish (*Odonus niger*) swim about. When approached each flees into a particular hole in the reef. Many damselfishes such as members of the genus *Chromis* hover above clumps of coral and flee between the stalks when alarmed. A diver can break off such a clump and bring the entire swarm to the surface, so strong is their attachment. Many fish have evolved special wedging devices which enable them to remain fixed in their holes, such as the velvet fish (*Caracanthus*),

Figure 162. Spine armored operculae of the velvet fish (*Caracanthus maculatus*) as an example
of a wedging mechanism for protection in corals. (From I. Eibl-Eibesfeldt [1964c].)

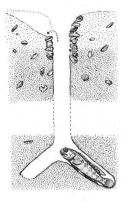

which lodge themselves in place with their spine-armored operculae
(Fig. 162). Fish that live above open sand have special escape
adaptations; they can withdraw quickly into burrows they have made
or bury themselves (I. Eibl-Eibesfeldt 1964c; W. Klausewitz and I.
Eibl-Eibesfeldt 1959) (Fig. 163).

Figure 163. Jawfish
(*Gnathypops rosenbergi*)
builds a vertical tube in
which it lives into the
sandy bottom. They
may be up to 1 meter
deep and have branches
at the bottom. The
rim is reinforced with
built-in pieces of coral,
clam shells, and sea
urchin pieces. (From I.
Eibl-Eibesfeldt and W.
Klausewitz [1961].)

The lizard fishes (*Synodus*), which rest on the sand, scoop sand
from beneath themselves with their pectoral and pelvic fins when
danger approaches so that within seconds they disappear into the
sand with only their eyes left showing. Garden eels (*Heterocongridae*)
remain in tubes in the sand which are held up by a gluelike substance secreted from their bodies (Figs. 164 and 165). They can
withdraw into the sand within seconds if one tries to pull them out.
Fish that live in the open water where there is no cover frequently
jump above the water surface and so from the pursuer's view. Many
of these fish, such as the *Mugilidae*, dive back into the water head

Figure 165. Garden eels (*Gorgasia maculata*) (Nicobar Islands). (Photograph: I. Eibl-
Eibesfeldt.)

Figure 164. Garden
eels (*Xarifana hassi*).

first, gather speed, and again jump into the air. A number of other fish return to the water tail first. Then they move the tail fin, whose lower part is extended, so rapidly that they are able to propel themselves along above the surface with the body pointing upward out of the water at an angle. As a result, the fish skims along the surface. Such surface skimmers were the evolutionary point of departure for the flying fish, as can clearly be shown by the increasingly differentiated forms within the *Synenthognathi* (K. Lorenz 1963b; see also Fig. 166). On one extreme is the garfish (*Belone belone*); at the other are various genera of flying fishes (Fig. 167), which can glide through the air with their pectoral fins that are considerably broadened into the shape of wings. They increase their speed on the water surface with their tail immersed in the water until they have attained enough speed to rise up into the air. They are able to sail along for many meters. Their head is higher than their tail and eventually dips back into the water as the fish slows down. The fish then can propel itself into the air again after having attained sufficient speed.

The South American hatchet fish (*Carnegiella vesca*), a freshwater fish, is said to be capable of an actual whirring flight. Breder and Eigenmann at the American Museum of Natural History observed that these fish raised themselves into the air when they were driven toward shore with a net (cited by K. Lorenz [1963b]; see Fig. 168).

Animals that are pursued often show irregular behavior patterns which make it difficult for a predator to predict their behavior and thus to maintain contact with them: Rabbits make sharp turns; some night moths fly zigzag courses or perform unpredictable loopings;

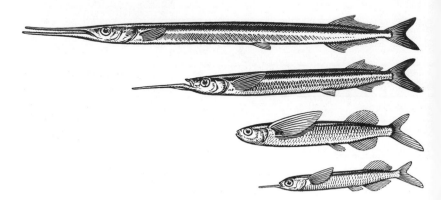

Figure 166. Differentiation of marine fish which skim along the surface. Top to bottom: Garfish (*Belone belone*), approximately 70 cm; halfbeak (*Hemiramphus*), 40 cm; *Oxyrhamphus micropterus* adult animal, 138 mm; young of the same species, 40 mm. None of these animals can glide through the air, but they are intermediaries that lead to the flying fish, which "fly" in that manner. (From K. Lorenz [1963b]; H. Kacher, artist.)

Figure 167. California flying fish (*Cypselurus californicus*) taking off from the surface and gliding. (From K. Lorenz [1965]; H. Kacher, artist.)

Figure 168. Freshwater fish capable of active wing beating. Left: *Carnegiella vesca*, the hatchet fish of South America; right: an intermediate form, *Triportheus elongatus*, that might possibly be capable of flight and lead into *Carnegiella*. (From K. Lorenz [1965]; H. Kacher, artist.)

pheasants scatter into all directions and then hide. Fish in a school react as a group. Sometimes there are deceptive maneuvers. A lizard that is caught by a predator often drops part of its tail, which continues to writhe on the ground, thus attracting attention while the lizard hides. These various deceptive behaviors have also been called "protean behavior" (M. R. Chance and W. M. S. Russell 1959), after Proteus of Greek mythology, who escaped his pursuers by assuming different forms.

Flight or escape reactions are often adapted to a specific category of predators. We have already mentioned that certain gastropods react to the odor of certain sea stars with escape reactions (p. 69). Domestic chickens show one set of behavior reactions to aerial predators and another to ground predators, each with its own specific warning calls. They take cover from birds of prey and fly up into trees before ground predators such as polecats and cats. In brain-stimulation experiments these behavior systems can be activated separately from different stimulus points (E. v. Holst and U. v. Saint-Paul 1960). If one category of predators is no longer present, one set of appropriate reactions against predators may drop out. On the Galápagos Islands, where predatory mammals are lacking, the hawk (*Buteo galapagoensis*) allows itself to be touched by humans (Fig. 169). Similarly tame are the marine iguanas (*Amblyrhynchŭs cristatus*) and the Galápagos penguins (*Spheniscus mendiculus*) when they are on land. In the water, where they are threatened by sharks, they flee even from a swimming man (I. Eibl-Eibesfeldt 1960b, 1964b). The Kittiwake gull (*Rissa tridactyla*) does not flee from humans when it is on the cliffs where it breeds, but it will flee when meeting man on land while gathering nesting material (E. Cullen 1957).

Figure 169. Example of "island tameness" (Duncan Island, Galápagos). The Galápagos hawk allows itself to be touched by man. (Photograph: I. Eibl-Eibesfeldt.)

Special adaptations to escape behavior is shown by the small crab *Dotilla*. After sifting sand pellets for food, it deposits them in a specific pattern, so that ring-shaped mounds are formed around the entrance to its hole. In addition, several radial "streets" are kept free of these pellets. In this way the crab maintains free escape routes on which it can reach its hole either directly or via a detour (H. Hass and I. Eibl-Eibesfeldt 1964; see also Fig. 170).

Escape behavior is certainly activated by external stimuli in most instances, but K. Lorenz (1943) has shown that this behavior may also be based on internal motivation. Ducks "escape dive" frequently

Figure 170. Sand pellets that were deposited in ringed walls around the hole entrance by the pellet crab (*Dotilla sulcata*) starting from left to right as the animal continues to feed. On the right the feeding crab. (From H. Hass and I. Eibl-Eibesfeldt [1964].)

in vacuo, and animals that have not been frightened for some time are inclined to flee in response to stimuli that would normally be ineffective as time goes on; they show a definite threshold reduction for escape reactions.

If an animal is prevented from escaping by driving it into a corner, it may attack once its assailant has reached a critical distance (H. Hediger 1942). A circus trainer must at all times be cognizant of this reaction. An animal will also attack if it is suddenly surprised and the critical distance has been inadvertantly violated. Frequently flight and defense are combined in behavior, for example, when an octopus ejects an inky substance when beginning its escape. The same occurs in an alarmed bombardier beetle (*Brachynus*), which secretes a volatile substance from its anal glands that quickly diffuses into the air.

Many animals seek protection near others that are better armed. We shall discuss this in Chapter 15. Animals that do not possess their own mechanical protection frequently obtain it by building containers which they carry with them. The tubelike structures made by caddisfly larvae are well known and their construction and repair has been described by C. Wesenberg-Lund (1943). The caterpillars of the *Psychidae* built a similar structure, an example of behavioral convergence. Empty snail shells are used by hermit crabs after careful inspection. The octopus (*Octopus aegina*) enters empty clam shells whose two halves it can open and close as needed (Fig. 171). The caterpillar of the wasp *Lygaeonematus compressicornis* surrounds its feeding place on an aspen leaf with a fence of foamy palisades, and sometimes it prevents access to a leaf by erecting such an obstruction at the stem. This foam is sticky and contains salicylic acid.

Many animals protect themselves through camouflage. Some are already adapted by their appearance to a specific background. However, they must also possess the appropriate behavior for select-

Figure 171. Octopus (*Octopus aegina*) in the clam shell in which it lived and protected its spawn. (Photograph: I. Eibl-Eibesfeldt.)

ing the correct background and to maintain specific postures if this protection is to be effective. In this way, caterpillars which are countershaded not only select the environment that fits this coloration, but they also assume the postures that make them least conspicuous (W. M. Herrebout and others 1963; H. B. Cott 1957). E. Curio (1966a) found one species of moth that had three different forms of caterpillars. Each form had a different color and showed the appropriate behavior in selecting the most fitting resting place (Fig. 172). Some looper moth caterpillars effectively mimic dead twigs. Many animals camouflage themselves by covering their bodies with foreign objects; thus sponge crabs (*Dromia*) use their modified third and fourth pair of legs to hold clams and sponges on their backs. In majids the carapace contains special bristles for attaching foreign objects. All these examples show that camouflage is the result of a complicated interaction between structure and behavior.

Widespread methods of defense against predators are those of warning and deception O. M. Reuter (1913). The forktail caterpillar (*Dicranura vinula*) will stop eating when it is disturbed and remain in a stretched-out position with the head slightly drawn in. In this position the green caterpillar is well camouflaged. If one touches the animal or moves the leaf on which it sits, the caterpillar raises its front end and turns a very conspicuous "face mask" toward the attacker. The brown head, framed by red and yellow edges, carries two dark pigment spots, which seem to be imitations of eyes. At the same time the animal ejects two red glandular filaments out of the last pair of modified abdominal legs, which twirl for several seconds before they are withdrawn (I. Eibl-Eibesfeldt 1966b; see also Plate IV). From a well-developed gland in the prothorax the caterpillar can eject a bad-smelling secretion.

T. Eisert and J. Meinwald (1966) reported on the chemical defenses of insects. Sometimes these defenses are neutralized by the

Figure 172. Resting places (branch or leaf) of three caterpillars of *Erinnyis ello* in all five stages (I–V). Arabic numerals, number of observations. (From F. Curio [1966a].)

special attack behavior of the predators. In this way the mouse *Onychomys torridus* deals with the beetles *Eleodes* and *Chlaenius,* which secrete a defensive substance from the posterior tip of their abdomen, by grasping the beetles and ramming them into the ground, abdomen first.

Often bad-tasting insects are very conspicuously marked. A good example is provided by the wasps. Once a songbird has been stung by a wasp it will remember this for months and avoid all conspicuously ringed objects from then on. This protection is also extended to a number of insects that are marked like wasps. Many bad-tasting butterflies are mimicked by those that are edible (Plate IV). They deceive their pursuers. Some unusual examples of mimicry have been collected by E. Curio (1966c), and we mentioned additional examples on page 149 (see also O. J. Sexton [1960]). Many animals are on the lookout for such surprises. All our native mammals raise their heads from time to time and look around, taking samples from various strata of air by sniffing while raising and lowering their head ("taking wind"). In addition, they interrupt other

Figure 173. Top: predatory *Mycteroperca olfax* about 1 meter in length chasing swarm fish *Xenocys jessiae* (Galápagos). The fish form a vacuole around the predators; bottom: *Xenocys jessiae* in flight just above the sea bottom. (Photographs: I. Eibl-Eibesfeldt.)

activities such as eating or digging with great regularity. A digging hamster repeatedly looks up and around. While analyzing films of people who were eating but were unaware of the observer, H. Hass (1968) discovered that they stopped at regular intervals and seemingly automatically looked up and around. Additional observations along these lines have convinced us that this looking up is a kind of reconnoitering.

A very conspicuous adaptation for protection from predators is the schooling of fish. First, it is easier for animals living in a swarm to detect danger; more eyes can see more. This, however, is by no means the most important factor favoring the aggregation. It can be shown that a single fish is much safer in the swarm, even when not warned, than if it is alone. Whereas a single fish can be easily fixated by a predator and then caught, it is much more difficult for the predator to focus and pursue one target out of many in a swarm. The aim of the predatory fish is confused and, of course, precise aiming is a prerequisite for successful hunting (I. Eibl-Eibesfeldt 1962b; G. v. Wahlert 1963). Predatory fish try to separate single fish from a swarm or lie in wait quietly until a fish comes within reach (Fig. 173). A falcon in pursuit of a flock of birds also tries to isolate an individual from the group by means of sham attacks (N. Tinbergen 1951). When a hawk chases a flock of pigeons it usually aims for the bird that is farthest from the group. In a flock of white pigeons it chases the single black one, and in a black flock it chases the white one.

A large number of brood-care behavior patterns for care of the young can be understood as adaptations to predators, such as the distraction display (p. 148), or the mouth-breeding of cichlids (p. 152 and Fig. 174). Black-headed gulls (*Larus ridibundus*) remove the egg

Figure 174. *Tilapia nilotica* female taking young into her mouth during the brood-care stage. (Photograph: R. Apfelbach.)

shells from the vicinity of the nest after the young have hatched. If they fail to do this a predator can more easily detect the nest (N. Tinbergen, G. J. Broekhuysen and others 1962). Adaptations to predators are also seen in the selection of the nesting place and the habit of breeding in colonies. Large gulls and crows can be driven off by many black-headed gulls. H. Kruuk (cited by N. Tinbergen [1965]) demonstrated that these attacks make it more difficult for predators to steal eggs from the center of the colony than from the periphery. The synchronization of egg laying, in black-headed gulls is a protection against predators. By this simultaneous egglaying the "market is flooded," so to speak, and this overabundance of eggs then becomes a protection (I. J. Patterson cited by N. Tinbergen [1965]).

To demonstrate what a variety of brood-care patterns evolved even in the quite homogeneous group of anurans, some examples of frogs that care for their brood will be mentioned (Fig. 175). Additional literature is in W. Klingelhöffer (1956) and R. Mertens (1959). In the central European midwife toad (*Alytes obstetricans*), the male takes up the strands of spawn around its legs until the larvae wiggle within the egg shells. Then the toad seeks out a pond in which the larvae hatch. In this way the spawn is protected from the numerous predators in the water. The Central American poison arrow frogs (*Dendrobates auratus*) deposit their eggs on a leaf outside the water. The male guards them and sits next to the newly hatched tadpoles, which then climb the back of their father. There they adhere by suction and are thus transported to the nearest puddle.

Figure 175. Frogs that care for their brood: Top right: midwife toad; below: Goldi's frog; left: *Hyla faber* in basin built into the mud which protects the spawn against predatory fish. (Drawn from photographs by H. Kacher.)

The frog (*Hyla faber*) of tropical South America builds small breeding basins at the edge of puddles by raising mud into a ringed wall. Spawning takes place in these small basins. Later the larvae are freed by the rising water. In the Chilean (Darwin's) frog (*Rhinoderma darwini*) the males guard the eggs, which are deposited on land. As soon as the larvae wiggle in the eggs they are snapped up by the males and are carried about in the vascularized throat pouch. In the South American toad (*Pipa americana*) the eggs develop in honeycomb-like pockets on the mother's back.

Marsupial frogs (*Nototrema*) protect their spawn in incubating pouches on their backs, and Goldi's frog (*Hyla goldii*) carries the spawn in a bowl-like depression on its back. Many frog species deposit their eggs in a frothy mass of air bubbles which they fasten to plants above the water, so that the hatching tadpoles fall into the water, for example, the grey treefrog of Africa (*Chiromantis xerampelina*), whose female clasps the foamy nest until the larvae have hatched. In this way the nest is protected against drying out too rapidly. The whistling frog (*Leptodactylus labialis*) builds its foam nests in caves it has dug into creek banks, and the larvae are freed when the water rises during the rainy season. A comparable multiplicity of brood-care behavior patterns can be observed in many other animal groups. We might think of the many fascinating adaptations of insects. Suffice it here to demonstrate this variability with the examples from the frogs. If such a uniform group of animals as frogs can show such a variety of brood-care behavior patterns, it is easy to imagine what diversity we can expect in other groups of animals.

The nonspecies member may also have an impact as a competitor, forcing the most varied adaptations in animals which are so threatened. On the Galápagos Islands the frigate birds (especially *Fregata minor*) have specialized in taking food away from other sea birds. They circle above a bay until they see a fishing booby or other sea bird. Then they rush at it and peck it with their bills until it regurgitates its prey, which they then skillfully catch. I once saw a tropic bird killed by a frigate bird. This formidable predatory competition may have caused the swallow-tailed gull (*Creagrus furcatus*) to fish only at night. The dusky gull (*Larus fuliginosus*), which fishes during the day, has gray feathers and is well camouflaged, which I interpret as an adaptation to evade the competing frigate birds. In this connection it is remarkable that many red-footed boobies (*Sula piscator websteri*) on the Galápagos possess a brown-feathered plumage which resembles that of juveniles.

Competition with other species finally demands behavior adaptations in the most varied areas. We already mentioned the food-getting behavior of the Darwin finches. As an example from another functional cycle: Blue-faced boobies (*Sula dactylatra granti*) and red-footed boobies (*Sula piscator websteri*), which occur sympatrically on the Galápagos Islands, have different breeding habits. The blue-faced booby breeds on the ground, the red-faced booby on trees (Fig. 176).

Special protective measures are required against parasites. Grass frogs, toads, and alpine salamanders free themselves of adhering leeches by sitting in the sun, which the leeches cannot tolerate. Fish that are plagued by parasites allow themselves to be cleaned by cleaners (p. 146). In the leaf-cutting ants (*Atta cephalotes*) the small minima workers protect the larger workers against the attacks of parasitic flies of the *Phoridae* group. While the larger workers cut leaves and are defenseless, the minima workers position themselves with open mandibles around them and snap at the approaching flies. They also ride along on the cut leaves as guards on the way back to the nest (I. and E. Eibl-Eibesfeldt 1967; see also Fig. 214).

Figure 176. Blue-faced boobies (ground nesters) and red-footed boobies (tree nesters) (Galápagos). (Photograph: I. Eibl-Eibesfeldt.)

Symbiotic relationships

We want to conclude the discussion of interspecific relationships with some illustrations of symbioses, which are of special attraction for the student of behavior. As we have said in Chapter 5, the problem of interspecific communication is raised in such partner relationships.

German biologists speak of symbiosis when two different species collaborate in some way to their mutual advantage. If only one species profits, without some disadvantage for the other, this is commensalism. In English, what we call symbiosis is called mutualism; the concept of symbiosis is a larger category which includes parasitism, commensalism, and mutualism.

The point of departure for parasitism as well as symbiotic partnership is probably in most instances some kind of commensalism. Many fish of the high seas seek protection in the vicinity of larger fish. I observed jackfish (*Caranx ruber*) accompanying barracudas (*Sphyraena barracuda*). They swam closely above their backs and performed each turn of the barracudas. Near the Cocos Islands I saw the jackfish (*Caranx chrysos*) accompanying sharks and rays, and in the Indian Ocean a closely related species of jackfish accompanied the large wrasse (*Cheilinus undulatus*). The blue runner (*Elagatis bipinnulatus*) sometimes swims with sharks and other large fish. They are safer there than alone in the open water. Sometimes they rub themselves on the skin of the sharks (Fig. 177), which, in turn, seem to derive no advantage from their companions. From these kinds of loose associations obligatory relationships evolved. Pilot fish (*Naucrates ductor*, L.) are almost never seen alone. According to our observations they behave differently swimming with predatory sharks than with the giant mantas (*Manta*) and harmless whale sharks (*Rhincodon*). When accompanying predatory species, the pilot fish usually swam near the ventral and dorsal fins and farther behind, and only for a short period of time in front of the mouth, such as when they passed the shark and swam toward the

Figure 177. Mackerels (*Elagatis bipinnulatus*) rubbing themselves on the back of a shark. (Maldive Islands). (Photograph: I. Eibl-Eibesfeldt.)

divers. Whale sharks and mantas were surrounded by pilot fish which fled into their mouths when danger approached (Fig. 178). H. Hass (1954) observed that the pilot fish cleaned the mouths of mantas. Here a partnership with mutual advantage for both has developed. *Echeneis* also clean sharks.

a   b

Figure 178. (a) Manta with pilot fish (Red Sea); (b) shark with pilot fish (Azores). (Photographs: H. Hass.)

Figure 179. Cleaner fish (*Labroides dimidiatus*) cleaning *Odonus niger.* (Photograph: I. Eibl-Eibesfeldt.)

In truc cases of mutualism simple signals develop between the partners in the service of interspecific interaction. These have been studied in some detail in the cleaning symbioses (I. Eibl-Eibesfeldt 1955). We referred earlier to the "cleaner dance" as well as the "inviting" and "rejection" postures of the customers, which even change color when they want to be cleaned. *Naso tapeinosoma,* for instance, turns light blue when it is being cleaned. This makes the parasites visible against the contrasting background. The cleaner rapidly taps its host with its fibrillating ventral fins keeping it informed as to where it is cleaning. The host acts accordingly: It stops moving the fins that are being cleaned (Fig. 179) and erects them, and it opens its mouth when the cleaner butts against its corners and allows the cleaner inside. If the host wants to breathe heavily, it signals its intention by partially closing its mouth and the cleaners leave the mouth cavity (p. 147). Other warning signals are the movement of the operculae and the shaking of the entire body by the host fish. These behavior patterns have convergently evolved in various species of fish.

Many different species of fish allow themselves to be cleaned by cleaners: Predatory fish and peaceful ones, reef dwellers as well as fish of the open seas. Sometimes we saw swarms of fish appear out of the deep blue water and stand head down above a cleaner station as if in response to a single command. They waited for the cleaners, which soon were busy with their task. After several minutes the fish school again disappeared in the depths of the sea.

Even manta rays visited the cleaner stations in the coral reef to be cleaned (Fig. 180). In a reef canal of the Addu Atoll (Maledive Islands) we observed how four mantas slowly circled a coral clump 15 meters below the surface. During this time they were being cleaned by numerous cleaner wrasses (*Labroides dimidiatus* and *Thalassoma* species). They opened their large gill covers and permitted the cleaners to enter.

Figure 180. Manta being cleaned by cleaner fish (*Thalassoma*). The manta opens the gill covers and permits the cleaners to enter. (Photograph: H. Hass.)

The cleaners play an important role in the life of the reef fishes, as was recently demonstrated by C. Limbaugh (1961). He captured all cleaners from two reefs in the Bahamas. A large number of reef fish left the area, and those that remained showed many skin and fin ailments 2 weeks later. Not until new cleaners migrated to this area did customers return again. Limbaugh also observed that one cleaner was visited by more than 300 customers of various species during a period of 6 hours.

We observed that at times the fish actually crowded around cleaner stations. Various species waited their turn, and as agonistic as these fish are at other places in the reef, they were peaceful here. The cleaner station was, so to speak, a barber shop in the reef, owned by all and therefore neutral ground.

Cleaning symbioses have been observed in various regions of the world, and it was found that all cleaners are not specialized in the same way to carry out this "trade." In the Indo-Pacific area it is mostly the cleaner fish of the genus *Labroides* that are active. In the tropical Atlantic there are many species of fish that regularly or occasionally clean others. In 1955 we observed near Bonaire (Caribbean) the following cleaner fish: *Elacatinus oceanops* (Gobiidae), *Gramma hemichrysos* (Hemichromidae), *Thalassoma bifasciatum* and *Bodianus rufus* (Labridae), and *Anisotremus virginicus* (Haemulidae); in the Bermudas we found *Chaetodon striatus* (Chaetodontidae) and *Abudefduf saxatilis* (Pomacentridae). It appears here as if the ecological niche of the cleaner has been occupied by not just one specialized cleaner. Various species are still competing for this position. Most specialized is the cleaner goby (*Elacatinus oceanops*), which has very similar colors to the cleaner wrasse (*Labroides dimidiatus*). A comprehensive summary of these cleaning symbioses can be found in H. M. Feder (1966). Near a Cuban island D. H. H. Kühlmann (1966) observed a toothed-carp (*Gambusia*) cleaning the mouth of a crocodile (*Crocodylus acutus*).

Many shrimp of the genus *Stenopus* and *Periclimenes* are known to be cleaners of fish. By waving their long antennae they attract the attention of their customers. *Periclimenes pedersoni* climbs on the fish and crawls under their gill covers or into their mouths with the fish exhibiting the appropriate "inviting" signals. When frightened the host fish spit out the cleaner prawns or warn them before swimming off. The cleaner prawns also remove parasites that are lodged below the skin (C. Limbaugh, H. Pederson, and F. A. Chace 1961).

On land comparable cleaning symbioses exist. The crocodile bird (*Pluvianus aegyptius*) had been cited by Herodotus as entering the mouth of crocodiles and eating leeches. This symbiosis has not been studied in detail. Very little is also known about the oxpeckers (*Buphagus africanus*), the relatives of starlings, which climb about on larger animals, catching insects, larvae, ticks, and other parasites. They specialize in this and are sometimes parasites themselves. I observed how they opened wounds on rhinoceroses and drank the blood. However, the benefits undoubtedly outweigh the disadvantages. Whether or not these species communicate with one another in the manner of the cleaner fish, is not known. One of my own chance observations points in this direction. An oxpecker that was working on a part of the skin of a rhinoceros, which was infested by the larvae of the botfly, was repeatedly interrupted in its task when the animal rolled on the ground. When the rhino again stood on its feet, the bird uttered rapid calls, before again alighting on it. The bird continued to call when it again sat on the rhinoceros before again approaching the wound. The rhinoceros did not roll on the ground after that. Zebras await oxpeckers by standing with spread legs, their tail raised and ears drooping, so that the birds can get at the ticks (H. Klingel 1967).

Oxpeckers are completely dependent on their hosts, However, there are also a number of more casual relationships which might serve as models for the development of such cleaning symbioses. Starlings (*Sturnus vulgaris*) can frequently be seen near grazing cattle, where they catch insects. White wagtails are frequently seen chasing insects on pigs. Cattle egrets (*Ardeola ibis*), which often ride on elephants, catch the insects that have been scared up by the grazing animals. The small ground finch (*Geospiza fuliginosa*) of the Galápagos Islands searches for ticks on marine iguanas. A meeting of two species in this manner provides the prerequisite for the development of an actual symbiotic relationship.

The partners may offer each other quite different advantages, and accordingly the symbioses vary. In the Red Sea and the tropical Indo-Pacific region live giant anemones (genus *Stoichactis, Radianthus, Discosoma*), between whose poisonous tentacles one often finds the anemone fish of the genera *Amphiprion* and *Premnas* (Fig. 181).

Figure 181. Anemone fish (*Amphiprion akallopisus*) between the tentacles of a *Radianthus* anemone. (Photograph: I. Eibl-Fibesfeldt.)

The fish are hardly ever encountered without an anemone, and the advantage of living together can be readily seen: The fish are well protected between the tentacles of the anemones. No predators can catch them there without being caught themselves.

In some instances it has been observed that the anemone fish clean their host: They carry off its waste products and swish the sand from its top side (I. Eibl-Eibesfeldt 1960a). In the aquarium some anemone fish feed their anemones, but it is not known if they also do this in the wild.

In any case the anemone fish seem to have the decided advantage. How is it, then, that they are not stung by the poisonous tentacles of the anemone? Here some opposing points of view exist, because various authors have studied species that differ in their behavior.

We have experimented with *Amphiprion akallopisus, A. xanthurus,* and *A. percula* and found that the fish are covered by a protective substance on their skin (I. Eibl-Eibesfeldt 1960a, 1964c). If one removes the mucus from the skin, the fish are stung and caught by the tentacles. An intact fish can be flung against the tentacles without harm resulting from this rough treatment. This holds true even when the fish is moved passively across the tentacles of the anemone in an atypical manner. This disproves the hypothesis of some who suggested that the anemone recognizes its fish by the type of movements they make. Investigations by D. Davenport and K. Norris (1958) and more recently by M. Blösch (1965) have also shown that anemone fish possess a protective substance. There exist anemones that tolerate all species of anemone fish, whereas others only tolerate a specific species of anemone fish and catch all other species. And finally Blösch found anemones that at first caught all kinds of anemone fish but gradually accepted them. But here, too, only anemone fish are able to accustom the anemone to their presence. Originally it was assumed that the protective mucus is secreted by the fish. D. Schlichter (1968), however, demonstrated that the anemone fish *Amphiprion bicinctus* impregnates itself actively with anemone mucus which makes the fish a part of the surface of the anemone itself. Whether this is true for all anemone fish needs investigation.

In the Mediterranean Sea E. Abel (1960a) observed the goby (*Gobius bucchichii*) as an "anemone fish" of *Anemonia sulcata*. Besides the typical anemone fish there exist a number of other fish which seek protection near the anemones but which avoid contact with the poisonous tentacles. Of special interest is the behavior of the damselfish (*Dascyllus trimaculatus*). Near the Maldive and Nicobar Islands we saw these fish, especially young ones, frequently near the giant anemones, without touching their tentacles. In the Red Sea, on the other hand, we observed small schools of 1- to 2-cm-long damselfish between the tentacles of anemones. Here one can observe the development to an anemone fish within one particular species.

Other animals have also joined up with anemones. It is well known that the hermit crabs of the genus *Eupagurus* have anemones on their snail shells in which they live and which protect them (Fig. 182). It has been observed that an octopus which attempted to catch such a hermit crab was stung by the anemones and retreated. However, the anemones probably also derive some benefit from this union, for instance by partaking of the meals of the crab. In any case they seem to be well adapted to this life with the hermit crabs. It is even possible that the initiative to live on snail shells came from the anemones. Near England the anemone (*Calliactis-[Sagartia] parasitica*) climbs without the aid of the hermit crab (*Eupagurus bernhardus*) onto the snail shell (D. M. Ross 1960). A

a   b

Figure 182. (a) Hermit crab *Pagurus arrosor* in symbiosis with *Calliactis parasitica;* (b) hermit crab *Eupagurus prideauxi* in a snail shell that is grown over with *Adamsia palliata.* (After L. Faurot from H. Füller [1958].)

detailed investigation showed that the classical conception of the stinging cells as independent effectors is not quite correct. If the anemone sits on the sea bottom and touches the horny outer surface of a snail shell, then the tentacles actively explore the shell, and many adhere to it by a discharge of the nettling cells. The anemone then releases its footing from the substratum and mounts the snail shell. Once in place its behavior changes. The tentacles no longer stick to the shell when they are touched with a piece of another shell. The threshold for the release of the nettling cells changes depending on the substratum on which the anemone sits (D. Davenport, D. M. Ross, and L. Sutton 1961).

In the Mediterranean the hermit crab (*Pagurus arrosor*) helps its anemone (*Calliactis parasitica*) to mount its shell. When the crab encounters an anemone fastened to the bottom, it at first attempts to remove it from the substratum by tapping it with its claws and the first pair of walking legs. The contracted anemone opens up as a result of this and releases its hold on the bottom. Without this "cooperation" of the anemone such a removal from the bottom is not possible, because the foot of the anemone adheres tightly. The released anemone then sticks to the snail shell with its tentacles and bends its body in a U shape, thus bringing its foot onto the shell. When the hermit crab chooses a larger shell, it will transfer its anemones to the new one (F. Brock 1927).

These two examples illustrate two developmental stages of this interspecific relationship, because in the first example of living together the crab is relatively uninvolved, whereas in the second one a clear adaptation is present in the behavior of the two symbiotic partners. It is possible, however, for each to get along without the other. In the crab *Eupagurus prideauxi* and the anemone *Adamsia palliata,* however, the relationship is so close that the anemone cannot live without the crab. The adult partners are never found alone. The anemone fastens itself below the mouth of the crab on the snail shell, which it gradually surrounds with its foot. Then it secretes a horny substance above the opening of the snail shell, and in this way achieves a growth (enlargement) of the snail shell. In this way it prevents frequent changes on the part of the crab, on which it depends as a supplier of food. Additional examples about symbioses with nettling animals can be found in H. Füller (1958).

In this way many protective alliances of various kinds are formed. A small coral, *Heteropsammia,* which lives in the sand, can only exist with the aid of a small sipunculoid worm (*Aspidosiphon*) which lives at the base of its lime skeleton. The worm moves the coral across the sand bottom, prevents it from sinking in, and straightens it up when it falls down. In return, the worm is protected by the coral (H. Feustel 1966). Prawns of the genus *Alpheus* live together with various gobies (*Cryptocentrus lutheri,* for example). The prawn digs a cave in the sand, and the goby will watch over it in return (W. Luther 1958; W. Klausewitz 1961; see also Fig. 183). Cardinal fish seek protection with sea urchins, *Siphamia versicolor* with *Diadema* sea urchins. The fish clean their hosts in return (I. Eibl-Eibesfeldt 1961c; see also Fig. 184). Additional examples will be found in E. Abel (1960b) and D. Magnus (1964).

The various symbioses known in insects are most fascinating. As an example we mention only the symbioses between ants and plant lice (*Aphids*). The plant lice secrete large amounts of sugary excrements and they are visited by the ants for this reason. The ants tap the aphids with their antennas and thus stimulate them to secrete

a   b

Figure 183. (a) and (b): While the goby (*Cryptocentrus lutheri*) watches at the entrance, the prawn *Alpheus djiboutensis* busily digs out sand from the cave. (Photographs: W. Luther.)

Figure 184. Cardinal fish (*Siphamia versicolor*) seek protection between the spines of a *Diadema* sea urchin (Nicobar Islands). (Photograph: I. Eibl-Eibesfeldt.)

their excrement. The behavior of the ant looks very similar to that which is shown when one ant begs for food from another, and it was thought that the plant lice imitated the head of an ant with their abdomen, especially because they raise their hind legs into the air, resembling feelers (Fig. 185). The relationship between the plant lice and the ants can be very close. The plant lice *Lachnus tacniato ides, Anuraphis farfarae, Pemphigus caerulescens,* and the *Stomachis* species can no longer remove their excrements from their bodies because hairs surround the anal region, which retains the excrements for the ants. These plant lice are not only protected against enemies by the ants, but they are cared for in a manner in which

Figure 185. Husbandry of the plant louse *Tramaradicis* on the roots of *Artemisia* by the ant *Lasius umbrabus*. (After A. Forel.)

man keeps valuable cattle. The ants make roofs for them out of earth and they bring the winter eggs into the lower part of their hills for safe keeping over the winter. In the spring the hatching larvae are taken to the plants for feeding, and on cold nights they are returned to the ant hill. Sometimes the ants also depend exclusively on plant lice. *Lasius brunneus* lives exclusively from the excrements of the *Stomachis* species. The symbioses between insects and flowers are mentioned at this time but will not be discussed furthcr.

Regardless of the kind of symbiotic relationship, they raise interesting questions for the ethologist in respect to their ecological significance, the way they originated, the methods of interspecific communication, and the development of signals.

Parasitism

The various relationships between parasites and their hosts raise a number of fascinating problems for the student of behavior, such as selection of the host, finding the host, and the defensive reaction of the host (G. Osche 1962, 1966). We already discussed how the parasitic widow birds mimic the host species and how the larvae of certain liver flukes reach their final host by means of deceptive signals. D. Davenport (1966) investigated the specific reaction of polychaetes which live on sea stars and hermit crabs to substances that are secreted by their hosts. Between symbionts, commensals, predators, and parasites there are transitional forms. This can be traced in ants and their visitors (E. Wasmann 1920, 1925; K. Escherich 1906; H. Bischoff 1927; R. Hesse and F. Doflein 1943). A number of these "guests" are true predators (*Synechtrans*) that damage the ant population by eating the brood and avoid pursuit by the ants by retreating into small crevices. Their hosts cannot follow them there. Others are heavily armored and can roll up into a ball so that the ants can get no hold on them. Besides these predators there are also harmless dwellers (*Synoeketes*) in an ant colony. They only eat waste products and are therefore tolerated. However, some of these have managed to obtain food during the social feeding among ants. In this way the silverfish *Atelura* (Fig. 186a) obtains nourishment. The ant cricket (*Myrmecophila acervorum*) robs food-carrying worker ants as well as larvae that have already been fed. The guest ant *Formicoxenus* begs from workers and is fed by them. True guests (*Symphyle*) provide special aromatic secretions for the host ants in return for food and protection. As long as no disadvantage derives from this alliance for the hosts one could speak of a symbiosis; frequently such symphiles cause extensive damage to their hosts. The short-winged *Lomechusa strumosa* live with the red wood ants *Formica rufa,* which care for them, feed them, and who in turn

Figure 186. (a) Silverfish *Atelura* takes part in the feeding of an ant; (b) the parasitic ant guest *Atemeles* is fed by an ant. (a) After C. Janet from K. Escherich [1906]; (b) after A. Forel from K. Escherich [1906].)

receive the aromatic gland secretion of the short-winged species (Fig. 186b). While caring for the other, however, they neglect their own brood, which is lost if there are too many symphiles. Furthermore, the guests and their larvae eat a large number of their hosts' larvae.

The mite *Antennophorus* becomes a parasite in a different way. The mites sit on the underside of the ants' heads and stroke the throat region and the side of the head with the first pair of legs so that the host ant regurgitates a drop of food, which is taken up by this ectoparasite (C. Janet 1897, cited by K. Escherich [1906]; see also Fig. 187a). The ants often try, in vain, to remove this living muzzle. On many ant larvae of the genus *Pachycondyla* there is

Figure 187. (a) A phorid larva is wrapped around the neck of the ant larva (*Pachycondyla*) like a living collar. The phorid larva is able to control to some extent the behavior of the larva (see the text). (b) The parasitic mite *Antennophorus,* which sits at the underside of the ant's head (*Lasius*) stimulates the ant with its legs, and thus releases feeding. (After C. Janet from K. Escherich [1906].)

wrapped around their necks, as a kind of living collar, a phorid larva. It eats when the ant larvae is fed. When all the food is eaten, the fly larvae manipulates the behavior of its host larva by pinching its skin so that it becomes restless, which in turn arouses the attention of feeding worker ants, which return to feed them (Fig. 187b).

Very unusual kinds of interspecific collaboration are found in the slave-making ants, where one ant species becomes the parasite on another ant. The females of the slaveowners no longer establish their own colonies. If the female of the red ant *Formica rufa* does not find a nest of its own species it will enter one of the related *Formica fusca*. There it will be adopted if no other queen is present. In the course of time this nest will become a *Formica rufa* state. The fertilized queen of *Formica sanguinea* always enters the nest of *Formica fusca,* robs some pupae, which it raises and defends against the host ants. The hatching workers care for the nonspecific queen and her brood, so that within the host nest a *Formica sanguinea* state develops. The host ants eventually kill their own queen, so the host population eventually dies out. When there is a lack of workers, the *Formica sanguinea* workers rob workers from nearby *Formica fusca* nests. The American *Formica rubicunda* ant, which lives in *Formica subserica* nests, collects new slaves when they are needed. The Amazone ants (*Polyergus*) are completely dependent on their slaves, because they are unable to break down food with their saberlike mandibles. They must be fed by their slaves and their main occupation is to go on slave-making raids. The small ant *Solenopsis fugax* builds its tunnels in the nests of *Formica rufa* so that it can rob their food. Similar prey robbers are also found among higher animals. We refer to the frigate birds mentioned earlier (for example, *Fregata minor* of the Galápagos Islands, which chases other birds and takes their food, (I. Eibl-Eibesfeldt 1964b). The European cuckoo possesses a supranormal gape releaser and throws its nest mates, the young of the host species, out of the nest. This behavior wanes after several days.

Many species live on others either for the duration of their life or for a shorter period only. Some species regularly use others as a means of transportation, for which the term *phoresis* has been coined. The marine snail (*Janthina*) is often used by the crab (*Planes minutus*) as a means of transportation. The book scorpion sometimes clasps the legs of flies and is transported to other localities in this way. Such phoresic relationships may lead to a symbiotic as well as a parasitic relationship. The larvae of *Meloë* climb the flowers of anemones and dandelions and wait there for bees which then carry them to their nests. There they eat the larvae and food stores. With these examples I want to close with the discussion of interspecific relationships, and we will now turn to intraspecific interactions.

Behavior toward species members

In some aspects there exists a surprising agreement in the social behavior of animal and human groups, so that one may be encouraged to hope that animal psychology could be useful in discovering laws that also govern the social life of human groups (D. Katz 1926:448).

The conspecific often plays the role of a partner as well as that of a rival in the ecology of an animal. As a result of the dual nature of this relationship, the species member frequently becomes the bearer of rejecting as well as attracting signals; some aspects of this duality have already been discussed under the heading of courtship and greeting ceremonies. In this section we want to demonstrate the selective advantage of these opposing contact-seeking and distance-maintaining mechanisms. We want to discuss advantages of forming groups with few or many individuals and the advantages of social intolerance. Only a few species are socially indifferent in the sense that they seek no contact. Such animals do not even meet for reproduction. Many marine animals discharge their sexual products into the surrounding water. Some of them do this in response to a chemical stimulus that results in synchronization, and when they signal to each other in this way, we are already dealing with contact behavior of a most simple kind. Many ground-dwelling male arthropods deposit spermatophores which are accidentally discovered by the females. The males of *Polyxenus* build a signal structure in the form of a double track made of filaments which leads the females to the spermatophores (F. Schaller 1962).

Special internal motivating mechanisms ensure that an animal will seek social contact with members of its species. Distance-maintaining behavior is based on the aggressive drive (p. 57), and there are special drive mechanisms which lead some animals to agonistic encounters with conspecifics. In the same way there exist drive mechanisms which are the basis for grouping behavior. The drive to sexual and parental-care behavior are examples of this. For the graylag goose H. Fischer (1965) demonstrated an additional drive which is independent of sexual behavior and caring for the young. Such drives to maintain bonds probably exist in other animals as well. A young fish that is separated from the swarm exhibits appetitive behavior in search of a swarm. It has not been possible to demonstrate an underlying sexual motivation, and it is probably appropriate, until the underlying mechanism is better known, to speak of a general drive to establish bonds as the basis for this appetitive behavior.

Intraspecific aggression

Territoriality

According to Rousseau the builder of the first fence was the founder of civilization. Since the work of H. E. Howard (1920) we know, however, that many animals defend a certain area of their habitat as a territory against members of their species and often mark it in a specific manner. The territory may be the possession of one individual which repels all conspecific animals or merely all species members of the same sex, but it can also be the possession of a group that repels only conspecifics that are not members of the group.

In hamsters males and females live solitarily, and they occupy a den only temporarily during the reproductive season. The females live with their young only for a relatively short time. In many birds and some mammals (gibbon, *Hylobates lar*; see J. O. Ellefson [1965]) the pair defends a territory, but many animals live in larger units (packs, herds, or clans) which occupy an area that they defend against conspecifics from different groups. This is the case with wolves, Hamadryas baboons, and rats, to give only a few examples. House mice and house and Norway rats (*Mus musculus, Rattus rattus,* and *R. norvegicus*) live in groups that develop out of the family unit as succeeding generations remain together. These animals defend their territories against strangers from other groups of their species. This intolerance, which is tied to a specific area, has received much attention recently, because certain parallels exist to the human attitudes and behavior with respect to property (p. 444).

Territorial behavior ensures a certain amount of living space or hiding places for an individual or a group of animals. Thus it is important if a songbird is to find sufficient food for its brood that no other species member breeds in the immediate vicinity of its nest. There may be competition for a suitable hiding place or nesting sites. Anemone fish do not defend their anemones as a feeding ground but as a hiding place, and the same is true for many other reef-dwelling fish. The animals distribute themselves equally through territorial behavior. Pressure is exerted on the neighbors and this results in an increase in the range of the animals in the final analysis. Finally, one of the results of territorial behavior is to prevent an overexploitation of the living space, for example through overgrazing (M. M. Nice 1941; N. Tinbergen 1957; V. C. Wynne-Edwards 1962; F. S. Tompa 1962). This principle holds whether individual animals, pairs, or larger groups oppose each other as intolerant units. When groups exert pressure on each other, then this also results in their dispersal. In free-living monkeys group territoriality is a widespread characteristic, and the analogies to human behavior are obvious according to C. R. Carpenter (1942). Members of different groups threaten and fight one another and actual fights between groups can develop. A. P. Wilson (1968) studied such fights in rhesus monkeys on the island of Cayo Santiago. In these conflicts the groups are lined up opposite each other. Several females of one group may rush forward in an attack; they fight briefly and retreat to their own line, while females that have waited until now rush forward and continue the fight. Fights between groups of Norway rats were described by F. Steiniger (1951).

Males and females may equally share in the defense of territory. Often, it is primarily if not exclusively the male that occupies the territory. This usually occurs during the reproductive period, outside of which the animals may be quite peaceful. Here possession of the female is the goal. The selective advantage of such rival fights is that the stronger and healthier animal will breed, and in some animals the stronger also assumes the role of protector of the brood. In the Uganda Kob (*Adenota kob thomasi*) there exist selected mating places or arenas which consist of a number of adjoining territories. Each territory is occupied by a male. In the center of these arenas, which have a diameter of 300 to 400 meters, the territories are most closely packed, 10 to 20 in number. The females seek out those males for mating which hold these center territories. There are also males in single, more widely dispersed, territories. The actual advantage of such arenas is not known. They have only been studied in this antelope (H. K. Buechner 1961; W. Leuthold 1966). F. R. Walther (1966) is of the opinion that this could be an adaptation against predators. When many neighbors are close together they can detect danger more easily. In addition, these otherwise very sociable animals can maintain social, if only visual, contact with their companions in spite of territorial separation.

To avoid misunderstandings it should be pointed out that territorial species do not defend all areas which they may visit against members of their own kind. In areas that are frequented regularly by an animal neutral areas exist. An area that is not defended by an animal may be called its home range. In the Galápagos sea lion (*Zalophus wollebaeki*) males defend a specific coastal strip on land as well as the water close to it. The fishing grounds in the sea, however, are not defended. The hamster (*Cricetus cricetus*) defends and marks the den and its immediate environment but retreats from other hamsters it may meet in the fields. There are then neutral areas that are not defended. In baboons (*Papio ursinus*) the troops each evening seek out a specific sleeping place on trees, which are defended against other troops. Each troop also has its own feeding grounds, where contact with other groups is avoided. At waterholes, however, they do come into contact with other groups without conflicts being the result (I. DeVore 1965). Sometimes an animal defends its entire living area, which then is usually small (Fig. 188). The territory is not necessarily a contained area with rigid boundaries. It can also be a system of paths with fixed points. Norway rats pursue strange rats from other territories only on paths which they have marked; house rats (*Rattus rattus*), on the other hand, defend the entire area that is crossed by their paths (H. J. Telle 1966).

Larger birds often have separate breeding and feeding territories. While the tree-nesting herons, such as the silk egret (*Egretta garzetta*), the rail heron (*Ardeola ralloides*), the night heron (*Nycticorax nycticorax*), and the gray heron (*Ardea cinerea*) breed in dense heronries in small forests on the bank of the Danube and Theiss with individual distances reduced to pecking distances, the birds distribute themselves on stagnant sidestreams and rice fields over wider areas, where they often hold larger, individual territories. The daily flights (feeding) between the two separate areas are still performed in dense masses; only on the feeding ground itself are distances kept, owing to increased aggression (A. Festetics 1959).

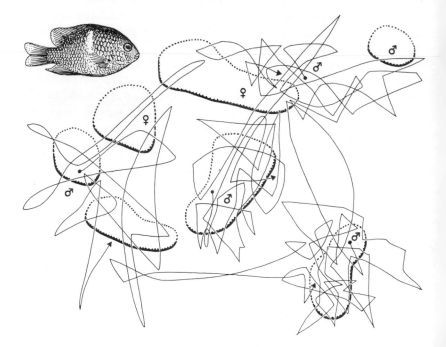

Figure 188. Territories of four male damselfishes *Abudefduf leucozona* (top left) can be recognized by the marked paths which the animal swims. It can be seen that the fish remain in a small area. Longer excursions came about during pursuit of entering neighbors. Each fish was observed for 5 minutes. (From I. Eibl-Eibesfeldt [1964c].)

It is an error to believe that territory-owning animals are in a state of continuous fighting with their neighbors. Such a preconception seems to have led R. Schenkel (1966) to conclude that the black rhinoceros is not territorial because he never observed fights between neighboring animals. He failed to consider that animals usually fight when the territories are first established and occasionally when a stranger trespasses but not with their neighbors. These know one another and respect each other's territories. One rarely observes fights between two neighboring Galápagos sea lions, but intense fights with intruders take place.

Schenkel seems to restrict the term territory to an area which the animal can continuously monitor and where there is a possibility of an intruder being confronted and driven off within a short time. In line with this reasoning a wolverine would be highly intolerant toward conspecifics of the same sex but would be unable to protect its large home range from intruders. Therefore, it would be wrong to speak of territoriality. I propose that any space-associated intolerance be called *territoriality*, where a "territory owner" is that animal before which another conspecific must retreat at a given time. In this connection P. Leyhausen's (1965b) observation that cats possess temporally defined territories deserves notice. Many male cats can use the same area, but at different, well-established times, and each is only a temporary owner of the territory—and is retreated from— during this time. This does not mean an equalization of territoriality with relative intolerance such as is observed in rank disputes. Intolerance connected with rank may lead to spatial avoidance from the side of the inferior but rarely leads to a spacing out. In territorial defense the competitors in rank often unite against foreign intruders. A low-ranking member of a group thus demonstrates that he is a member of the group, occupying and defending the group territory. Rank disputes are not linked to territorial claims; they might be observed even in a migrating flock and the rank relationship does not change regularly with the time of the day, whereas in temporally defined territories the temporary owner gains superiority.

Considering this addition we basically agree with Schenkel that a territory is an area in which an individual or group does not tolerate particular members of the same species either repulsing all strangers or only those of the same sex. The territory ownership might be restricted to established periods of time only.

Natural landmarks are often used as territory boundaries. In the three-spined stickleback one can experimentally move the territorial boundary. A newly planted row of *Elodea* will be accepted by a cichlid fish as its territorial boundary, even if it reduces the size of the original territory. Likewise, a row of bicycle spokes, 3 to 4 cm apart will be accepted (J. v. Iersel 1958). However, the new border is only accepted if it borders the territory of a neighbor and when the row is not closer than 30 cm to the nest.

An area that is occupied by an animal or a group is often marked. Many mammals place scent marks by depositing gland secretions, urine, or excrements at certain places around the territory. The method of marking differs from species to species. The hamster smears the secretion of its flank glands onto the walls of its den and on clumps of grass and stones in the vicinity of its den. Badgers and martens mark objects with a secretion from a gland pocket under the base of their tails (Fig. 189). Antelopes place secretions from their preorbital glands on bushes and tips of grasses (H. Hediger 1949; F. R. Walther 1965). The tenrec (*Echinops telfairi*) puts saliva on the object to be marked and transfers its body odor by alternately scratching itself with a foot and then rubbing it in the saliva (Fig. 190a). The giant galago and the Senegal galago (*Galago crassicaudatus* and *G. senegalensis*) urinate on the palms of their hands and rub it into the soles of their feet. When climbing about they leave behind obvious scent marks which are also visible as dark spots. House mice and rats mark their paths with urine and follow these trails like trains on a track (I. Eibl-Eibesfeldt 1950c, 1953c, 1965c). These odor trails can also be used by strange mice and rats. When one mouse population, which lived in a wooden barracks, was exchanged with another, the new mice quickly found their way about by utilizing the scent-marked paths of the previous owners (I. Eibl-Eibesfeldt 1950c). H. J. Telle (1966) poisoned one of two adjoining rat populations, each of which had its own marked trails that were separated only by some narrow objects. When he introduced new rats in the freed area the newcomers used all the available trails, including those of the adjoining population. They soon learned that they would be attacked there, however, and thereafter they restricted their activities to the area that became available when the previous population had been poisoned. Male rabbits mark their territory with chin and anal glands. The chin glands in males are

Figure 189. Tame badger marking the shoe of the caretaker. (Photograph: I. Eibl-Eibesfeldt.)

Figure 190. (a) and (b) Olfactory marking of the tenrec (*Echinops telfairi*). The animal is marking the head of the caretaker. (Photographs: I. Eibl-Eibesfeldt.)

larger than in females, and this difference in size increases with the approach of sexual maturity. They are more strongly developed in higher-ranking males but with no direct relationship to body size. Sometimes a smaller but more dominant male can have a larger chin gland than a heavier, sexually inactive animal. The secretion of these chin glands is rubbed on the ground, branches and stones, and on females. High-ranking animals mark more frequently than low-ranking ones (R. Mykytowycz 1955). The secretion of the anal glands adds a particular smell to the dung pellets. Those dung pellets that are deposited for the purpose of territory markings on especially dug up earth mounds have a stronger odor than those scattered all over the ground during feeding (R. Mykytowycz 1966).

The scent marks are chemical property signs (F. Goethe 1938). They aid the territory owner, first of all, as signs of recognition. They help in orientation and make the area familiar. A badger that becomes agitated or frightened in a strange environment can be calmed by letting it sniff an object that it had marked previously (I. Eibl-Eibesfeldt 1950). A male hamster that enters the territory of a female during the mating season will mark this strange territory before it actually begins to court. It is probable that this also has a repelling function for others. Strange scent marks have an aggression-releasing effect in hamsters, which show threat behavior when sniffing strange scent marks (I. Eibl-Eibesfeldt 1953). Rabbit dung that is introduced into the territory of an established animal is highly arousing. A male rabbit marks with its chin glands and defecates significantly more than when some fresh earth is placed into its territory. Aroused in such a way male rabbits will also begin to attack members of their group, but will stop when they have come close enough to recognize them. It appears as if a rabbit in this mood considers everyone a potential intruder (R. Mykytowycz 1966). The scent marks of the flying marsupials (*Petaurista*) have no repelling effect on strangers but they increase the aggressivity of the territory owners while they lower the aggression of the strangers (T. Schultze-Westrum and B. Braun 1967).

In addition to scent markings there are other possibilities to indicate territorial ownership. For example, it can be advertised by calls and conspicuous behavior. The male sea lion calls continuously when swimming back and forth before its part of the beach. Near the territorial boundary it occasionally climbs out of the water and calls toward its neighbor, who reacts in the same way, without fighting. Male fur seals (*Callorhinus ursinus*) who own territories move toward the neighbor, throw themselves on their bellies, and glide toward each other until they bump noses near the territorial boundary. In this manner they indicate their boundaries without engaging in fights (G. A. Bartholomew 1953). Groups of howling monkeys (*Alouatta palliata*) mark their territories by howling displays, especially during the morning hours (C. R. Carpenter 1965). The territorial song of many songbirds is well known.

Many animals display themselves conspicuously within their territory and are frequently conspicuously colored. Genital displays of some primates can be interpreted as a visual marking of territory (p. 428).

Ownership of territory is frequently a prerequisite for the occurrence of aggressive behavior. Sticklebacks swim peacefully together in a swarm without reproductive coloration until they have found a suitable location for the establishment of a territory. As soon as one fish has found a place, its belly turns red; other males are attacked when they approach too closely. Its readiness to attack is reduced the farther it gets from its own territory. This can be demonstrated easily if two neighbors are each placed in small glass containers. They can then be moved about at various distances from each other. If male *a* remains in its territory and *b* is placed close to it, it will attempt to attack *b* through the glass wall, while *b* attempts to flee. If both are placed into the territory of *b*, then the change in behavior can be observed: *b* attacks and *a* tries to escape (Fig. 191).

More aggressive individuals generally conquer more favorable and larger territories. If a territory-owning red grouse (*Lagopus*) is implanted with a pellet of male sex hormone under its skin, then the aggressivity of the male is increased and it expands its territory substantially at the expense of its neighbors (A. Watson 1966).

Sometimes the defended areas are quite small. Many birds that breed in colonies nest just out of range of the pecking distance of their neighbors (Fig. 192).

Outside of their territory many animals are willing to attack a conspecific if it comes too close. They are surrounded with a small inviolable space (Fig. 193). This "individual distance," which if

Figure 191. Two stickleback males (a) and (b), which live in two territories, A and B, respectively, were moved about between the two territories in glass tubes. When *a* is in its territory it attacks *b*, which tries to escape. If both are moved into *b*'s territory, he attacks and *a* tries to escape. (From N. Tinbergen [1951].)

Figure 192. Flightless cormorants nesting side by side just out of range of reach (peck distance) of the next nest. (Photograph: I. Eibl-Eibesfeldt.)

Figure 193. Individual distance of gulls on Lake Zurich. (Photograph: H. Hediger.)

crossed by a member of the same species will release fighting, has been measured precisely by P. R. Marler (1956a) for the chaffinch. Males permitted females to approach closer than they did other males, where the males' plumage is the distinguishing mark. Females whose undersides have been artificially dyed red are not permitted as close as normal-colored ones and are attacked at a larger distance, as if they were males.

Intraspecific fighting behavior

In rival fights and in fighting for territories we observe attack behavior, which is exclusively released by the appearance of conspecifics even before any physical contact has been made. This aggressive behavior has been the subject of many discussions and there are many contradictory views. These contradictions are especially prone to arise in discussion of the degree to which phylogenetic adaptations determine the behavior and especially whether internal drive mechanisms are the cause for a spontaneous aggressive drive.

Intraspecific fighting behavior has several remarkable characteristics and in most species easily can be distinguished from interspecific fighting behavior. An Oryx antelope will never use its horns to gore another oryx but fights according to strictly observed rules (Fig. 194).

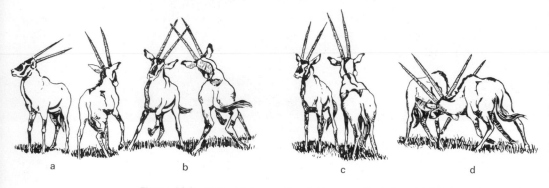

a   b   c   d

Figure 194. Fight of two bulls of *Oryx gazella beisa*. (a) Initial posture (head-up threats); (b) the first clash, where the horns touch in the upper third; (c) pause; (d) second thrust, which leads to the head-to-head pushing contest. (After F. Walther [1958].)

It does, however, stab lions in this manner (F. R. Walther 1958). A giraffe uses its short horns to fight rivals, but uses its hoofs in defense against predators (D. Backhaus 1961). A predator fights differently with a species member than with a prey and by electrical brain stimulation it could be shown in cats that these two types of behavior have different neural substrates in the brain. From areas in the lateral hypothalamus stimulation evokes eating and prey-killing responses. From areas in the ventral and medial hypothalamus stimulation elicits intraspecific ("emotional type") aggression (B.

Kaada 1967). These differences between inter- and intraspecific aggression must be emphasized, because they are not always clearly recognized. Thus R. Ardrey (1962) traces the aggressive behavior of man back to the predatory ways of his australopithecine ancestors. He overlooks the fact that there is no necessary connection between aggression and predation. After all, plant eaters are no more peaceful against their own kind. Bulls fight no less intensely than do rabbits, sparrows, hamsters, or cats. Z. Y. Kuo (1960–1961) also treats intra- and interspecific aggression as if it were the same. We do not want to assert that there are absolutely no connections between aggression and food getting, but only that a predatory way of life does not necessarily lead to an increase of intraspecific aggression. W. Wickler (1961a) and H. Albrecht (1966a) have shown that fighting behavior can frequently be derived from feeding behavior. Fish that feed on algae which they pull from the substratum fight and threaten with the same behavior patterns. Others, which capture larger prey, threaten with movements and postures which they assume before striking at prey. Furthermore, the readiness to feed and to fight are often positively correlated: Stimuli that release fighting facilitate eating movements; eating increases the readiness to fight. The phenomenon of competition for food, which we can also observe in mammals, is facilitated in this way.

In this connection we can point out that the manner of fighting of a species is also naturally determined by a number of other factors which in themselves have little to do with aggressive behavior.

a

R. Apfelbach's (1967a) investigations on mouth- and substrate-breeding cichlids of the genus *Tilapia* revealed obvious differences in the mouth fights. Substrate-breeding fish hold on tightly to each other's lips and engage in pushing-pulling contests until a winner emerges. Mouthbreeders never bite each other in this way; instead they butt each other with opened mouths. This mouth butting becomes more and more ritualized as the mouthbreeding itself becomes more specialized. Apparently mouthbreeding requires a sensitive mouth, which is incompatible with the rough ways of push-pull fighting (Fig. 195 a and b).

b

According to a widely held opinion aggressive behavior ultimately aims at the destruction of the opponent. This can easily be disproved. Where one species possesses very dangerous weapons such as teeth or claws, which could easily kill an opponent if they were used, special inhibiting mechanisms have usually evolved which prevent killing of the species member; often the entire fight has become transformed into a tournament (K. Lorenz 1943, 1963a).[1] Only rarely do well-armed animals use their weapons against a conspecific

Figure 195. (a) The cichlid *Tilapia zilli* is a substrate breeder and grasps the opponent firmly with the jaws during fighting; (b) the mouthbreeding *Tilapia nilotica,* on the other hand, fights by pushing with the mouth. (Photographs: R. Apfelbach.)

[1] In German this is called *Kommentkampf* [*comment* = French, *wie* (how)] and refers to the rules of duelling by students. In English: tournament or ritualized fights.

without any inhibition. This is true for some rodents, for example, hamsters (*Cricetus*), which can get away from one another quickly after a short exchange of bites. This ability to escape protects the pursued, and under natural conditions one hamster rarely kills another. Why lions in some areas of East Africa kill animals belonging to other prides without inhibition (R. Schenkel 1966) remains to be studied in more detail.

A large number of invertebrates fight without doing damage to each other. The best-studied forms are the tournaments of some fiddler crabs (R. Altevogt 1957; J. Crane 1966). During low-intensity fighting the animals merely butt each other with their large claws, which are only slightly opened. Crane reports that small protuberances on the front side of the claws prevent them from slipping off one another. Only with an increase in the intensity of fighting do the crabs hold on with their claws, but here again other morphological adaptations ensure that the animals grasp each other in a particular way.

Ritualized fights can be observed in many fish. In cichlids (*Cichlidae*) rivals threaten each other either frontally or by lateral displays, in which they spread the fins and especially the gill covers and gill membranes (Fig. 196a–f). At the same time the combatants display a colorful appearance. Before the fight begins they may circle each other. Then one animal beats its tail against the other, and the strength of the created pressure wave gives an indication to the other as to the strength of its opponent. The cichlid *Apistogramma wickleri*

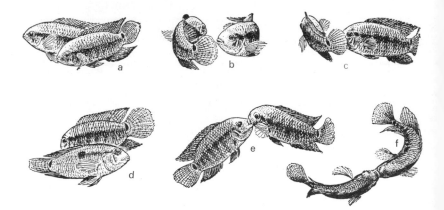

Figure 196. Fighting scenes from *Aequidens pulcher* (= *A. latifrons*). (a) Lateral display threats and light tailbeats, uneven fin spread; (b) circling, threatening swimming about, lower part of the mouth lowered; (c) tailbeat against the head of the opponent; (d) the front animal gives up (coloration, fin position); (e) rearing up in front of the other and grasping the mouth; (f) mouth pulling. (From W. Wickler [1962b].)

specializes in tail beating. The tail beat aims below the opponent and creates a drag so that the other fish is pulled downward (W. Wickler 1962b). After an exchange of tail beating most species begin mouth pulling or mouth pushing by grasping each other at the upper or lower mandible. Finally one of them gives up, folds its fins, and swims off. If the fish cannot withdraw, then it is continually attacked with ramming thrusts by the opponent, which attacks without any inhibition. The sides and fins of the opponent are damaged and it is quickly killed. This occurs, however, only when the fish are kept in an aquarium.

Grunts (*Haemulon*) fight by attempting to push each other mouth to mouth from their places. Anemone fish (*Amphiprion percula*) also have tournament fights in which they parry thrusts of the opponent with their pectoral fins (I. Eibl-Eibesfeldt 1960a, 1965a; see also Fig. 197). Butterfly fishes (*Chaetodon, Chelmon, Heniochus*) fight by head butting (D. Zumpe 1965; see also Fig. 198). The clinid fish *Emblemaria pandionis* grasps its opponent by the head after an initial threat display and attempts to retreat into its own cave while holding on to the other. If this succeeds, the other will give up soon, because it is exposed helplessly before the entrance (W. Wickler 1964e).

Figure 197. Sparring technique of fighting anemone fish (*Amphiprion percula*). (From I. Eibl-Eibesfeldt [1960a].)

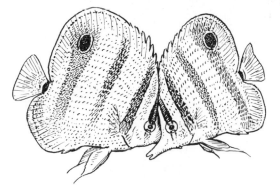

Figure 198. Fighting pipe-fishes (*Chelmon rostratus*). (After D. Zumpe [1964].)

In poisonous snakes rival males fight according to strict rules which differ slightly from species to species (E. Thomas 1961; C. E. Shaw 1948). Rattlesnakes (*Crotalus ruber*) wind themselves around each other's tails and raise the anterior third of their bodies. In this position each tries to hit the head of the other with its own (Fig. 199). This they do alternately until one of them is so fatigued that it gives up. In the marine iguanas (*Amblyrhynchus cristatus*) the males fight by rushing at each other after an initial threat display, and butting the tops of their heads together. Each attempts to dislodge the other. If one of them realizes that it is losing, it will give up and lay down flat on the ground before the victor, who in turn respects this submissive gesture of the loser, and waits in a threat posture until the

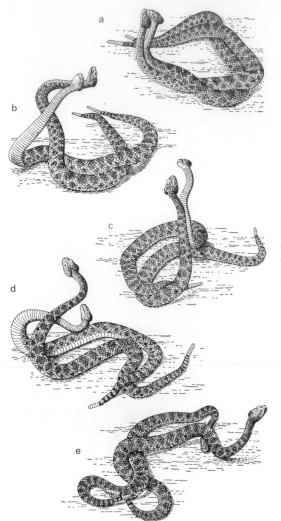

Figure 199. Fighting male rattlesnakes (*Crotalus ruber*): The rivals beat one another with their heads. The loser is pushed against the ground with the body. The rivals never bite one another. (After C. E. Shaw [1948].)

other retreats (Fig. 200). In this way the stronger males do not kill their weaker, often younger rivals with their powerful teeth and jaws, and the species does not lose its reserve of growing males (I. Eibl-Eibesfeldt 1955).

Fighting males of the lava lizards (*Tropidurus*) whip each other with their tails (Fig. 201). In special circumstances damaging fights will occur in these lizards. One *Tropidurus* male which had lost part of its tail in some way tried at first in vain to defend itself with tail whipping against its opponent. It seemed to be unaware that it did not possess a complete tail. Finally it successfully defended itself with biting (I. Eibl-Eibesfeldt 1966; see also Fig. 202). Male marine iguanas bite and shake a rival if another animal is abruptly placed

a

b

c

Figure 200. Ritualized fight of the marine iguana. (a) Territory of a male (center) with several females; (b) two fighting marine iguanas (head pushing); (c) the weaker yields and lies flat before the victor, who ceases to fight as a result. (Photographs: I. Eibl-Eibesfeldt.)

Figure 201. Lava lizards fighting by beating with their tails (*Tropidurus delanonis*). (From I. Eibl-Eibesfeldt [1964a, 1966a].)

into their territories. The unwilling intruder does not have an opportunity in this case to show the threat display which initiates the usual encounters. This seems to be the reason an immediate attack occurs.

In fence lizards (*Lacerta agilis*) one of the combatants will permit the other to grasp its neck. This they do alternately until one seems to realize that the other is stronger. This it can recognize by the strength of the grip of its opponent. Sometimes a smaller animal will give up when it realizes, while biting the other, that it has encountered an especially large opponent. The loser prostrates itself on the ground, treads, and runs off (G. Kitzler 1942).

Turkeys try to drive off rivals with threatening calls or by jumping at them or beating them with their legs, which have large spurs. If one of the opponents stands up to fight, a tournament-like struggle

Figure 202. (a–c): Damaging fight of lava lizards (*Tropidurus albemarlensis*) after one lost its tail. The animal at first tried unsuccessfully to fight the attacker with tail beats. This did not work, so at first it fled from its territory when the neighbor appeared. After several days the animal defended itself by grasping the opponent at its tail and winning the fight. (Photographs: I. Eibl-Eibesfeldt.)

ensues wherein each bird attempts to grasp the other by the conspicuously colored red neck and head skin, to push or pull the other and to press it to the ground. The strong skin resists this rough treatment much better than feathers could, and according to W. M. Schleidt (1966) the significance of the red signal color is that it draws the pecks of the opponent so that the feathers remain undamaged. Hens and vanquished males that show no red skin are not engaged in fights. Domestic roosters peck and kick each other after an initial display. The loser finally submits.

Many mammals change to a damaging fight after an initial threat display. Attacking Norway rats erect their hair, hump their backs, gnash their teeth, and approach their opponent broadside (Fig. 203). They avoid all sudden movements that could release defensive biting and begin to push their opponent from its place. Until this time the fight does not result in damage, and it can end if the threatened animal gives up and leaves by jumping up and away. After this beginning rats often box each other with their front paws while standing opposite each other on their hind legs, and they may kick with a hind leg. If the opponent falls down and the attacker lies on top of it, both become rigid and they make threats by gnashing their teeth and by squeaking. At last, the damaging part of the fight commences, where both animals will bite each other and hold on (Figs. 204 and 205). After a few seconds one of them will give up and flee (I. Eibl-Eibesfeldt 1958, 1963).

Wolves also bite each other in a fight until one becomes submissive either by behavior identical to the food begging of a puppy or

Figure 203. Fight of two male Norway rats (wild form). (a) and (h) The attacker approaches by exposing its lateral side to the opponent; (c) the opponents fight standing up in front of each other and the stronger attacker throws its opponent on its back and kicks with its hind leg; (d), (e), and (f) the opponents rigidly maintain their "wrestling" position. (From Scientific Film F1.31, I. Eibl-Eibesfeldt [1957a, b].)

Figure 204. Damaging fight of Norway rats. The animals are biting each other and hold on.

a   b

Figure 205. Submissive behavior in wolves. (a) Appeasing by food begging; (b) passive submission by rolling on the back. (After R. Schenkel [1967].)

by rolling on its back and remaining still. The latter posture seems to be derived from the behavior repertoire of the puppy and may be interpreted as a ritualized presentation for cleaning of the anal region. Quite often the submissive wolf urinates, thus releasing actual cleaning by its opponent. Dogs behave similarly (K. Lorenz 1963a; R. Schenkel 1967). Submissive postures by presentation of the neck region was also described by D. Backhaus (1960) in the zebra and by H. Kummer (1968) in the baboon.

Quite extensively ritualized are the fights of horn- and antler-bearing hoofed animals. Bighorn sheep walk toward each other raised up on their hind legs and beat their heads together from above (H. Bruhin 1953). In Nilgau bulls (*Boselaphus tragocamelus*) F. R. Walther (1958) described a ritualized neck fight which occurs in addition to the usual head-against-head butting (Fig. 206). Horn-like organs have developed repeatedly and convergently in the service of intraspecific aggression (V. Geist 1966a). The multiplicity of forms in horn development within horn-bearing hoofed animals shows

Figure 206. Fighting ritual of various antelopes. (a) Two Nilgau bulls (*Bosephalus tragocamelus*) during the neck fight. The animals try to press one another to the ground. This species also fights by thrusts with the horns. (b) In *Taurotragus oryx* the thrusts are always directed against the head of the opponent. Other antelopes with well-developed horns do the same.

clearly that these organs primarily function in the service of intraspecific conflicts and that they are adapted to the specific style of fighting of a particular species. If they were, in fact, weapons against predators, then most of them would probably be dagger- or saberlike and certainly not coiled backward as in bighorn sheep. In those animals that ram their opponents the forehead is provided with massive bases for the horns, which are reminiscent of armored plate, as, for example, in the buffalo. In "wrestlers" the horns are shaped in such a way that they can hook or fork into each other in some way (Kudu and Impala). "Fencers" beat each others' horns with the long side, where the middle part of the horns is especially subject to wear. In this case they are then curved and cross-ridges of horn prevent a sliding off (*Capra ibex* and *Hippotragus niger*) (F. R. Walther 1966).

In species with sideways-protruding horns the flanks are covered, preventing side attacks. The opponent is only able to attack from the front. In such cases the rules of tournament are determined by the very structure of the weapons, however, rarely so. The Thomson's gazelle has no protruding horns; a side attack would be easy and, indeed, the rivals sometimes stand so to each other that the observer would expect it. In spite of this, in more than a thousand fights, F. R. Walther never saw a single flank attack. The rivals used exclusively frontal attacks, directed against the head, especially the horns. The rule and regulations of the fight must be based upon specific central nervous system structures in such cases.

Among the antelopes we find, besides these ritualized fighting methods, submissive, attack-inhibiting postures whereby the head is turned, the horns held backward, or the animals turn aside all together. In all cases the horns are turned from the partner and the appeasing animal makes itself appear smaller (F. R. Walther 1966).

In a comparable manner we find that antlers are adapted to the style of fighting of their bearers. Here we also find a change in the style of fighting because each year the antlers are dropped. After the antlers have been dropped and the new ones are still covered with velvet, elk fight with their hoofs, without attempting to fight with their heads first.

Antlers and horns constitute visual signs of rank in some species (p. 353).

The forehead-butting technique of the horn- and antler-bearing hoofed animals probably evolved out of biting (G. Tembrock 1961). One could suppose that the bite attack became inhibited in social encounters, where the animals then lowered their head and butted together with their heads as they were carried forward by the initial momentum of the thrust. This hypothesis is supported by the observation that hornless females still perform snapping movements when they butt their heads together, as well as by the biting-threat posture of elk (p. 192). The hornless females butt their opponents in the sides in 50 percent of the cases, something that horn-bearing males and females never do (F. R. Walther 1961; see also Fig. 207).

Figure 207. Hornless females—here Nilgau antelopes—butt in 50 percent of the cases with the forehead against the flank of the opponent, an act that the males and females of horn-bearing species rarely do. (After F. Walther [1961].)

The tournaments of marine iguanas could also have evolved in this way. Howling monkeys defend the territory of their troop by loud calls. C. R. Carpenter (1942) and C. H. Southwick (1963) talk of actual vocal battles which prevent a more bloody encounter. In man we find song duels. For additional examples of tournament fights the reader is referred to B. Oehlert (1958), W. M. Schleidt and M. Schleidt (1962), K. Fiedler (1964), D. Ohm (1964), I. Eibl-Eibesfeldt (1961b), and J. Crane (1966).

In submissive postures, which terminate a fight, fight-releasing signals are turned away, as was discussed in respect to appeasing greeting gestures. The exact opposite of threat behavior is shown by making oneself smaller, the principle of antithesis, which was discussed in detail by C. Darwin (Fig. 208).

a   b

Figure 208. Principle of antithesis discovered by Darwin. (a) Threat posture of the dog; (b) submissive posture. (From C. Darwin [1872].)

Many fish collapse their fins and change their coloration (Figure 209). The cichlid *Tilapia mariae,* for example, changes from conspicuous coloration into the juvenile colors. The cross-bars provide excellent camouflage between plants (H. Albrecht 1966). Often infantile behavior will appease another animal. A dog that lies on its back and urinates frequently may be licked by the attacker as a mother would her young. We have already pointed out that many animals (for example, hamsters) appease during courtship by showing infantile behavior (p. 123). Sometimes the female presenting movement (baboons, p. 112) is used as a submissive posture. The aggression-inhibiting behavior patterns of man will be discussed in Chapter 18 (p. 454).

Figure 209. Threat posture (top) and submissive posture (below) of *Blennius fluviatilis.* (Photographs: W. Wickler.)

The existence of tournament fights indicates how strong a selection pressure in favor of aggressive behavior exists. Otherwise counterselection would have bred out aggressive behavior in species that can do damage to a conspecific. Instead, the most complex fighting techniques have evolved in order to allow fighting to occur as a spacing-out mechanism.

The dynamic instinct concept of aggression

Because aggressive behavior in human society plays a large role, there have been several attempts to understand the determinants of this behavior through the study of aggressive behavior in animals. J. P. Scott (1960) demonstrated that aggressive behavior can be decisively influenced by experience. He was able to produce very aggressive male mice by enabling them to win fights repeatedly, and he could produce peaceful males by raising them with females and by picking them up daily by their tails and stroking them gently. Aggressive puppies that were lifted off the ground so that they lost their footing also became docile. B. Ginsburgh and W. Allee (1942), J. P. Scott and E. Fredericson (1951), M. W. Kahn (1951), and J. Uhrich (1938) proved experimentally that success in fighting makes mice aggressive and that losses dampen their aggressiveness.

J. P. Scott came to the conclusion that aggressive behavior is learned. The habit of attack is said to emerge gradually during the course of development, when the animals experienced pain in the competition for food and during play fighting.

From a more general viewpoint the experiments with mice show us that aggression has to be learned. Defensive fighting can be stimulated by the pain of an attack, but aggression in the strict sense of an unprovoked attack can only be produced by training (J. P. Scott 1960:20).

Z. Y. Kuo (1960–1961) also is of the opinion that aggressive behavior is an acquired habit, and a bad one.

J. Dollard and others (1939) developed the hypothesis that aggressive behavior is always the result of frustration; frustration they considered to be any impediment, or external influence, which interrupts striving toward a goal. In this view aggression is primarily of a reactive nature and not the expression of a spontaneous drive. L. Berkowitz (1962) and M. F. A. Montagu (1962) agree with this hypothesis, which is also reflected in educational programs. Thus J. P. Scott suggests that children be raised in an environment which is devoid of all stimuli that could release aggressive behavior through frustration. W. Craig (1928) holds that there is no appetitive behavior for fighting; an animal merely defends its interests. In short, aggression is reactive behavior in his view. A similar position is taken by P. R. Marler (1957), although he emphasizes that endogenous influences (for example, hormones) sensitize an animal to specific categories of external stimuli.

This concept of aggression is opposed by the conception of aggression as a dynamic instinct by S. Freud and K. Lorenz (see also A. Mitscherlich [1957, 1959]). Freud, however, postulated a mystical death wish, a concept from which many modern psychoanalysts have departed (H. Hartmann and others 1949). K. Lorenz (1943, 1963) recognized, on the other hand, the instinctive basis of aggression as well as its species-preserving function. According to his reasoning, aggression is a true instinct with its own endogenous excitatory potential and the appropriate appetitive behavior. This concept of aggression is opposed to the earlier discussed hypotheses but has been supported by experimental results with animals. There are strong indications that the dynamic instinct concept of aggression holds true for man too, but we have to admit that a clear proof for an inborn aggressive drive in man has not yet been presented.

A number of investigations have shown that even rats and mice that were raised in isolation attacked conspecifics that were introduced into their cages, and that they exhibited all the species-typical behavior patterns of threat and fighting (E. M. Banks 1962; J. A. King and N. L. Gurney 1954; I. Eibl-Eibesfeldt 1963). Mice that were raised in isolation by Banks were even more aggressive than those raised with companions. Mice that were raised in isolation by King and Gurney attacked another mouse less readily, perhaps because their innate aggressive behavior was initially more suppressed by the number of new stimuli than was true for animals raised with others. In contrast, G. A. Hudgens, V. H. Denenberg, and M. X. Zarrow (1968) report that mice who were deprived of the opportunity to play with their litter mates from weaning on were more ready to fight than mice that grew up together. The authors presume that the mice learn during play to live peacefully together. Their fighting plays do not lead to injuries because their jaws and teeth are not fully developed. According to K. Lagerspetz and S. Talo (1967) aggressive behavior in the albino mice matures spontaneously around day 28. This maturation can be postponed by punishing them with painful stimuli and can be facilitated by withholding food. Experience with other mice in a group is not necessary. Genetic control of aggressive behavior was demonstrated by K. Lagerspetz and K. Worinen (1965) when they exchanged litters of mothers from aggressive and nonaggressive mouse populations. The young from the aggressive line that were raised by the docile mothers were clearly more aggressive than the animals that came from the docile strain and were raised by aggressive mothers (K. Lagerspetz 1964). J. P. Kruijt (1964) raised jungle fowl cocks in isolation, and they were more aggressive than those raised in a group and they exhibited the behavior patterns of threat and fighting. When left in isolation they even fought against their own tails, clearly demonstrating an appetite for fighting. Siamese fighting fish (*Betta splendens*) raised in isolation attacked conspecifics or their own mirror image with species-typical behavior patterns (H. Laudien 1966).

Of course, various experiences with conspecifics have an effect on aggressiveness (E. McNeil 1959; F. Merz 1965). Rhesus monkeys that grew up in the exclusive company of their mothers are later more withdrawn as well as aggressive when placed with playmates of the same age. In general, they adjust readily, however. The monkeys that were raised in this way are quite normal with respect to all other behavior (B. K. Alexander [in press]). J. P. Scott (1963, 1964) repeatedly pointed out the importance of early social experiences for the development of friendly social relations in mammals. Nevertheless, the fact that aggressive behavior can be so greatly influenced does not permit the conclusion that it is completely learned.

For aggressive behavior to be shown an animal must generally be in its familiar territory, or it will flee rather than attack (N. Tinbergen 1951; see also p. 313). An animal trainer who enters the cage first and only then allows the lions to enter utilizes this knowledge. In this way he is the territory owner and the lions are inhibited in their aggression from the start.

We have already discussed those key stimuli that release fighting behavior (p. 72). These are often of a simple nature. Fence lizard males (*Sceloporus undulatus*) attack females whose abdominal sides have been painted blue, and they ignore males whose abdomens were painted grey (G. K. Noble and H. T. Bradley 1933).

Similarly unexperienced male sticklebacks attack simple decoys with red undersides. It is evident that fixed action patterns, releasing mechanisms, and releasing signals have evolved in the "service" of intraspecific aggression. The neural substrate for aggressive behavior is fairly well known in a number of instances (B. Kaada 1967; J. M. K. Delgado 1967; W. R. Hess 1954). In their readiness to react aggressively animals show clearly fluctuations that are not necessarily related to corresponding fluctuations of the environment. In vertebrates the male sexual hormone plays a decisive role in inducing the specific readiness for aggression in the adult animal, as well as in the organization of the neural structures during early ontogeny (A. B. Rothballer 1967). Appetitive behavior for fighting develops in isolates, as was demonstrated with cocks in the experiments just mentioned. There are strong indications, therefore, that inborn drive mechanisms underlie aggression.

By means of electrical brain stimulation it is possible to release a true appetitive behavior for fighting in chickens (E. v. Holst and U. v. Saint-Paul 1960). The previously discussed experiments of A. Rasa (p. 57) prove finally that aggression can be dammed up and discharged. N. H. Azrin, R. R. Hutchinson, and R. McLaughlin (1965) released aggression by electric punishing stimuli in squirrel monkeys. The animals learned a task following a shock, when they were rewarded by being allowed to attack a ball for a short time. In the latter case we are not dealing with spontaneous aggression. The experiments demonstrate, however, that by provocation a physiological state is achieved that results in an appetite for attacking which is probably similar to the physiological state responsible for the spontaneous urge to fight, as observed in the isolated jungle fowl cocks and cichlids mentioned above. That this internal aggressive urge built up by provocation can be discharged, leading to a reduction of tension, was demonstrated in experiments with humans. J. E. Hokanson and S. Shetler (1961) had an experimenter induce anger in student subjects and as a result their blood pressure rose. One group of angry subjects was then given a chance to administer electric shock to this experimenter, whenever he made an error in his task. Another group could inform him of his errors by flashing a light. In those who believed they were shocking the experimenter the blood pressure dropped rapidly, while it remained much higher in the other group. The possibility of administering verbal insults also resulted in a discharge (J. E. Hokanson and M. Burgess 1962). The experiments of S. Feshbach (1961) and J. W. Thibaut and J. T. Cowles (1952) show that aggressive impulses can be discharged. However, the release of tension is only of short duration, as is the case also with other instinctive behavior patterns. In the long run, the possibility of discharging aggressive impulses constitutes a kind of training for aggression. The animal becomes more aggressive. In the same way an aggressive drive can atrophy when an animal does not have an opportunity to discharge it for some time (W. Heiligenberg 1964). We want to emphasize this point because sometimes the view is expressed that a child should have the opportunity to discharge its aggressive impulses so that it will be all the more peaceful as an adult. This possibility needs to be investigated in humans directly. There is no evidence available, to my knowledge, of such a long-lasting cathartic effect. That aggressive behavior can also be rewarding was indicated by the experiments of T. I. Thompson (1963, 1964). His fighting fish (*Betta splendens*) and fighting cocks learned a task when they were rewarded by the presentation of stimuli that released fighting and threat behavior.

There are many facts that argue for the dynamic instinct concept of aggression. D. E. Davis (1962), who investigated the behavior of gangs, pointed out that rank and territory are the objects of aggression and he concludes from his observations on humans:

A wide variety of observations suggest that fighting for rank and territory has innate features. . . . Thus contrary to the conclusion of some authors, it seems that aggression is heavily dependent on genetics. Probably only the means of fighting and the objects of attack are learned.

In R. Ardrey's (1966) book, evidence is presented which supports the Freudian-Lorenzian instinct concept of aggression quite clearly. Resistance to this concept is primarily based on philosophical convictions. Thus L. Berkowtiz (1962:4) writes:

But aside from its theoretical significance Freud's hypothesis has some important implications for human conduct. An innate aggressive drive cannot be abolished by social reforms or the alleviation of frustration. Neither complete parental permissiveness nor the fulfillment of every desire will eliminate interpersonal conflict entirely, according to this view. Its lessons for social policy are obvious: Civilization and moral order ultimately must be based upon force, not love and charity.

Is this conclusion actually compelling? We are in a position to deny this. Already the observation of aggressive animal species shows that the gregarious forms are definitely able to neutralize their aggression, which in fact is a prerequisite for the formation of groups (p. 342). Individual acquaintance generally inhibits aggression. The lions, which according to R. Schenkel (1966) show no social inhibitions about killing other lions as such, have an absolute inhibition against biting their own pride members. This is true for many other animals and on the whole also for man, as shown by the necessity for forbidding fraternization with the enemy during war. When the Patasiwa tribe in western Seran still engaged in headhunting activities, it was their custom to attack their victims from behind in order to kill them. To attack a man from the front to take his head was considered murder. Only as long as the headhunter cannot look his victim in the eye can it be considered prey with whom there exists no personal bond. This bond is at once established, however, when one man looks the other in the eye. To kill in that circumstance was considered a crime (O. D. Tauern 1918).

In other cases the bond is established by sharing a meal. In certain New Guinea headhunting tribes even strangers can not then be killed.

The inhibitions to kill are nevertheless graded. They are stronger toward members of the individualized group than toward strangers. Women and children, especially small children, are more protected than men. Although true reports exist that children were killed in wartime, it is described as an outrage, a deviation from the normal. The strong inhibition to kill children is exemplified by the custom in several cultures to use a child for the establishment of bonds with strangers. The Massai of East Africa often pushed a little child to the fore, hands held open, to beg for sweets. H. Basedow (1906) reports how aborigines in Australia approached Europeans in a formal manner, with one or two high-ranking men pushing a little child in front of them, their hands on his shoulders. They were sure one would not harm a child. The same author relates how in Central Australia a woman who was suddenly surprised gripped her breasts to spray the intruders with milk. Later asked why, she explained that she did it to show that she was a mother, hoping they would then leave her unharmed. Thus the existence of an innate killing inhibition may be presumed, especially as we find, all over the world, as subjective correlation, the emotion of pity. In this sense, innate and thus binding norms of ethical behavior seem to be programmed into man.

As an additional safeguard against the possible release of aggression in a member of the group, animals and humans possess a repertoire of behavior patterns which buffer aggression (greeting ceremonies and other appeasement gestures, p. 125). When animals form larger groups, whose members can no longer recognize each other individually, they develop signals which unite them, for example, group odors. The familiar odor of the group inhibits aggression against a member. Man is also equipped with this additional capacity to identify with someone with whom one is not personally acquainted, and as this often involves abstract ideas and uniting symbols, it would be in the realm of the possible to create symbols that unite all mankind (p. 451).

To control aggression one must promote those behavior mechanisms in man that appease aggression and facilitate the formation of bonds between members of a group—an idea advanced by S. Freud. Freud also considered it hopeless to attempt to abolish aggression, but he thought that the disruptions of life in human societies might be overcome by the promotion of "libidinal" forces, by the activation of all forces that are capable of facilitating emotional ties among people. Freud wrote in 1932:

If the readiness to make war is a channel for the discharge of a drive towards destruction, then it seems logical to call upon its great opponent eros to curb it. Everything that establishes emotional ties between people must work against war. These ties can be of two kinds: first, the relationship towards a love object, albeit without sexual goals. Psychoanalysis need not be ashamed to speak of love in this context, because religion says the same: Love thy neighbor as thyself. This is easy to demand but difficult to fulfill.

The other kind of emotional tie is by identification. Everything that establishes meaningful similarities between people calls forth such feelings of communality identifications. A good part of the structure of human societies is based on them (S. Freud 1950, vol. 16:20).

Sometimes objections are raised which claim that man needs an enemy to discharge his aggressive drive, but experiments which show that aggression can be discharged without the performance of aggressive acts indicate that this is not so (p. 329). It is not even necessary that verbal insults and similar behavior be engaged in. Athletic competition is just as helpful as the passive viewing of a motion picture. S. Feshbach (1961) presented angry and nonangry college students with either a 10-minute boxing film or neutral film. The angry students were less angry after viewing the boxing film than after viewing the neutral film. There was no significant difference between the groups among the nonangry students. The film experiment indicates that the viewers can benefit in the sense of a release of tension by seeing films with aggressive content. This needs to be studied more and further research will undoubtedly reveal additional channels for releasing aggression. In primitive people customs that serve as a kind of safety valve have been described. It has been reported that some Australian tribes come together at certain times to insult one another and to fight according to specific rules. Eskimos settle many of their disputes by song duels. Additional examples of these customs are given by P. Bohannan (1966). K. Lorenz (1943) proposes that certain combative sports may be possible safety valves for aggression.

Another possible way to control aggression would be by radical counterconditioning. E. McNeil (1959), however, raised the question as to what degree this kind of training would result in a loss of general initiative. It is most certainly dangerous to conduct educational experiments before the characteristics that are correlated with aggression are known. We speak of attacking a problem, and there are many other indications that a general tendency to explore is positively correlated with aggression. This needs to be studied before one thinks of curing man of his aggression. The same applies to eugenic attempts to eradicate aggression.

Aggression definitely has positive aspects, which K. Lorenz has emphatically pointed out. Bloodless competition is an important driving force toward cultural development. I. Kant (1784) also recognized this positive aspect of aggression:

Without those not especially loveable characteristics of unsociability, from which resistance is derived, which each person must necessarily encounter in his own striving, all talent would forever remain hidden in its buds as in the arcadian life of the shepherd in complete harmony, satisfaction, and mutual love: People, docile as the sheep which they herd, would impart no greater value to their lives than their domesticated animals possess: they would not fill the empty spaces left for them by creation in line with their purpose as rational creatures. Therefore we owe our gratitude to nature for the quarrelsomeness, for the envious, competitive vanity, for the never satisfied desire to possess or even to rule. Without these all the excellent native potential of humanity would slumber eternally. Man desires harmony, but nature knows better what is good for his kind: it wants discord. He wants to live at ease and happy; nature, however, desires to stir him from his laxity and unproductive contentedness into work and hardship, so that he may again discover means to cleverly extricate himself from them. The natural drives to accomplish this are unsociability and universal opposition, from which much unhappiness comes, but which also motivate to new efforts and hence to a further development of natural potential. Thus they betray, it seems, the plan of a wise creator and not the hand of an evil spirit who meddled in his majestic plan or spoiled it because of envy (I. Kant 1960, vol. 6:30).

However, we must be aware that aggression can be trained to be excessive and destructive, so that the natural counterforces of love fail to curb it. We should furthermore not take the one-sided view of accepting our aggressive disposition as an excuse for reckless competition. To a great extent the struggle for life consists of co-operation. Our aggressive impulses are counteracted by our bonding impulses and they indeed are so strong that, for example, in trench warfare soldiers have to be shifted from time to time in order to prevent bonding over the lines by exchanging cigarettes. Indeed, any war propaganda has artificially to build up barriers against communication and bonding. And all efforts aim at making the members of their own group believe that the others are not real human beings. It is my personal feeling that this capacity of man made man more murderous than the invention of armament. Control of aggression requires the diligent pursuit of friendly, altruistic behavior. Only then is peaceful competition possible. If uncontrolled, aggression will lead to further murderous strife between peoples, which will endanger our very existence. The understanding of the causal relationships involved should help us in the control of our aggressive impulses. A criticism of the Freudian–Lorenzian instinct theory of aggression which is sometimes heard—that it is designed to exonerate us, and should therefore be rejected—is nonsense. Whoever argues along these lines has not read carefully what these authors have said. Actually there are surprisingly numerous misinterpretations presented in the discussion of human aggression. A. Plack (1968) imputes that Lorenz declared aggressivity to be the basic drive of all life, although he never said such a thing. Sociability, readiness to cooperate, and altruism are as much part of human nature as man's occasional incompatibility.

One remarkable form of aggression which has been little studied is the expulsion reaction, which is directed not against strangers but against a member of their own group. T. Schjelderup-Ebbe (1922) found that chickens attack one member of the group and even kill it, if it deviates from the norm, whether it be different because of weakness or a physical handicap. He could release this reaction when he marked the comb of a chicken with paint or tied it down into another direction. C. Kearton (1935) described how three slightly different colored penguins were continuously attacked by their own kind. Young herring gulls attacked a sibling that had a caked cloaca (F. Goethe 1939). Humans also tend to expel group members who deviate from the norm (K. Schlosser 1952a–c). In a milder form this behavior can be seen in school classes or in the military. Someone who is fat, cross-eyed, or has some other deviant habit is teased, laughed at, or even mistreated. This aggression against deviating group members undoubtedly results in preserving homogeneity, which under the archaic conditions of life in very small groups may be of selective value. In man a mild form of this expulsion reaction, teasing, can be considered as a kind of educational devise to bring the outsider into line, in that deviating "asocial" habits are suppressed. Where this is not possible, a more radical expulsion reaction may result. In these instances aggression tends to be more cruel and stronger than when it is directed against enemies who are less well known individually. This is perhaps because in addition to everything else, those common characteristics that unite the group with the outsider must also be destroyed. This norm-preserving function of the expulsion reaction is today not advantageous in human society; "outsiders" are frequently highly talented and valuable persons. Here a phylogenetic behavioral adaptation proves to be a historic burden, a handicap comparable to the appendix. We need by the use of insight to curb our intolerant impulses.

Living in groups (contact behavior)

The selective advantages of living in groups

Most animals—but by no means all of them—come together, at least temporarily, with another species member for the purpose of mating. This is apparently the best method to ensure fertilization and an exchange of genetic codes on which further evolution is based. Animals also come together for other reasons to form permanent or temporary groups. Alpine salamanders (*Salamandra atra*) gather in the fall in cavities under rocks and sleep there during the winter. Land isopods, which are normally solitary, bunch up into balls during dryness and thus protect themselves against excessive loss of moisture (W. C. Allee 1926). Some species of harvestmen in Mexico gather in tight clumps during the dry period in favorable locations and so prevent dessication. Such an aggregation of harvestmen (*Leiobunum cactorum*) was discovered by H. O. Wagner (1954) in the lowest fork of the branches of the candelabra cactus (Fig. 210). He estimated that approximately 70,000 animals had gathered there. The legs of these animals were folded over their backs and pointing outward, giving the mass the appearance of a piece of fur. Like a pelt it retained the moisture given off by the cactus. A pheromone which is secreted from a pair of glands at the edge of the head and thoracic segment attracts additional species members. Animals that were forcibly removed attempted to reach the aggregation from as far as 30 meters away. Here the aggregation is a protection against climatic conditions (see also p. 272). The animals are not only attracted by a favorable location but also by each other and coordinate their behavior to a certain extent.

For protection against predators many fish and birds collect in swarms or flocks. Large mammals living in savannah, which affords little protection, gather in herds. These protective aggregations can

Figure 210. Collection of approximately 70,000 harvestmen (*Leiobunum cactorum*) in the branches of the large Mexican candelabra cactus. (a) Side view; (b) cross-cut through the cactus and the collection; (c) view of the whole cactus. (From H. O. Wagner [1954].)

be temporary, such as migratory groups and swarms of young fish, or they can be lifelong, as with herring swarms. In fish swarms the individual is protected by the confusion effect (p. 289). Birds of a breeding colony actively assist each other in mutual protection in the same way as members of families or herds, when they are threatened by a predator. Rhesus monkeys will even attack their keeper when he catches an animal from the group and it utters the alarm call. Jackdaws attack anyone who holds a conspecific in his hand, including the caretaker, when he is holding a tame jackdaw (K. Lorenz 1935). The reaction is released whenever something black is dangling, even a black pair of swimming trunks. Porpoises aid wounded species members and raise them to the surface so that they can breathe. They circle females giving birth and thus protect them against sharks (J. B. Siebenaler and D. K. Caldwell 1956; see

Figure 211. Porpoises carry an injured conspecific to the surface. (After J. B. Siebenaler and D. K. Caldwell [1956].)

also Fig. 211). Various predators form hunting packs. Jackfishes (*Caranx*) circle swarms of fish (p. 289), wolves run down game in packs; some pass an intended victim and try to cut it off, while the other pack members chase it (A. Murie 1944). Cape-hunting dogs (*Lycaon pictus*) hunt their prey in groups. At first each hunting dog chases the gazelle nearest it, but continually watches the other pack members. If they observe that another is gaining on a gazelle more than it is, it will come to the aid of its fellow (W. Kühme 1965).

Family groups are often formed to care for young (p. 289). If only the father rears the young, paternal families result, as in sticklebacks, pipefishes, sea horses, labyrinth fishes, and some birds, such as phalaropes (*Phalaropus*) and quail (*Turnices*) (A. Remane 1960; N. Tinbergen 1951; D. Morris 1954; K. Fiedler 1954). In maternal families the females alone take care of the young. This is true in many cichlids (*Tilapia macrochir*) and other mouthbreeders, wolfspiders, reeves, hummingbirds, and many mammals (polar bear, hamster, squirrels) to give only a few examples. Parental families in which males and females care for the young are gibbons (*Hylobates*), crested bull-faced tamarin (*Oedipomidas*), songbirds, and many cichlids (*Hemichromis*). In such instances there is a certain division of labor in that the male usually undertakes the defense of territory, and sometimes even carries the young, as in *Oedipomidas* (H. Wendt 1964). This is usually done by females, which in mammals also take care of the feeding and cleaning, but male wolves and foxes also bring food. In many birds both parents usually incubate and feed.

Frequently one male protects several females and young, for instance in sea lions. In the Congo cichlid (*Lamprologus congolensis*) each male has a large territory with several hiding places into which it entices females, one after another. Each of these females has her own subterritory and defends it against others. First, each newly acquired female is attacked by all the others, but because the male sides with her she eventually obtains one part of the territory. In this way an upper limit to the size of the male's harem is established because he can help out against only a certain maximum number of females (W. Wickler 1965d).

The division of labor becomes possible only when animals come together. In insect states this division of labor has reached its ultimate form; we are reminded of the various castes in termites, which in addition to sexual animals contain workers and soldiers with highly specialized tasks. In some species of the family Termitidae the anterior portion of the heads of soldiers has been elongated into a long proboscis. From the tip of this nose the "nasuti" secrete a sticky and perhaps poisonous substance which they use for defense. Their mandibles have retrogressed to such a degree that they can no longer feed without the aid of others. They have to be fed. In the tropical ant *Colobopsis* one caste protects the entrances. Their heads are flattened and they possess a head protrusion that is colored like the bark of the plant into which they dig their tunnels. By placing their heads into the openings to this tunnel they can effectively close them. If a worker ant wants to leave or return to the nest, the guard must be notified by special signals (A. Forel, cited by K. Escherich [1906]; see also Figs. 212 and 213). The various

a

Figure 212. Examples of specializations of various caste-forming insects. (a) Division of labor and cooperation in the weaver ant (*Oecophylla longinoda*): While one group of workers pulls the edges of the leaves together, another group sews them together by pressing the spinning glands of their larvae against the edges and weaving back and forth with them. (b) Guard of *Colobopsis* closing the entrance with its head. (a) After F. Doflein [1905]; (b) after A. Forel from K. Escherich [1906].)

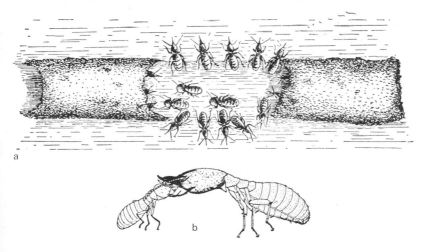

a

b

Figure 213. (a) Termite soldiers (*Eutermes*), the so-called "nasuti," guard the opening of a damaged passage way, while workers begin with repairs; (b) workers of the termite *Bellicositermes natalensis* feed soldiers. ((a) After P. Grassé; (b) after Beaumont from E. Hegh [1922].)

Figure 214. Defense against parsites by the caste of minima workers in the leaf-cutting ant (*Atta cephalotes*). (a) Leaf-cutting ant: The minima worker wards off the attack of a fly (*Phoridae*). (b) Transporting the leaf: The minima worker guards the carrier. (From I. Eibl-Eibesfeldt [1967].)

castes of the leaf-cutting ants each have specific tasks to perform (p. 292 and Fig. 214).

In workers of the honeybee the activities differ at various ages. From days 1 to 10 after hatching a bee cleans the hive. It cleans the cells and keeps the brood cells warm. After several days they develop nurse glands and the bees take care of the larvae. Toward the end of this first stage of its life the bee undertakes short exploratory flights outside the hive. In the second stage (10 to 20 days of life) the nurse glands regress and the wax glands develop strongly. The bee is now a building bee and receives nectar from other workers and stores it in the storage cells and also cleans the hive. Toward the end of the second stage some of these bees are guards near the entrance. From day 20 until death a bee is active as a pollen and nectar collector. These dates, which were determined by G. A. Rösch (1930), are accurate on the average. The thorough investigations of M. Lindauer (1952) show, however, that most of the activities above are not as strictly separated in the sequence; there is considerable overlap between them, and nurse and wax glands are often simultaneously functional in one bee. Only the transition to foraging is fairly close to day 21.

The castes of many animals have their parallels in the physical constitutions of man up to a certain point. I. Schwidetzky (1950), among others, has pointed out that in humans there exist inherited body forms which must be considered to be adaptations to the environment. Herdsmen and warrior-herdsmen are long-legged and tall; planter types are more stocky and short-legged. It is noticeable that in sedentary, city-building peoples several constitutional types occur simultaneously and that preferences for various jobs seem to be correlated with certain constitutional types (examples can be found in Schwidetzky). Perhaps this rather conspicuous polymorphism is of selective advantage by providing a kind of predestined suitability for diversity of jobs. Additional advantages of living in groups, especially higher vertebrates, are the possibility that experiences and inventions of individuals can be passed on to other members. They can then spread faster than if transmission from one generation to the next were the only available means (Japanese macaques, p. 222).

When living together partners influence one another in various ways. The social facilitation of moods, for example, results in consumption of more food in chickens that are kept in a flock than if they are kept singly. The same holds for rats, fish, and many other animals as well (J. C. Welty 1934; H. F. Harlow 1932). We have already noted that the presence of a male can stimulate the development of the gonads in females (partner effect, p. 51). Cockroaches (*Blatella germanica*) grow better in groups than alone. The effect is transmitted via the sense organs on the antennae, because animals in which these have been amputated grow like those kept in isolation. In addition, there exists a nutritional effect, because the addition of pulverized feces to the food of isolated animals results initially in a higher growth rate and, with high concentrations, in an inhibition of growth (R. Chauvin 1952). House mice females show a regular estrous cycle when excrements of familiar males are present. The odor of strange males, on the other hand, results in resorption of embryos or premature birth (H. M. Bruce 1961). One result of population density is reduced fertility; for instance, flower beetles (*Tribolium confusum*) eat their eggs when the population density is too high.

In mammals overpopulation results in a kind of stress which eventually leads to a population crash long before there is a food shortage. Fourteen miles from Cambridge, Maryland, is an island of 280 acres (James Island) where, in 1916, 4 or 5 Sika deer (*Cervus hippon*) were released. In 1955 there were 300 healthy animals. In 1958 about half of them died, although the food supply was adequate, and the population continued to decrease to 80 animals during the following years. The animals that were studied during the years of the decline showed histological changes in the adrenal glands which indicated that the stress caused by the overpopulation led to the decline (J. J. Christian 1959, 1963).

In the tree shrew (*Tupaja belangeri*) density-dependent stress symptoms have been studied in detail (H. Autrum and D. v. Holst 1968; D. v. Holst 1969). Stress causes a delay in the development of the young and numerous changes in the behavior and physiology of the adult. Females under stress produce less milk or none at all. The sternal gland ceases secretion and the females therefore cannot mark their young olfactorily as usual. Without this protection the young get eaten by the cage mates or even by their mother. Under strong stress females do not reproduce any more and they show masculine behavior, by mounting cage mates. In young males the *descensus testiculorum* is delayed, and under extreme stress the testes, even of older males, recede into the body cavity. Stress is mainly but not solely caused by aggressive interactions. This leads to an activation of the sympathetic nervous system and the adrenal chord. Tree shrews under stress fluff the hairs of their tails conspicuously. The time of tail-hair fluffing, expressed as a percentage of the total daily activity time, gives us a means to define the degree of stress. The interdependence of tail-hair fluffing and other changes from crowding can be seen in Fig. 215.

If a group of rats is kept in a limited space with adequate food supplies, their number will increase up to a certain point, and then they develop abnormal behavior. They no longer care properly for their young and do not build adequate nests; as a result the death

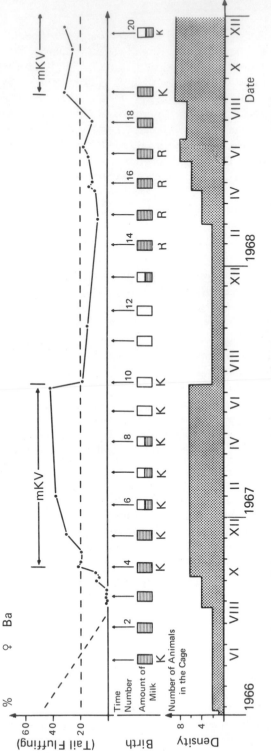

Figure 215. Relation between the number of animals present in one cage, tail fluffing (TF) value as expressed in the percentage of total activity time, propagation, and male copulatory behavior. Abbreviations: mKv, male copulatory behavior; K, cannibalism (eating of the young); R, rhythm of nursing disturbed. Hatched squares, young fed as normal; half-hatched squares, young inadequately fed; empty squares, young not fed at all. (After D. v. Holst [1969].)

rate among young increases to such a level that no additional increase occurs in the population, although theoretically there would have been more space available for them. The animals continuously disrupt each other's activities. (J. B. Calhoun 1962). In field mice F. Frank (1953) found similar conditions. The animals increase in number when the food supply is adequate until the optimal density has been passed. As a result of continuous conflicts with conspecifics various disturbances occur, including those of an endocrine nature, which eventually result in death of the animals. Under unfavorable social conditions the embryos of the wild rabbit are reabsorbed, and more so in low- than high-ranking animals (R. Mykytowicz 1960). Lemmings, on the other hand, reproduce without inhibition during summers in which the food supply is ample. They are finally forced into mass migrations, which for most of the animals ends in catastrophe. Here there is no mechanism of limiting births. In most instances, however, adaptations in the social behavior have been found which serve to prevent overpopulation of a given area (V. C. Wynne-Edwards 1962). This is one of the functions of territorial behavior. In Scotland defeated willow grouse males (*Lagopus lagopus scoticus*) survive for a time hidden in the territories of the victors. However, they hardly obtain enough food and usually die in the latter months of winter. If death of a territory owner opens up a territory one of these animals can take over and survive. On the average about 60 percent of the male offspring die as a result of this social mechanism. They are the tribute which the species pays for the preservation of its kind. Catastrophic population crashes are prevented (D. Jenkins and others 1967; A. Watson 1966). In some mammals, these mechanisms, which would lead to a density-dependent reduction in the rate of increase and so prevent a breakdown of the social system, do not function adequately. This most likely also applies to man, as H. Autrum (1966), E. T. Hall (1966), and T. Schultze-Westrum (1967) emphasize.

It is not the danger of hunger but the danger of a breakdown of the supporting and order-preserving social structures as a result of overpopulation which threatens our future (H. Autrum 1966).

We discussed in this chapter a number of behavior patterns that are of advantage to the group but not to the individual. The jackdaw that attacks an enemy is endangered, and superficially seen it might appear as if an animal that does not behave so altruistically would have a better chance to pass on its hereditary material. W. Wickler (1967c) has discussed the phylogenesis of altruistic behavior in great detail and has shown that it is possible to discuss it in terms of a Darwinistic point of view. The fact that a genome prevails in a population means that its bearer distributes it better within a population than a competitor which has different characteristics. Even if the carrier of a certain genotype of parental care and defense of siblings should die, the population in which this altruistic behavior is contained survives better than one in which it is lacking. The individual that regularly kills group members of its species certainly prevails quickly; however, compared with a population with inhibitions to kill the former is clearly at a disadvantage. If one bird would throw its siblings out of the nest it would survive alone. However, if a mutation would occur so that a bird would not eject its nest mates, the female would raise correspondingly more young, and the population with an "altruistic" inhibition to throw out siblings would prevail. From what has been said it is clear that mutations for altruistic behavior can best prevail in closed groups such as families or clans—a decided advantage for the formation of closed groups.

Mechanisms of group cohesion

Socially indifferent species sometimes form aggregations when they are brought together by a common goal such as a favorable roosting place for the night. However, not until they are socially attracted by other species members can they be said to constitute a true association. In general, it is enough that they possess one signal that attracts the partner, such as an odor, song, or a visual releaser. Such simple signals keep the fish swarm together (p. 71), bring harvestmen to a gathering place (p. 334), or attract the sexual partner (p. 116), to recall only a few examples.

If the species are not especially aggressive by nature, no further obstacle exists to coming together. It is different with aggressive animals, but even they have evolved ways to form groups. This can come about, for example, by a periodic suppression of aggressive impulses. In the spring the three-spined stickleback moves peacefully in a swarm to the shallow breeding grounds. There it establishes a territory and only then assumes sexual colors and becomes pugnacious. In other species the agressiveness may be restricted to a specific category of conspecifics. Male fence lizards (*Sceloporus undulatus*) attack only males of their own species, which they recognize by the blue stripes on their flanks (p. 72). The females lack this inciting signal, they are tolerated unless one paints blue stripes on their sides. Many young animals are especially protected by baby characteristics, but this is not always so. If one hamster (*Cricetus cricetus*) meets another in its territory, they will attack each other regardless of sex. Only when a female is in estrous will it temporarily tolerate a male. What then inhibits her aggression in this case? The house mouse (*Mus musculus*), which lives in clans, attacks any mouse that does not belong to its group which may have strayed into the territory. Within the group, however, they get along very well. They clean one another and do not even compete for the favor of an estrous female. Why then does not aggression break through within the group? Graylag geese and herring gulls allow contacts by their mates and young, but strange young or adults are attacked (N. Tinbergen 1963). A sea lion female takes care of her own young; the same is true for the domesticated sheep or the mountain sheep (*Ovis aries musimon*) (B. Tschanz 1962).

In the last-quoted cases aggression of the female is always inhibited by individual acquaintanceship with the young. This close bond is established immediately after birth or hatching of the young, and during this time it is possible to have strange young adopted. Later this is possible only rarely (H. Blauvelt 1964; P. H. Klopfer, and J. Gamble 1966; W. Leuthold 1967). The bond of individual acquaintanceship also unites siblings. This can also be established artificially between members of different species. I raised polecats with young rats. Both species got along well until the rats died a natural death when they were 2 years old. They cleaned one another and engaged in play fighting. If the polecats became too rough, the rats squeeked and this inhibited further attacks.

Strange rats, which were later introduced to the polecats, were sniffed thoroughly, and were not harmed. The deciding factor permitting this peaceful cohabitation between rats and polecats was probably the possession of similar bonding and aggression-inhibiting mechanisms so that they "understood" each other's expressive behavior. Polecats squeek when a conspecific nips them, and rats do likewise; and in both species further aggression is inhibited in this way. Both species understand grooming as a friendly gesture. When animals know one another individually, the aggression that is released by the partner frequently becomes redirected against other objects and this may even get transformed into a group-uniting ritual (K. Lorenz 1963a; see also p. 129).

Within such a group aggression is often neutralized by the establishment of a rank order. This prevents continuous conflict with members of the group. Not until there is a disruption of this order can one observe at times intense outbreaks of aggression within a previously harmonious group. This also holds, as revolutions teach us, for human societies. J. P. Scott repeatedly stated that "social disorganization" is the cause of aggressive behavior. This statement should be made more precise by saying that social disorder releases aggression against members of the group. On the other hand, we know that the group consciousness of a well-organized group, especially increases aggressiveness against strangers (H. D. Schmidt 1960).

If a group grows into a larger association, so that individual recognition of members is no longer possible, then the recognition of members (the familiarity effect) is brought about by other means. Rats and mice mark each other with odors (p. 310), and they recognize group members by this odor. If a rat is removed only for a few days from the group, it looses the group odor and it will be attacked by all former group members (I. Eibl-Eibesfeldt 1950c). Symbol identification in man is discussed later (see p. 45).

In addition to such cues for recognition which tie together aggressive animals that live in groups, they possess a number of appeasement ceremonies which serve as buffers against aggression. We discussed these behavior patterns in Chapter 6 and refer here only to the greeting ceremonies of the flightless cormorant, whose appeasing function has also been experimentally supported (p. 126).

In the parrot *Agapornis personata* a brief interlocking of beaks is used as a gesture of appeasement, in greeting, to strengthen the bond, when danger threatens, and as an expression of tenderness. Similarities with preening and feeding suggest possible origins of this pattern. The question remains open until we know more about related species (R. A. Stamm 1960, 1962). The appeasement gestures do not in themselves keep the group together, but they enable animals to remain together in groups.

Display behavior patterns in which the weapon is turned away have an appeasing function (p. 123) and so do a number of behavior patterns of young animals, such as begging movements and other infantilisms (p. 122), especially behavior patterns of care for young ritualized feeding (p. 122), social grooming (p. 94), which appease as well as establish ties because they are rewarding. The significance of sexual behavior patterns as appeasement gestures have been discussed in more detail by W. Wickler (1965e, 1967a,b) (p. 112). The equivalent group-uniting mechanisms in man will be discussed later (p. 425).

In groups whose members know each other individually, the highest ranking animal often has an important group-uniting function (p. 351). Sea lion bulls settle disputes between females with special appeasement ceremonies (p. 106). In some animals the young inadvertantly promote the cohesiveness of the group, as in lemurs (*Propithecus verreauxi* and *Lemur catta*), where the attraction toward the young keeps the adults together. When an infant is born in a group of *Propithecus,* the adults groom each other four times as much as they normally would. They discharge the activated brood-care behavior of social grooming on adult group members (A. Jolly 1966).

All gregarious animals display an obvious desire for contact. Separated from their group and kept forcibly in isolation they do not do well. A fish that has been separated from its group swims back and forth rapidly and seeks contact with the group. Gregarious mammals become apathetic when they are kept alone; they suffer from "loneliness" (M. Meyer-Holzapfel 1958). In higher animals group members attempt to bring back into the group those members which are in danger of becoming separated from the group. This has been observed by K. Lorenz (1931) in jackdaws. In 1929 a large flock of migrating crows and jackdaws settled near his colony of tame jackdaws. The young birds of that and the previous year had mingled with the strangers and the possibility existed they would be carried away in the excitement when the flock took wing to migrate—which is a powerful flight-releasing stimulus. This undoubtedly would have taken place had not two old and experienced males of the tame colony brought the youngsters back individually. They flew from the house to the meadow into the flock, searched out the young of their flock, and called them away from the strange flock by flying closely above them from behind with their tails spread and giving the flight-call note. In this way they returned all but two of the young during the course of 2 hours.

In cichlids the cohesion of the family is achieved by the following reaction of the young as well as by the brood-care behavior of the parents. The mother leads the young as soon as they are free swimming, and takes into her mouth those who swim too far away from the swarm and spits them back among the others (E. Kuenzer and P. Kuenzer 1962).

In spite of this desire to be with others an individual animal may avoid bodily contact. This can be seen in many social birds. Starlings (*Sturnus vulgaris*) have great attraction for each other they form large flocks outside the reproductive season—but they avoid body contact with flock members. When they perch on electric wires they maintain a certain distance between each other. Swallows (*Hirundo rustica*) behave similarly. The long-tailed tit (*Aegithalos caudatus*) and the gold crest (*Regulus regulus*), on the other hand, keep together in a family after fledgling of the young and sit huddled together at night in the closest possible contact. Alone they would probably freeze. Animals can then be divided into gregarious and solitary animals, and the gregarious ones can be grouped into *contact* and *distance* animals. It should be obvious that the nongregarious animals are also extreme distance animals.

A contact animal displays a definite appetitive behavior for bodily contact. This is true in chimpanzees, gorillas, and many other primates, which if kept alone often deteriorate, unless their keeper permits them to make contact, plays with them, and scratches or strokes them. This need for contact seems to have its roots in the drive for close contact on the part of young animals, and the

Figure 216. The need for contact has its roots possibly in the drive to be with the mother. (a) Rhesus mother with an infant; (b) with an older infant (Cayo Santiago, Puerto Rico); (c) two Sonjo children clasping each other in fright. (Photographs: I. Eibl-Eibesfeldt.)

behavior of contact-seeking adults shows clear similarities with parental care and infantile behavior (Fig. 216). Chimpanzees put their arms around each other, and even high-ranking animals, when frightened, will clasp a lower-ranking animal for reassurance. In general, however, other animals flee to the high-ranking one for protection. Baboons even flee to the high-ranking animals that has just mistreated them. Bodily contact has a calming effect, and low-ranking chimpanzees beg for it (p. 127 and Fig. 217). Young gorillas and adult females seek body contact with old males when they rest (G. B. Schaller 1963; see also Fig. 218) just as the young of most mammals seek contact with their mother.

In contact animals social grooming is a widespread behavior. We find it, for example, in Norway rats, agoutis, house mice, vervet

Figure 217. For contact animals bodily contact has a calming effect. (a) An approximately 4-year-old female under the hand of an old chimpanzee male; (b) a human couple. (Photographs: (a) Baron and Baroness van Lawick-Goodall, with permission of *National Geographic;* (b) I. Eibl-Eibesfeldt.)

Figure 218. Resting positions of a female gorilla and her young who sought contact with a silver-backed male. (a) Female and young sit next to the resting male whose hand rests on the shoulder of the young; (b) a young gorilla rests with its head on the hand of a male; (c) female and young sit close to a sitting male; (d) two young rest by leaning against the legs of a male. (From G. B. Schaller [1963].)

monkeys, chimpanzees, and many other primates (Figs. 219 and 220). This behavior certainly seems to have its own strong motivation, because caged animals invite the keeper to groom them, for example, agoutis by licking and combing the skin with their teeth. Tame vervet and other monkeys "delouse" the hair of their keeper and demand in return that the same be done to them, even when they are free of parasites. In this way one can make friends with

Figure 219. Female lions licking each other. The animals clean those parts of the body which the other cannot reach. (Photographs: W. Kühme [1966].)

Figure 220. Social grooming in the vervet monkey (*Cercopithecus aethiops*) and in humans (Bali). (Photographs: I. Eibl-Eibesfeldt.)

shy contact animals (p. 94). Such behavior patterns facilitate group ties via the reinforcement mechanism.

A bond can also be established and strengthened via aggression. A pair of graylag geese is bonded as a combat unit, and their greeting rituals derived from aggressive threat (K. Lorenz 1963a; see also p. 129). Fighting together establishes a bond in the rhesus monkey and other macaques (p. 354), and this is also true for man. In everyday life one can observe how laughing at someone unites those that join in (p. 132). Lorenz even expresses the opinion that love—defined as the personal bond—evolved in many instances from intraspecific aggression. This thesis I am reluctant to fully accept. I agree that aggression is older than the individualized bond, but I doubt that it derived from aggression. There are many indications that the individualized bond evolved from the parent–child bond. The majority of animals that live in closed groups show parental care and bond with patterns derived from this relationship. Exceptions are some butterfly fish, but there is no indication that their bond is derived from aggressive behavior. (A detailed discussion of the origin of bonds by the author is in preparation.)

Various groups of animals differ markedly in their social potentials. In reptiles one does not find groups whose members collaborate in some way. Marine iguanas seem to be gregarious; one sees them rest on rocks on the shore by the hundreds, crowded together and sometimes one on top of the other (Fig. 221). They tolerate each other but exhibit no altruistic "friendly" behavior patterns. They do not clean or feed one another. Their social behavior is limited to the repertoire of fighting and threat behavior, with which even their courtship behavior has much in common. This is a conspicuous difference from most birds, mammals, and bony fishes, which often form associations in which they cooperate in an altruistic manner; they may take different roles in hunting or defending a territory together. These animals also have a rich repertoire of friendly gestures such as grooming behavior, feeding ceremonies, and greeting ceremonies derived from them. These behavior patterns usually

a   b

Figure 221. (a) Aggregation of marine iguanas on Narborough (Galápagos); (b) despite close body contact these animals show no altruistic behavior patterns. (Photographs: I. Eibl-Eibesfeldt.)

evolved from behavior associated with care of young, and some are derived from infantile patterns (p. 122).

Apparently, during the course of the evolution of parental behavior, many behavior patterns became available which are also very suitable as indicators of readiness for social contact as well as for interactions of adult animals, all of which make it possible for altruistic cooperation to occur. W. Wickler (1967b) points out in this connection that only those insects that have highly developed brood-care behavior form states, and that their brood-care behavior also ties the adults together. Bees and ants feed not only their own young but also each other.

The fact that behavior patterns derived from the brood-care system are found to subserve the general function of keeping groups together indicates, among other things, that the larger, individualized and most exclusive groups can be traced back to the mother-child association.

Types of groups

Aggregations

At times aggregations of animals of one or several species occur where the sole reason for coming together lies in the attraction of some environmental factor. Butterflies may congregate at water places. If no social attraction exists, one speaks of *aggregations*.

Anonymous groups

When animals are brought together by social attraction for one another, but subsequently do not develop a bond based on individual recognition, one speaks of *anonymous groups* (G. Kramer 1950). They can be open or closed to others.

In an *open anonymous group* new species members may join at will. Individual animals are freely interchangeable as far as the group is concerned. One such example is a swarm of fish (I. Eibl-Eibesfeldt 1962b). The swarm is kept together by means of simple species-specific signals. Swarms of minnows (*Phoxinus*) accept new conspecifics only if they do not deviate more than 1 cm from the average length of the swarm members (Berwein, cited by A. Remane [1960]). Individuals that have become separated from the swarm show a definite appetitive behavior to rejoin a swarm of their own species.

Anonymous groups can consist of subgroups in which members know one another individually. This is true in the breeding colonies of many bird species, where many breeding pairs form a larger association. They collectively attack predators and are clearly grouped together because of social attraction and not because of environmental factors. The mates know each other individually; each pair marks off its own nesting area and tolerates only its partner and later its own young in the immediate vicinity.

In *closed anonymous groups* individual members do not know each other individually, but they recognize from other cues whether or not an animal belongs. Only group members are tolerated; strangers are vigorously attacked. In this category fall rats and mice, which have already been mentioned. The bond of individual familiarity is lacking, but a collective odor based on a mutual marking with urine identifies members of the group. They do not know one another, so there is no rank order. Males mate with females without any rivalry among the males. Conflicts about food are not bloody. The animals groom one another; sometimes an especially large individual may dominate others.

If one of two house mouse males, which have until then lived together peacefully, is marked with the urine of another mouse, aggressive behavior in the other will be released. On the contrary, one can also reduce the intensity of conflict between two strange males by rubbing one of the males with the urine of a mouse that is known to the other (J. H. Mackintosh and E. C. Grant 1966). The members of a bee colony also know one another by a hive-specific odor, but they recognize each other as belonging to one colony only after they have exchanged food. If worker bees of one hive have been separated by a double screen, they fight each other even though they have been exposed to the same hive odor (J. Lecomte 1961).

Individualized groups

If a group of animals is kept together by bonds of individual acquaintanceships, we speak of an *individualized group*. Its social organization can be quite complicated with the establishment of a *social hierarchy* that develops as a result of occasional fights. Each group member learns from the repeated conflicts who is superior and who is inferior and behaves accordingly. Once the matter of rank has been settled, fights are rare and usually a brief threat by a high-ranking animal is sufficient to keep a lower-ranking one in its place. The high-ranking animal not only has a number of special advantages, such as being the first at the feeding place or obtaining the best sleeping place, but it may also assume the responsibility for protecting the group against predators or of one group member against another. The dominant animal may ensure the cohesion of the group by breaking up fights; it may assume the function of leadership in certain respects, such as determining the time for moving on and by giving the direction during migrations. The role of the protector makes the highest-ranking animal a focus around which the group gathers.

A social hierarchy or rank order not only presumes that some members of the group will fight for authority either by fighting for it or by some other special achievement, but also that the lower-ranking animals accept this order. Only the capacity and readiness to submit makes the formation of stable societies possible. This often does not become apparent until one attempts to raise a higher, solitary mammal. My own quite intelligent badger completely lacked the ability to submit. He remained self-willed and accepted no reprimands. If punished for some misdeed by as much as a light slap, he at once became seriously aggressive. A dog, however, will readily submit and adjust his behavior accordingly. It is a group animal by nature.

The presence of a high-ranking animal influences a low-ranking one in many ways. E. Diebschlag (1940) has shown that low-ranking pigeons had more difficulty in learning a color and position discrimination in the presence of a higher-ranking animal. He also kept such low-ranking pigeons in individual cages where they became used to models that retreated whenever the birds showed aggressive display. They were finally able to win over the previously dominant bird when they were again placed into the old environment. This bird was then also clearly superior in its learning performance, when both were present in a training session.

The phenomenon of rank order was first studied by T. Schjelderup-Ebbe (1922a, 1922b, 1935) in chickens. At a particular feeding place some hens have certain privileges. They are first at the feeding place and peck at other lower-ranking hens which were already present or which come too close to them. Who may peck whom is well established. Chicken *a* may peck chickens *b, c, d,* and *e;* chicken *b* pecks all others except *a; c* all others except *a* and *b;* and so on. The lowest-ranking hen is pecked by all the others, but is generally left alone by the higher-ranking ones, because these are more attentive to the next lower-ranking hens, which are their most serious rivals. When strange chickens are put together, they will at first fight intensively. Each animal will fight every other animal and victory or loss will determine its future standing. A chicken that has lost a fight will remember the victor and avoid it in the future. The victor usually is the strongest animal, but agility, perseverance, and aggressiveness are also of importance. It is also possible that a high-ranking hen *a,* which was victorious over *b* and *c,* loses to *d,* which had lost to both *b* and *c,* perhaps because hen *a* had just been weakened in a fight or had been frightened by something.

Then *a* is still dominant over *b* and *c,* but is below *d* in the rank order, although *d* is subordinate to *b* and *c.* Thus there are, in addition to simple linear rank orders, more complex *triangular* relationships as well.

The subordinate animal will be pursued by the winner for only a few days; after that it will usually be left alone. Once the peck order of a flock of chickens has been established, everything is peaceful. If necessary, the higher-ranking animal will assert itself with a short threat. Roosters in general are dominant over hens, but they must fight their way up through the ranks, including the hens. Several roosters within a flock also have their own peck order. In jackdaw colonies there is also a strict rank order. Very high ranking animals are quite peaceful toward very low ranking ones, but they are very aggressive toward those who are just below them. They may also become involved in fights between two lower-ranking birds, and then they always attack the lower ranking of the two. Male jackdaws will only mate with females that are of lower rank than they (K. Lorenz 1931, 1935).

The colorful plumage of males plays an important role in the fights for rank. Female chaffinches whose undersides were colored red to resemble those of males dominated the social hierarchy when they were placed together with normal-colored females. They also dominated over the others in fights and usually won, which further demonstrates the intimidating function of the colorful male plumage on the other sex. Very low ranking females from a group that had lived together for a long time could be raised in rank by artifically coloring them. Even hand-raised females which had never seen a male retreated before red-breasted females. This reaction, then, is not acquired as a result of social experiences (P. R. Marler 1955a, 1955b). It was not possible to teach the females to avoid green-breasted birds. Even when they had been placed together for a long time with green-breasted males, they henceforth did not avoid artificially green-colored females.

In many hoofed animals that bear antlers or horns, conspecifics judge the fighting qualities of other species members by the size of the antlers or horns. In the red deer only animals with approximately equally well developed antlers will fight together. After dropping their antlers high-ranking deer rapidly drop in rank, and the attacks of lower-ranking animals take place always immediately after the others loose their antlers. This occurs even though just prior to dropping them, the antlers are already useless in fighting because of the onset of osteolysis. This shows that their possessors were only protected by the symbolic significance of their antlers (H. Hediger 1954; see also A. Bubenik 1968). Wild sheep estimate the strength of conspecifics according to the size of their horns, and strange animals that join the group can readily fit into the existing rank order without fights (V. Geist 1966b).

A particular rank order is by no means stable. Small changes occur continuously. A young hen with chicks advances in her standing; even higher ranking hens will tolerate her. Low-ranking baboon females rise in rank when they are in estrous or when they have small young (I. De Vore 1965), and the same is true for many other animals. In animals that have permanent mates the rank position of the female may change dramatically as a pair bond is established. A low-ranking jackdaw female at once advances in the hierarchy when she mates with a high-ranking male and she changes her behavior accordingly (K. Lorenz 1935). She is aware of her new position, which is based on protection by her higher-ranking mate.

One of the free-living chimpanzees that was observed by J. van Lawick-Goodall (1965) utilized an accidental discovery and substantially improved his standing within the group. Chimpanzees are afraid of loud noises. The low-ranking Mike had discovered that one can make loud noises with empty kerosene cans by dragging them over the ground or throwing them.

Mike often walked to the tent while a group of chimps was resting peacefully nearby, selected a can from the veranda, and carried it outside. Suddenly he would begin to rock slightly from side to side, uttering low hoots. As soon as the hooting rose to a crescendo, he was off, hurling his can in front of him. He could keep as many as three cans in play, one after the other (p. 813).

J. van Lawick-Goodall later hid the cans, but by then Mike was no longer in need of them. Whenever he approached other chimps of his group, they bowed low to the ground, acknowledging his dominance.

It is therefore not physical strength alone, but intelligence as well, which contributes to the rank of an individual animal. In free-living rhesus monkeys the rank of a male does not only depend on his physical strength but also on his ability to form friendships (alliances) with others. In so doing males sometimes change groups. An animal that seeks contact with another male stays near him and tries to entice him to join him in a common aggressive action by sham attacks against other males or the observer, on the principle that he who fights alongside another is also his friend (p. 129). Females, on the other hand, remain in their groups (A. P. Wilson, personal communication). In contrast to the males, the females have no close, friendly relationships toward nonrelatives of the same sex. In the Japanese monkey low-ranking individuals seek the friendship of the high ranking by assisting them in attacks. When the superior makes an attack the inferior individual joins it from the side, leading the attack and roaring. These "attacks for flattery" (S. Kawamura 1963) help to strengthen the bond and raise the confederate's status," for to be the friend of the most dominant animal is the surest and shortest way of promoting oneself (S. Kawamura 1963:203).

In baboons the strongest male is usually the highest-ranking one. Sometimes, however, two or three older males join together, where each alone would be subdued by single, younger males in the group. By being allied they rule the troop. Within the central hierarchical group there is a certain ranking. Males outside of this central group may be the highest-ranking individual animal. In one of the instances described by I. DeVore (1965) the central hierarchy consisted of the males Dano, Pua, and Kovu. The highest-ranking individual, however, was the fourth male, Kula, which was dominant over each individual of the group when they were encountered alone. Kula had to move aside only when he encountered all three together, who then usually acted in concert. With respect to the choice of sleeping place, the direction of migration and in situations of danger the high-ranking animals determine the behavior of the group. The relationship to the highest-ranking animal finally determines also the rank of an individual. If a baboon secures the friendship of a high-ranking animal, he raises his status in relation to others. Lorenz found similar conditions in the graylag geese.

Most revealing observations about the function of high-ranking animals were made by L. Williams (1967). When he released his colony of woolly monkeys (*Lagothrix*) for the first time into an open area, only the alpha male climbed into the trees and carefully inspected all possible routes for climbing. He broke off dead branches. Group members that attempted to follow him were at first chased off. He permitted them access to the trees only after 2 days of careful exploration.

There is a study by D. W. Ploog (1963; see also Fig. 222) of the development of the social relationships within a group of squirrel monkeys during the course of 1 year. The fusion of two groups of squirrel monkeys began in a collective discharge of aggression, in the course of which the highest-ranking females engaged in screaming and biting fights. The males performed display duels. The vanquished group finally withdrew into a corner, and a new rank order then developed among the males, with one male of the losing group becoming the whipping boy. The aggression of all others was directed toward this one male, and this seemed to release sexual behavior that had been inhibited previously by the collective display of aggression. As social contacts increased, the groups fused more and more into one and the aggression against the whipping boy decreased gradually (R. Castell and D. W. Ploog 1967). R. M. Yerkes (1948) described cyclical changes in the ranking of chimpanzee females

Legend

⊂⊃ Mutual Friendship ——► Heavy Attacks

⊕ One-Way Friendship —► Light Attacks

         ·······► Ambivalent Relationship

Figure 222. Example of the development of social relationships in a group of squirrel monkeys. (From D. W. Ploog, J. Blitz, and F. Ploog [1963].)

which paralleled their estrous cycle. During the estrous period the females are superior to the males in the rank order, but they are subordinate to them at other times.

The rank position of an animal is sometimes determined by the standing of its mother. Then it is transmitted by tradition and has been demonstrated in rhesus monkeys and Japanese macaques (M. Kawai 1958; C. B. Koford 1963a; D. S. Sade 1967; S. A. Kaufmann 1967).

Newborn young are introduced to other animals in an individualized group, a practice that undoubtedly protects them against aggression. The chimpanzee mother is alone when she gives birth to her young. When she returns to the group she approaches other group members with an open hand, palm up, and shows her child at the same time. She behaves in an obviously apprehensive manner but is at once reassured when the other animal reaches out with its hand (J. v. Lawick-Goodall 1965). The gorilla mother Achilla of the Basel Zoo presented her newborn young within the first few days to familiar persons when they came near her cage.

She pressed "Jambo" against her body with her left arm, held his left arm with her right hand and stuck it through the cage bars while holding her own right hand next to his. She had learned to present her hand, and it seemed as if she also wanted to include her infant into this manner of indicating contact readiness. She seemed quite contented after one had given one's hand to her and her child (R. Schenkel 1964:243).

It is remarkable that she reached out her hand in the manner of a chimpanzee mother, a gesture which Schenkel considered learned but which is possibly homologous to that of the chimpanzee. Achilla occasionally tried to push her child under the bars to the outside. She was taught to refrain from this simply by breaking off contact with her. Lion mothers also introduce their young to other members of the pride (R. Schenkel 1966).

Individualized groups can be found expressing various degrees of recognition. The partners of one pair frequently know one another individually (K. Lorenz 1963a). This is true for many fishes, especially in the cichlid group (*Cichlidae*) and damselfishes (*Pomacentridae*), where the parents know each other individually but the young are known only as a group (W. Wickler 1967). In many birds, graylag geese, for example, the parents know each other as well as their young. In mammals a monogamous, individualized family is rarer, but it does occur (for example, in the gibbon). Usually mammals form maternal families or larger packs or clans which are maintained on the basis of individual acquaintanceship.

The flying marsupial (*Petaurus breviceps*) lives in such clans and it was possible to demonstrate the existence of a clan odor as well as individual odors (T. Schultze-Westrum 1965). These animals keep together in a group and have rank orders within the group. In this species group members mark one another and fight against outsiders. If one wants to get strange animals used to each other one can place them into a nest box together, separated by a screen.

The prairie dog (*Cynomys ludovicianus*) lives in exclusive family groups which consist of one male, several females, and the young of the present and previous year. The family members know one another individually (J. A. King 1955). By constant control the dominant male makes certain that no stranger intrudes. During the breeding season the young males of these groups become quarrelsome and emigrate. They found new colonies and females at first move freely back and forth between them until they finally choose a specific male.

Cape hunting dogs (*Lycaon pictus*) live in packs with a highly developed social life. All pack members are of equal rank and feed each other; each may beg from any other and receive food. In this way all food is equally divided within the pack, so that the less skillful hunters also obtain food (W. Kühme 1965). The lack of rank order in the cape hunting dog is in this case a specialization.

During the hunt and when encountering predators one animal can still be seen in the forefront, but on the next occasion another animal may assume this position. If it were otherwise, according to Kühme, this would result in failure to adapt. If there were a division of labor according to rank, the pack would often be involved in internal rank-order fights, whenever an alpha animal became slightly injured; in cape hunting dogs injuries are quite frequent.

In wolf packs that are otherwise quite similar there is a clear rank order (A. Murie 1944; L. Crisler 1962; R. Schenkel 1947). In red deer and the wapiti we find groups of females that are led by an older female. The leading position is always held by a female with a calf; she is the most alert animal in the group. The other females with older young form subgroups that are driven off temporarily when a new calf is born. The adult males form loose groups outside the rutting season without any clear leadership being in evidence, and they disperse with the onset of the breeding season. Then a male will join a female group, but will not assume leadership, which is still provided by the leading cow; she warns when danger approaches, and the male keeps the herd together by circling it (F. F. Darling 1937; M. Altmann 1952).

In horses one finds large herds which are led throughout the year by a stallion. He lives in the center of the herd and the younger stallions live on the periphery. When these are 4 to 5 years old they separate themselves from the group with a few mares each. Among the mares there is also a rank order. According to H. Ebhardt (1958) there are social organizations among horses in which a lead stallion lives with only a few mares and their foals in a family group. Zebras (*Equus quagga*) lives in permanent families and in stallion groups. The families consist of an adult stallion and one to six mares and their foals. The old mares remain in the family for the remainder of their days, but sick and old stallions are replaced. During their first estrous young mares often leave the herd to accompany strange stallions. Between the first and fourth year of life stallions leave the herd on their own to associate with stallion groups. Within the zebra family an older mare leads the group. The highest ranking stallion follows it (H. Klingel 1967).

In eared seals the males rule over a stretch of coast as their territory where they gather their females. There they tolerate no other males. The Galápagos sea lion bulls participate in the care of young; they drive back young animals that swim too far out and protect them from sharks this way. The females in general are quite peaceable; any conflicts that might arise are stopped by the bull.

Prosimians (*Lemur catta* and *Propithecus verreauxi*) live in exclusive groups. As with many primates behavior patterns of social grooming and the attraction toward the young are the most important binding forces which keep the group together. There is a rank order between group members, and A. Jolly (1966) developed the interesting hypothesis that this complex social life facilitated the development of primate intelligence. During the mating season there are often great disruptions of the social structure, as intense fights break out among the males which often lead to injuries.

The sociology of rhesus monkeys is known from intensive observations made on the Island of Cayo Santiago off Puerto Rico, where several groups live in freedom (A. P. Wilson 1968). They, too, form exclusive groups which have their own rank order. If one high-ranking group approaches a feeding place, the lower-ranking group will move away, even if the approaching animals, within their own group, have only a low-ranking position. The ranking within a group depends, with females, on their ancestry. High-ranking females usually produce high-ranking daughters, and the animals that belong together as a result of common descent form subgroups consisting of several generations within the larger troop which the females never leave (D. S. Sade 1967). Males, however, do on occasion change over into other groups. They initially associate with the strangers by remaining near the periphery of the group they wish to join, and attempt to groom another male on occasion. They begin their life in the new troop with a low rank but they can rise higher. The rank of a male depends on his ability to establish an alliance (A. P. Wilson, personal communication; see also p. 354). High-ranking animals settle disputes between group members. In group fights they leave the fighting to the next lower ranking animals, and if the strange group is victorious, they are the first to retreat. There is no negative effect on the cohesion of the group as a result of the breeding season. The animals are promiscuous, but usually only half of the males will mate. The lower-ranking males copulate rarely or not at all, apparently because of social inhibitions (C. B. Koford 1963). I. DeVore (1965) observed that low-ranking baboons do not mate. This is reminiscent of some instances of psychically caused impotence in humans. Otherwise, sexual activity seems to cement the rhesus group together, in contrast to the lemurs. According to C. R. Carpenter (1942) this is also the case in howling monkeys, baboons, chimpanzees, and gibbons. In rhesus monkeys the estrous period during which females permit copulation is relatively long. They take up 9.2 days per sexual cycle of the female, which is about one third of the total cycle. Carpenter interpreted this as an adaptation in the service of group cohesion.

In baboons (*Papio ursinus*) groups consist of numerous infants, juveniles, adult females, and males. One troop of 80 included 54 young, 18 adult females, and 8 adult males. Of these males the strongest is usually dominant, and between the others there is a graded hierarchy. It happens also that two or three older males unite to form a central hierarchy group which rules the troop (I. DeVore 1965; see also p. 355).

The highest-ranking males advance toward predators and strange baboon males whenever there is danger. Otherwise they remain in the center of the troop and the females with very small young gather around them and are thus protected against encroachments by other group members. They copulate exclusively with females in full estrous, while the younger, lower-ranking males can only mate with females in partial estrous. They also have access to younger females. When two low-ranking animals fight, one of them may flee to the proximity of a high-ranking one, present his rear to him in an appeasing gesture, and threaten against the enemy. In these instances the high-ranking animal will take sides with the one that has fled to him and chase away the pursuer (H. Kummer 1957). In migrating troops a certain marching order is evident. An advance group of strong adult males is followed by childless females and young males. Then follows a group of dominant males, including the highest-ranking ones and the females with small infants. Other young adult males bring up the rear (Fig. 223). Young baboons seek protection initially with their mother, later with the alpha male, even when he is the cause of the fear. Until they are 2 years old young baboons are not subject to the harsh rules of adult life. They

Figure 223. Marching order of a troop of baboons. The dominant males accompany the females with infants in the center of the troop. A group of juveniles is seen at the lower part of the picture. Other males and females precede and follow the center group. Two females in estrous (swelling shown by dark markings) are each accompanied by a male. (From I. DeVore [1965], with permission of Holt, Rinehart and Winston, Inc.)

become only gradually involved in the social tensions that exist (H. Kummer 1957).

Chimpanzees in the wild live in larger, loose groups of animals who know one another personally. Females with their young form subgroups and young animals live in more or less close contact with their mother and siblings. Females with newborn infants introduce them to other group members, and the young are very interested in their newborn siblings and are permitted to hold them after several weeks have passed (J. Goodall 1963, 1965; J. Van Lawick-Goodall 1965). Among adult animals a rank order exists; the animals, males included, get along quite well. They also do not fight over females. If one female is in estrous males that are present mate with her without fighting about her.

G. Schaller has reported on clans of gorillas whose family life is more similar to that of humans. Here one group usually consists of several females and males. An especially large silver-backed male, as the dominant animal, determines the time and direction of moving and settles disputes between females. The males are tolerant toward others, but it is an exception to find two silver-backed males in one group. The bloodless rank-order fights are carried out by chest beating. (G. B. Schaller 1963). Members of different groups avoid contact. They threaten each other at a distance. Occasionally fights between males of different groups have been observed. M. Kawai and H. Mizuhara (1959) report a fight during which one male throttled another. In general, however, gorillas have peaceful habits in the wild. And on the whole this is true for primates, where violent aggression rarely occurs in nature except under situations of stress and overcrowding (C. Russell and W. M. S. Russell 1968).

Family groups dissolve frequently when the young grow up and no longer get along with one another, as in polecats and hamsters, or because the mother drives off the young. Squirrel mothers become very unsociable shortly after weaning and reject approaching young actively by threat calls and by biting or pushing with the feet. If such behavior mechanisms for breaking up the family group have not evolved, the groups grow into a larger clan association. If this group reaches a certain size, the members can no longer recognize each other individually. If such a group does not dissolve, we have an anonymous, closed group (p. 351).

In some vertebrates strong inhibitions against pair formation between parents and their young as well as between litter-mates can be observed. This is true, for example, for the graylag geese. In the Japanese macaque a mother-son "incest taboo" seems to exist and J. van Lawick-Goodall (1968) reports that in the two observed cases where a female had her swellings all males of the group copulated with her, except her two grown-up sons. A young female allowed her brother to mate when she had her first but incomplete swellings. She refused him, however, when she got her first real swellings. Whether the incest taboo in man has a biological basis or is purely culturally traditioned is a matter of controversy (F. David and others 1963; K. Kortmulder 1968; F. B. Livingstone 1969).

The structure and the size of animal groups changes from species to species, or within a species according to the time of the year (compare with stickleback, p. 163). Convergencies are numerous. There are monogamous birds and mammals as well as monogamous fish. Animals that live in open and exposed areas such as plains or the high seas tend to form large groups. Those who occupy biotopes with adequate cover live in smaller groups, but there are always exceptions—the house mouse, for example, which lives in superfamilies despite adequate cover. The multiplicity of social groupings has been discussed in detail by P. Deegener (1918). Excellent reviews can be found in W. C. Allee (1938), E. A. Armstrong (1947), G. P. Baerends (1950), F. Bourliere (1950), I. DeVore (1965), J. F. Eisenberg (1965), R. F. Ewer (1968), W. Goetsch (1940), F. E. Lehmann (1958), G. Le Masne (1950), H. M. Peters (1956), A. Portmann (1953), A. Remane (1960), E. Stresemann (1934), and N. Tinbergen (1953).

16 ORIENTATION IN SPACE

Each living organism is in continuous control of its environment through its sense organs and each is programmed in such a way as to avoid unfavorable conditions and to stay in a favorable environment. A daphnia swims close to the water surface when the water contains much carbon dioxide, which is appropriate because the water contains more oxygen there. Two stimuli play a role in this reaction: carbon dioxide as a releasing stimulus and light as a directing stimulus. If one illuminates the water from below the daphnia will swim downward as soon as carbon dioxide is added (A. Kühn 1919).

Yet this retainment in a favorable surrounding can be achieved by simpler means, such as by moving faster in the unfavorable environment and by slowing down in congenial surrounding, without necessarily altering the direction of moving. This principle, which is called kinesis, suffices to keep an animal for a longer period in an environment favorable to its survival.

In most animals of higher organization environmental stimuli are processed in such a way that an angle-controlled change of the direction of movement is achieved. Such topical reactions represent, of course, definite progress against the kinesic reactions. The orientation movements (taxis) and their mode of operation shall now be considered. I want to restrict myself to a short summary of the attempts to classify orientation processes, concluding with a catalogue. This is followed by the more detailed discussion of selected examples, to demonstrate the main problems and the experimental approach involved.

As already discussed (p. 18), orientation movements are dependent on directing stimuli. According to the resulting position, or the involved mechanisms, taxes have been variously named. A. Kühn (1919) distinguished phototaxis or alarm movements and four kinds of topotaxes:

1. Tropotaxis: With the help of paired receptors the animal assumes a position in which each receptor is equally stimulated. If one receptor is destroyed, the animal turns in circles.
2. Menotaxis (compass orientations): nonsymmetric orientation toward an orienting stimulus, for example, by keeping a constant angle toward light rays.
3. Telotaxis: goal-directed orientation. During this process the goal is fixated.
4. Mnemotaxis: orientation based on memory.

In accordance with the kind of stimulus, one refers to phototaxis, rheotaxis, phototaxis, geotaxis, chemotaxis, and galvanotaxis. Helpful summaries are presented in O. Koehler (1950), G. S. Fraenkel and D. S. Gunn (1961), and M. Lindauer (1963).

One species may possess several orienting mechanisms. The grayling butterfly (*Eumenis semele*) will escape from an enemy by flying toward the sun. It flies in circles if blinded in one eye; this is a tropotaxis. The males fly toward passing females in response to optical stimuli; this they can do even when blinded in one eye, hence their orientation in this functional cycle is a telotaxis (N. Tinbergen and others [1943]).

Orienting processes are not governed by strict stimulus-response relationships. E. v. Holst (1950a) demonstrated that the specific physiological condition of an organism—its drive state (p. 44)—exerts a decisive influence. Many fishes orient simultaneously to gravity and to the light. In horizontal light from the side, the fish is turned 90° in response by the dorsal-light reaction, while the static receptors attempt to hold the fish in equilibrium in his normal position. The fish then comes to rest in an in-between position which can be measured precisely in high-backed forms (for example, *Pterophyllum*, angelfish). The stronger the intensity of the light, the more the fish will orient toward the light. If the weight is increased on the static organs by placing the entire experimental apparatus into a centrifuge, then the influence of the light is reduced. Up to this point it appears as if the fish is suspended between two arms of a scale— one representing the light components, the other the static components, but there is another factor at work: Should the hungry fish perceive a prey, the orientation to the light exerts more influence, that is, optical stimuli are now stronger. This example supports the contention that orienting reactions can be dependent upon an inner readiness to act, such as hunger (E. v. Holst 1950a).

According to their function, orienting processes can be classified into three main groups:
1. Space orientation, which encompasses the orientation in spatially oriented reference systems, where the stimulus sources provide the coordinates.
2. Stabilization of posture and movement.
3. Object orientation, which encompasses all orientations toward directly perceived objects, which are at the same time goals.

Into these three main categories it is possible to fit all the known orienting capacities that are known to us, as can be seen from H. Schöne's (1965–1966) compilation:

A. Orientation in fixed-space reference systems: Stimulus sources provide the coordinates and/or serve to maintain the normal position, and they are *not* the goal of the orientation.
  1. Orientation with gravity receptors.
     a. Maintenance of physical balance against the pull of gravity and compensatory eye movements (postural reflexes).
     b. Control of position and direction of action in space.
        b_1.Of the body (directional control of free-swimming fishes, octopuses, crustaceans, and so on).
        b_2.Of body appendages via proprioceptive control and of other objects by means of optical, tactile perception (recognition and distinction of horizontal and vertical structures).
  2. Orientation with the eyes.
     a. Orientation related to light: orientation of many water animals; not quite correctly called "dorsal-light reaction".
     b. Horizontal orientation: compass and sun-compass orientation, orientation to landmarks.
  3. Tactile, kinesthetic (registration and repetition of the own movements) orientation.
  4. Electrical and magnetic orientation (electric fishes).
  5. Orientation in currents (air, water).
B. Stabilization of posture and movement (locomotion) independent of locality.
  1. Via mechanical rotational sense organs (semicircular canal structures).
  2. With the eyes [optomotoric (kinetic) control].
C. Object orientation: Stimulus sources are the objects of orientation; frequently locating processes (goal *and* direction is considered).
  1. Optical target orientation (mantis, chameleon).
  2. Acoustical location (echo location of bats, porpoises).
  3. Tactile location (dragonfly larvae, clawed frogs).
  4. Chemical orientation.

5. Thermal orientation (infrared organs of pit vipers, temperature preferences).

In the following sections we want to demonstrate the way in which orienting mechanisms work by discussing some selected examples. We shall pay special attention also to the active role of the organism in the orienting process. Comprehensive reviews on this topic have been published by M. Lindauer (1963), S. Gerlach (1965), and B. Hassenstein (1966). The phylogeny of orienting capacities, especially the light orientations of arthropods, were discussed by R. Jander (1966a, 1966b).

The control of body position and movement in space

Many fishes and water beetle larvae (*Dytiscidae*) orient themselves with their eyes toward light, which can be easily demonstrated with an experiment (Fig. 224). Water beetle larvae swim up to the surface to get air. If an aquarium is illuminated from below, the animals will swim to the bottom and turn their back toward it as if it were the surface. Unless the illumination is reversed the animals will suffocate there.

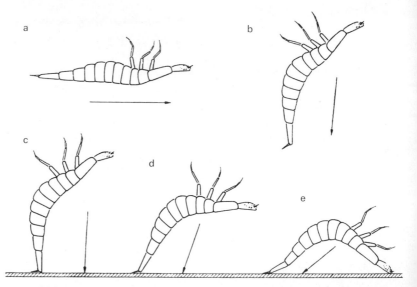

Figure 224. When the light comes from below, *Acilius* larvae swim with their backs down. Horizontal swimming and attempts to get air can be seen. (From H. Schöne [1962].)

However, it is not as if the light stimulus is coupled, so to speak, with a rigid, obligatory reaction. The animal may orient its direction of movement toward the light, but which particular direction with respect to the light source is followed depends on its internal state. If the animal needs to get air it will swim upward; afterward it will swim away from it. Depending on the internal state different orders go to the orientating mechanism so that the animal can choose between various directions and orient itself accordingly. There is no fixed reflexlike stimulus–response relationship; the organism can actively change the reference direction. This was elegantly demonstrated by H. Schöne (1962).

Dytiscid larvae have six stemmata on each side from which information is integrated in a way that enables the animal to move in a specific direction with respect to the angle of light incidence. The angles of light stimulation can be controlled by covering individual eyes and by diffusely illuminating the aquarium from all sides. The animal will then swim with its dorsal or ventral side up in circles in the case of symmetric blinding, in an attempt to maintain a reference position with respect to the direction of illumination. If the angle of light incidence coincides with the angle for the reference position (Soll-Lage) the larvae stop circling. The turning tendencies, which can be computed from the circles that the animal performs, can be represented on a curve where the orientation of the body depends on the angle of light incidence, so that one obtains a sinusoidal curve from which the strength of the turning tendency for a given expected position can be read.

A normal larva shows a number of different behavior patterns in which it can assume various reference positions in space (Fig. 225). After breathing at the surface it swims downward at an angle and continues horizontally when hunting, or it waits motionless for prey. Then it swims upward again, at an angle, toward the surface. Shortly before reaching the surface the larva changes the direction of movement and swims backward until the tip of its abdomen breaks the surface. For each of these behavior patterns there is a different reference value. H. Schöne (1962) determined the curves that indicate the turning movements in degrees of magnitude for various reference positions and found that with each change of the reference position the curve that represents the strength of the turning tendency is shifted up or down on the ordinate. The position of the upper and lower values on the curve remain unchanged (Fig. 226).

Postural orientation based on statolith organs has been studied in detail by E. v. Holst (1950) and H. Schöne (1959). The gravity detectors of crustaceans and vertebrates consist of statocysts with a sensory epithelium. On the hairs of the sensory cells rest the stato-

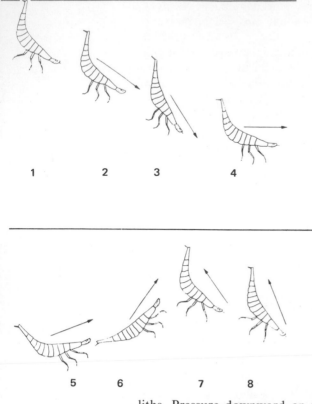

1   2   3   4

Figure 225. Characteristic positions of eight behavior patterns in *Acilius* larvae when the light comes from above. (From H. Schöne [1962].)

5   6   7   8

liths. Pressure downward on the statoliths and pulling into the opposite direction elicit no excitation, but a parallel shearing force does. If the hair receptors of the left statocyst are bent outward in a crustacean, a turning tendency to the right with respect to the longitudinal axis is released. If bent inward they cause turning in the opposite direction. With unilateral removal of the entire statocyst fish and crustaceans turn about their longitudinal axis in the direction of the injured side. E. v. Holst thought this was due to the continuous activity of the sensory cells in the statocysts. Normally these continuous discharges of impulses in the left and right statocysts cancel each other; after elimination of the statocyst on one side the turning tendency in the opposite direction becomes manifest. This interpretation is supported by the results of experiments in which sharks, whose statoliths were removed without damaging the sensory epithelium, showed no tendency to turn (S. S. Maxwell 1923). Crustaceans without statoliths, in which one of the empty statocysts had been removed, turned toward the injured side (H. Schöne 1959).

The discharges of resting potentials of the sensory epithelium on both sides cause, according to these findings, opposing turning tendencies which cancel each other. The shearing forces in the stato-

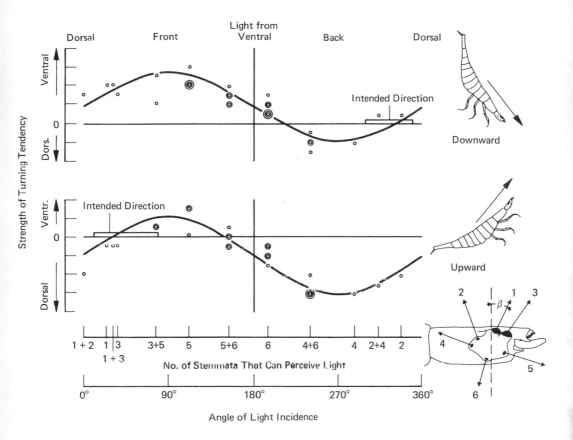

Figure 226. Example of an orientation with changeable reference (− intended) positions; Up and down swimming in larvae of the water beetle *Acilius sulcatus*. Experimental technique: The incidence of light in each case is coupled to the subject; All stemmata are covered, with the exception of those pointing into the desired direction toward the light. The larvae is observed in a container that is illuminated equally from all sides. When the "set" direction of light incidence coincides with the reference position, the animal swims straight; otherwise it turns toward its back or ventral side. The diameter of the performed circles serves as the measure for the turning tendency. Resul : The values for the strength of the turning tendencies obtained in this way are plotted in relation to the direction of light incidence: in the upper curve for reference direction downward, in the lower curve for reference direction upward (swimming into the breathing position and leaving the breathing position). Conclusion: The curve for the downward direction is shifted parallel up against the curve for the upward direction. The reference values for the positions are changed by central nervous system processes affecting the turning tendencies in correspondence to the direction of light incidence. All turning tendencies are increased by the same value in the same direction. (From H. Schöne [1962].)

cysts, which emanate from the statoliths when the animal turns, increase the rate of discharge of impulses in one direction and decrease it in the other according to the principle represented in Fig. 227.

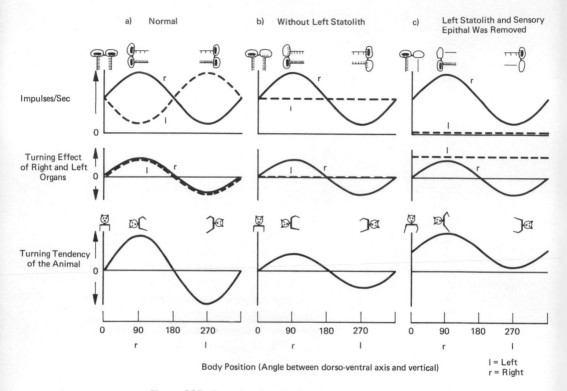

a)   Normal   b)   Without Left Statolith   c)   Left Statolith and Sensory Epithal Was Removed

Impulses/Sec

Turning Effect of Right and Left Organs

Turning Tendency of the Animal

Body Position (Angle between dorso-ventral axis and vertical)

l = Left
r = Right

Figure 227. Example of a simple orientation process. The principle of postural orientation with the aid of the statolith apparatus in vertebrates. The normal position (alpha = 0°) is the expected position; all turning tendencies seek to return to this zero position. The diagrams of the top row (1) show the excitatory values in the statolith organs, as they can be determined in electrophysiological experiments from recordings of single fibers; the figures above the curves symbolize the two statolith organs and the sensory epithelia, statolith bodies, and the flow of excitatory currents. The shearing force of the statolith mass modifies the rate of discharge of the impulses which emanate from the sensory cells. In the center row (2) the hypothetical possibility of the relationship between the excitation of both individual organs and the resulting turning tendency is portrayed. The turning tendencies of both sides *add up* to the turning tendency of the entire animal (diagram 3). The figures above the curves illustrate the appropriate position reactions, which correspond to the turning tendencies in the example of a standing mammal. The diagrams (a) represent a normal animal; (b) represent an animal that lacks its left statolith mass, but whose sensory epithelium is still intact; (c) represent an animal whose left sensory epithelium has also been removed. The curves show the relationships immediately after the operation. Later-appearing compensating processes result in a shifting of the turning tendency curve downward; it then cuts across the zero line. The Initially continuously occurring rotations eventually come to a rest in this position. (From H. Schöne [1965–1966].)

Many animals also orient themselves according to the direction of light incidence. In fish E. v. Holst investigated this integration of two orienting mechanisms. Fish turn their dorsal side to the light and assume a slanted position toward the light when it comes from the side. This slanted position is the resultant of stimulation received from the eyes and the statolith organs. Changes of the endogenous condition can lead to a differential evaluation of the optical components, which is then expressed in a different postural angle (Fig. 228).

Figure 228. *Myripristis murdjan* in a cave of the Miladummadulu atoll (depth approximately 30 meters). Some of the fish swim with their backs down. They orient themselves with the aid of the dorsal light reflex toward the light, which is reflected from the sand bottom of the cave. (Photograph: I. Eibl-Eibesfeldt.)

Distance orientation and migration

Distance orientation has always been of special interest (G. Kramer 1961). Many animals are capable of finding a goal which they cannot directly perceive, and their achievements during such migrations are at times astounding. The golden plover (*Pluvialis dominicus*), which breeds along the northern shore of Alaska, migrates during the fall via Labrador to Argentina with a flight over the open ocean from Nova Scotia to Guiana. The return flight is across land, over Central America, up along the Mississippi River toward the north. The long axis of this migratory ellipse is 11,000 km long. The Mongolian plover migrates from Siberia to Australia and South Africa. Figure 229 presents an overview of the extraordinary migratory capacities of some American migratory birds. E. Stresemann (1934) computed the work performed by such a migratory bird, on the assumption that a medium-sized wading bird such as the golden plover

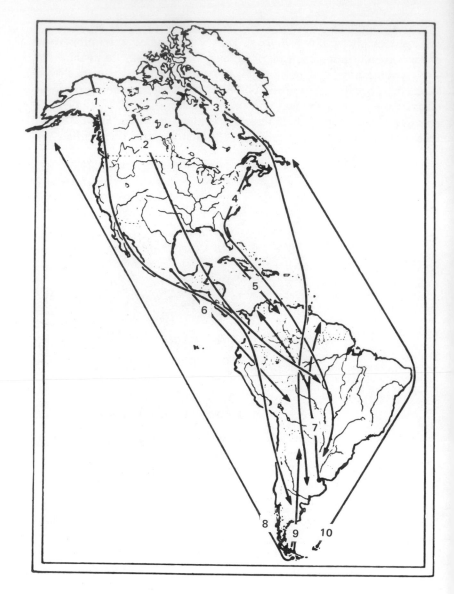

Figure 229. Migrations of some American birds after the breeding season. (1) Knot (*Calidris alba*) from northern Alaska to southern Argentina; (2) red-eyed vireo (*Vireo olivaceus*) from Makenzie to Mato Grosso; (3) American golden plover (*Pluvialis dominicus*) from Melville Peninsula to Argentina; (4) bobolink (*Dolichnyx oryzivorus*) from Maine to Brazil; (5) gray kingbird (*Tyrannus dominicensis sequax*) from Cuba to Venezuela; (6) sulphur-bellied flycatcher (*Myiodynastes luteiventris luteiventris*) from southern Mexico to Bolivia; (7) swallow (*Phaeoprogne tapera fusca*) from Argentina to British Guiana; (8) sooty shearwater (*Puffinus griseus*) from the Magellan Islands to the coast of Alaska; (9) *Lessonia rufa* from Tierra del Fuego Islands to New Argentina; (10) Wilson's petrel (*Oceanites oceanicus*) from the Falkland Islands to Newfoundland. (After van Tyne and Berger [1959] from G. Diesselhorst [1965].)

flies 26 meters per second and makes two wing beats per second. The distance of 3300 km, which is the shortest distance from the Aleutians to Hawaii, would require 35 hours, during which the bird would move its wings up and down 252,000 times. The American golden plover requires about 48 hours for its nonstop flight from Nova Scotia to South America.

European storks migrate in two groups. Storks that live west of a line extending from Leiden (Holland) to Giessen, Würzburg, and Kempten (Germany) move westward over Gibraltar to Africa. The eastern storks fly across the Bosporus (Turkey), the Jordan Valley, and the Gulf of Suez to tropical Africa. Knowledge of the general migratory direction is innate; East Prussian storks that had been displaced to West Germany moved toward the southeast when released, that is, in the direction they would have taken from their original home location, in order to reach the Bosporus and East Africa. On the other hand, when young East Prussian storks made contact with the resident West German populations, they flew with them in a southwesterly direction. In this case they followed the group. The Baltic starlings winter in England and northern France. To reach these areas they must migrate in a southwesterly direction. This general direction is inborn, because young starlings that were displaced to the latitude of Genoa (Italy) migrated to Spain. If the same experiment was made with experienced starlings which had once migrated from the Baltic to England, they corrected their course to compensate for the displacement and flew northward from Genoa to England (A. C. Perdeck 1958b; E. Schüz 1952). Graylag geese learn the migratory route south from their parents, whom they accompany. Without their lead they remain where they were raised.

Other groups of animals also perform remarkable migratory feats. Salmon return from the sea into the small rivers in which they spawned. To return they swim against strong current for hundreds of miles. The Atlantic green turtles (*Chelone mydas*) regularly visit small islands in various places of the world to bury their eggs in the sands. Marking experiments showed that turtles which feed near the Brasilian coast migrate to Ascension Island, 2000 km out in the Atlantic (A. Carr 1965).

In all these examples the question arises as to how these animals orient during their travels. Some migratory birds orient themselves by utilizing landmarks, but many can do without these. They maintain a specific direction by orienting themselves with reference to stars or, as more recent work indicates, by using the earth's magnetic field. The various kinds of compass orientations will be discussed.

Many migrating animals use the sun as an aid in navigation. G. Kramer (1952, 1957, 1959) was the first to demonstrate the use of the sun as a compass by migratory birds. His caged birds, which showed migratory restlessness, fluttered in the same direction independent of optical landmarks or earth magnetic influences: in the fall toward the south and in the spring toward the north. They oriented themselves according to the position of the sun. When Kramer deflected the sun rays with mirrors by a specific angular value in the horizontal plane the birds compensated their directional tendency by the same value (Fig. 230). Starlings are also capable

a)9^{25}–9^{45} b)10^{15}–10^{25} c)10^{40}–11^{00}

Figure 230. Mirror experiment of G. Kramer [1952]. The dots represent individual observations that were made at equal time intervals.

of compensating for the slow movement of the sun during the course of the day. Starlings that had learned to obtain food at the same time of day in a specific direction in a cage chose the same direction at other times of the day when the sun was in a different position. Three of these starlings that were trained to respond to a specific direction were tested by Kramer in a cellar under a fixed artificial sun. In these conditions the direction of choice changed in a lawful manner over time. The birds behaved in accordance with their central compensating mechanism as if the sun had moved by a specific amount (Fig. 231).

This ability presupposes a precise internal clock, and the experiments of K. Hoffmann (1954, 1960) show that the mechanism that underlies circadian periodicity is involved. This internal clock, which is so important for orientation, can be reset by an artificial light–dark alternation schedule (p. 394). In a starling that had been trained to seek food in a specific direction, the choice direction changed accordingly. If a bird had been trained to a southerly direction before its internal clock was reset, then it chose an angle of about 45° to the right of the sun at 9:00 in the morning, and at 3:00 in the afternoon it selected the same angle to the left of the sun. When the day had been moved in such a way that it began 6 hours later than usual for the bird, then the early afternoon became the subjective morning (Fig. 232). The results of Hoffmann's experiments cor-

Figure 231. Series of choices made by a starling that was trained to the west with an artificial sun. Arrows, incidence of the light from the artificial sun; filled dots, choices that resulted in eating food; empty circles, choices without food intake; blocks with cross near the periphery, baited cork in the expected direction; two points connected by an arrow, choice corrected. (From G. Kramer [1952].)

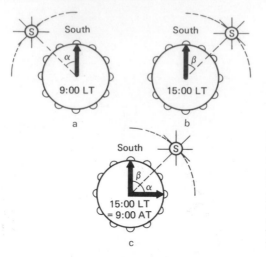

Figure 232. Expected results after changes in the internal clock. S, sun; filled arrows, direction of training; empty arrows, expected choice direction after the change. The circles symbolize the choice presentation, the half-circles the feeding cups. WOZ, true local time; KOZ, artificial local time.

responded exactly to this prediction in experiments in which the 6-hour-deviation from the normal day was used.

When the artificial day is again brought in line with the actual time the starlings also made their choice according to the existing situation (Fig. 233). This demonstrates that the compensation for the movement of the sun depends upon an endogenous physiological clock which is synchronized with local time through the day–night cycle. Under constant conditions of continuous illumination this orienting mechanism continues to function. It is therefore not set in motion by an external zeitgeber (internal clock) working on the hour-glass principle. This clock shows its own circadian frequency under constant conditions (p. 392), which deviates somewhat from the time based on the rotation of the earth. By waiting long enough one obtains deviations which are similar to those achieved by artificial resetting of the day–night cycle (Fig. 234).

That birds utilize a sun compass during their migrations can be indirectly deduced from translocation experiments, which were successful only during clear weather. Under covered skies the birds were unable to orient themselves (G. Kramer 1961; G. V. T. Matthews 1955; J. D. Carthy 1956). According to the findings of A. D. Hasler and H. O. Schwassmann (1960), H. F. Winn and others (1964) and W. Braemer (1960), fish are also capable of using the sun as a compass. Lizards (*Lacerta viridis*) (K. Fischer 1961) and land and water turtles (E. Gould 1957; K. Fischer 1963) are also capable of this orientation. Salmon can also utilize the sun as a directional cue. An interesting, innately programmed control of orienting behavior was recently described by C. Groot (1965) who investigated the migration of salmon (*Oncorhynchus nerka*) from their spawning grounds in Babine Lake (British Columbia) to the sea.

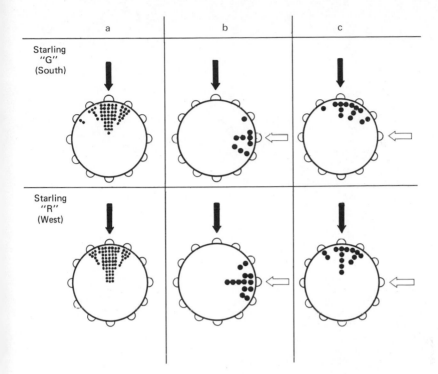

Figure 233. Result of experiments in which the internal clock was changed in starlings trained to respond in a certain direction. The starlings were trained to search for food in the south or west. (a) Choices during training under natural daylight conditions; (b) choices after 12 to 18 artificial days, which were shifted 6 hours from the actual time of day. The direction of choice shifted in (b) in the expected direction and in (c) returned again into the original direction. The large circles symbolize the choice situation, the half-circles the food cups (which were empty during the tests). Each dot represents one choice. The filled arrows show the original training direction, the empty arrows show the expected training direction after the artificial shifting of the day. (From K. Hoffmann [1965].)

The lake system has many branches so that the young which were born in various parts of the lake initially migrate in different directions until they reach the common river which leads to the sea (Fig. 235). They orient not by the currents but by the sun and other still-unknown cues when the sky is overcast. One group from one of the arms of the lake (Morrison Lake) has to migrate south southeast to reach the exit from this lake, then turn 180° in a north northwest direction, while other groups from the lake can simply maintain a general northwest migratory direction. Young fish that were caught in various parts of the lake at the beginning of migration and were kept in round or octagonal aquaria without a view of the horizon under the open sky oriented themselves according to the direction

Figure 234. Comparison of the rhythms of locomotor activity (left) and finding of direction with the aid of the sun after maintenance under constant light and temperature conditions. A: The starling N is represented by an open circle, starling W by a dot with respect to the onset of activity. When the onset of activity coincides for both, a line is drawn through the dot. Crosshatched: Times of darkness. Black squares indicate the time of the choice tests, which are shown in (b) and (c). B: Choice test during the time of training in a normal day (a), after 10 to 11 days under constant conditions (b) and during an artificial day, which was synchronized with the natural day (c). LL, constant light; LD, artificial light–dark alternation, which corresponds to the normal day–night cycle. Centripetal arrows show the direction of training, centrifugal arrows the average direction of choices made. Additional explanations are in Fig. 233. It can be seen from this figure that the activity rhythm occurs ½ hour sooner each day and that the direction of choice shifts accordingly. (From K. Hoffmann [1965].)

of migration which their fellows showed in the various parts of the lake; for example, those from Morrison Lake in a southeasterly direction, those from the main lake in a northwesterly direction. As time passes the orienting direction of the fish from Morrison Lake changes. A similar change from the preferred direction was also observed by C. Groot in salmon from another lake as the migrating season progressed. Here there was also a correspondence of the change in direction to the natural migratory route, which none of the fish had previously traversed. The way home is found by salmon by recognizing the odor of the waters in which they were born (A. D. Hasler 1960).

Bees orient with the aid of the sun compass. They learn the direction of flight or obtain the information from conspecifics (p. 136). In various species of wolf spiders, ants, and water boatmen

Figure 235. Skeena River system. The arrows indicate the migration direction of the salmon toward Babine River and the sea. (From C. Groot [1965].)

a sun-compass reaction could also be demonstrated (R. Jander 1966; F. Papi 1959; G. Birukow and E. Busch 1957).

When the sun is covered bees are able to orient themselves according to the pattern of polarization of a part of the exposed blue sky (K. v. Frisch 1950), a capacity that has meanwhile been demonstrated in other arthropods as well (K. v. Frisch, M. Lindauer, and K. Daumer 1960).

Quite remarkable is the orientation of the beach hopper (*Talitrus saltator*), which lives in and around the waterline of beaches. When displaced inland the animal escapes toward the sea using compass orientation with respect to the sun and compensating for the movement of the sun. The appropriate compass direction in which to escape is inborn in the respective populations that live in different locations (L. Pardi 1960). These Beach hoppers' offspring who were raised in the laboratory under artificial light and whose parents came originally from various different locations, and took different escape directions in their specific localities, spontaneously chose the flight direction of their parent populations when placed under the open sky. At night they oriented themselves by using the moon (F. Papi and L. Pardi 1959; J. T. Enright 1961). Some migratory birds that migrate at night, such as several species of warblers and lesser grey shrikes, orient themselves by the stars. They are disoriented when the sky is covered with clouds, but they are able to maintain their migratory direction in a planetarium under an artificial starry sky. They apparently orient themselves to the patterns of the fixed stars. Moonlight disrupts their orientation (F. Sauer and E. Sauer 1955, 1960).

By means of photoperiod manipulation St. T. Emlen (1969) induced the physiological states of spring and autumn migratory readiness in indigo buntings (*Passerina cyanea*). He tested the birds simultaneously under an artificial spring planetarium sky and found that birds in spring condition oriented northward; those in autumnal condition, southward. The results suggest that the seasonal reversal of the orientative tendencies are brought about by changes of the physiological state of the birds rather than by differences in the external stimulus situation.

Whether some animals can achieve compass orientation by utilizing the earth's magnetic field is still disputed. The direction of crawling of the snail (*Nassalia obsoleta*) can be influenced by the intensity of a magnetic field (A. W. A. Brown and H. M. Webb 1960). By changing the alignment between a magnetic and an electrical field F. Schneider (1961) was able to influence the activity of the cockchafer (*Melolontha vulgaris*). According to G. Becker (1965) diptera reacted to changes in the magnetic field. That some birds may utilize the earth's magnetic field to find the direction when they migrate, which had long been suspected, has been shown probable by the recent experiments of F. W. Merkel and W. Wiltschko (1965) and W. Wiltschko (1968). They were able to influence the direction of migration in the European robin in an artificial magnetic field.

Special problems are posed by true *navigation,* that is, the ability to find home from an unknown place. Homing pigeons which have been displaced for hundreds of miles and which have no sensory contact with their home region find their way home to their loft, but only when the sun can be seen (G. Kramer 1952, 1957; H. G. Wallraff 1960b). Pigeons that have been raised in cages until they were tested can do this provided the cage was in an open area and was constructed of wire mesh (H. G. Wallraff 1967). Just how the animals determine the geographic position of their home location and then compare it with the location at which they are released is unexplained to date. There are several hypotheses, but according to H. G. Wallraff (1959, 1960a, 1960b) they are not convincing. G. V. T. Matthews (1955) developed the sun-navigation hypothesis according to which the bird knows the path of the sun's movement at its home location. When circling over the releasing point the bird is said to take cognizance of a part of the sun's path from which it can determine the apex of the sun in that locality. By comparing the highest position of the sun at both places the bird would then be able to compute the geographical latitude of its position. The bird is also said to remember the time at its home locality, so it could also compute the distance of the displacement. With some modifications C. J. Pennycuick (1960) supported this hypothesis, which presupposes the existence of an extremely precise internal clock; it would seem to be improbable for this reason.

According to F. Sauer (1956, 1957, 1961) lesser whitethroats, blackcaps, and garden warblers can navigate by the stars, a conclusion that Wallraff is unwilling to accept on the basis of a statistical analysis of Sauer's data. The more recent investigations of F. Sauer (1961) however, further support astronavigation. According to these findings star orientation involves more than a simple compass orientation. Although we are relatively well informed about the compass orientation of animals, we know very little about the mechanisms underlying true navigation—the ability to find home from an unknown location.

Frequently more than one mechanism is involved in the orientation of an animal. Salmon can use a sun compass, but also stimuli from currents and finally the odor of the home river (A. D. Hasler 1954, 1956, 1960). Eel larvae flow into river estuaries with the rising tides. With falling tides they sink to the bottom and are thus protected against being washed out to sea again. They react innately to the odor of specific substances in the inland water, and it has been demonstrated that the saline concentration is not a factor. This odor keeps them at the bottom until seawater returns, when they continue their movements (F. Creutzberg 1961).

Proximity orientation during locomotion

During their daily excursions animals orient themselves by various means. Many insects and mammals leave odor trails when they move about, which serve as their orientation (p. 310).

Acoustical orientation exists in the oilbirds (*Steatornis*), many bats, porpoises, and probably Wedell seals (*Leptonychotes*) (D. R. Griffin 1958, 1962; F. P. Möhres 1953; W. E. Schevill 1955; W. N. Kellogg 1961; J. Schwartzkopff 1960; C. Ray 1966). Electrical orientation with the aid of self-produced electrical fields is known to exist in some fish. The Nile pike (*Gymnarchus niloticus*), which lives in cloudy waters, sends out an almost continuous series of electrical impulses, approximately 300 per second with a voltage of 3 to 7 volts. The tail becomes negatively charged with respect to the head during a discharge. The fish are very sensitive to the changes in potential in the surrounding water. A drop in potential of as little as 0.04 millivolt per centimeter is still responded to. The fish is able to detect and localize the sound of disruptions in the electrical field with the aid of its electrical sense (H. W. Lissmann and K. G. Machin 1958; F. P. Möhres 1961).

Change in the reference value during active movement ("reafference principle")

Oriented behavior in space is only possible when an animal can actively take up a variety of postures. This requires special mechanisms. If a passive fish resting in its normal position is tilted, say, to the right, the increasing shearing force of the statolith on the sensory surface of the labyrinth on this side results in increased activity in the postural center of the central nervous system, and this sets in motion the motor movements that bring the fish back to its normal position. Under such experimental conditions this postural reflex works untiringly. During the normal activity of the fish, as in the search for food, however, spontaneous tilting movements frequently take place. Why, in this case, does the postural reflex fail to pull the fish back automatically to its normal position? According to the reflex theory, it was assumed that the postural reflex is blocked during spontaneous movement. It can be shown by experiment that this is not the case. By placing the fish in a constant centrifugal field, it is possible to double the shearing force that the statolith exerts on the sensory surface of the labyrinth. Now, if one measures the frequent, spontaneous tilting movements which the free-swimming fish makes, one finds that they have decreased in magnitude and that the tilting movements become less the heavier the statoliths are made. Thus the righting reflex is not blocked during spontaneous movements but is dependent upon or regulated by the afferent feedback which they cause (E. v. Holst and others 1950).

Likewise it was assumed that optomotor reflexes are blocked during active locomotion. Again it could be shown that this is not the case. The fly *Eristalis* orients optically by fixating its visual field when in a resting position. If a cylinder painted with vertical stripes is rotated around the fly, it will turn as a result of the optomotor reflex in the direction of the moving stripes. If the fly moves spontaneously in the resting cylinder it can make many turns without being forced to return to its original position, although there is a displacement of retinal images from the environment. The opto-motor reflex in this case does not interfere with the fly's movements. H. Mittelstaedt (1954) turned the head of an *Eristalis* 180 degrees and fixed it in this position to the thorax (Fig. 236). The sequence of visual elements is now reversed, so the animal turned to the left whenever the striped cylinder turned to the right. If active move-ment would result in inhibition of the optomotor reflexes the fly with the reversed head should be able to move in the resting stripe cylinder. Actually, however, when the fly begins to move spon-taneously, it now moves alternately to the right and left in small circles and finally stops in a bent posture.

Figure 236. Schema of a fly with normal head position (a) and the head turned 180° about the longi-tudinal axis (A-A). (b) Six sectors for the eyes are shown (1–3) for the left eye (L) and (4–6) for the right eye (R). When a striped wall is turned to the right about the animal, it will turn to the right when its head is in the normal position (a). When the head is inverted it will turn to the left (b) with respect to the striped wall. (Additional explanation in the text.) (From E. v. Holst and H. Mittelstaedt [1950].)

During spontaneous movements, therefore, the stimuli, which otherwise release postural reflexes, are not inactivated but must become neutralized in another way. Additional experiments and considerations lead to an hypothesis called the *principle of reafference* [E. v. Holst and H. Mittelstaedt (1950)].

This functional organization can be represented as a feedback loop. We differentiate in this schema *afferences,* which flow toward the central nervous system, and *efferences,* which lead from the cen-tral nervous system to the motor areas. Afferences in turn can be subdivided into receptor excitation caused by internal changes in the muscular system (*reafference*) and those which are produced pas-sively by external energy changes from the external environment (*exafference*). Reafference and exafference are integrated in some higher center. E. v. Holst and H. Mittelstaedt postulate that with each voluntary movement of the organism a copy of the motoric efferent impulse branches off as an efferent copy which is stored in a sub-ordinate lower center Z_1. The efferent impulse goes to the effector and the sense organs report the result of the movement as a reafference. Centrally this reafference is matched against the efferent copy and is canceled. If the total afference is too large or too small as the result of external stimulation, then there remains a plus or minus value in the center Z_1. This is reported to the higher center and the initial command is correspondingly strengthened or weakened.

E. v. Holst and H. Mittelstaedt have illustrated this principle with the example of "space constancy" (Figs. 237 and 238). This is the observation that we recognize nonmoving objects as stationary and moving objects as being in motion regardless of whether or not we ourselves or parts of our bodies move. For example, if we glance along a train that is beginning to move, its image passes across our retinas in the same manner as if we actively pass our glance along a standing train. In either case, however, we know when the train stands still and when it is moving. Three simple experiments help us to understand the mechanism that underlies this capacity.

If we fixate an object with one eye and then move it passively to the left by pressing lightly against the eyeball with a finger, we have the impression that the object moves to the right. In this case an intentional command is missing and as a result the efferent copy present during an active movement is lacking. The retinal displacement of the image is reported on to a higher center, and we draw the erroneous conclusion that the object is moving.

Next we may temporarily paralyze the eye muscles by a drug and ask the subject to look to the left. This movement cannot be carried out, but interestingly enough the subject experiences a movement. He sees the object moving to the left. In this case, then, a movement perception takes place although there is no retinal displacement of the image. This is again in agreement with the reafferentation principle, because the intentional command produces an efferent copy that passes to the higher centers without being canceled by afferent impulses.

Finally, we can combine both experiments by asking a subject to move the previously paralyzed eye which is focused on an object to the left. At the same time we move the eye passively in the same direction by means of a clamp ring. If this experiment is carried out properly, the subject will perceive no movement. Both commands, reafference and the efferent copy, which in the previous experiments led to erroneous perceptions, now cancel each other. This is the reason we perceive our environment as stationary although we actively look about. The functioning of the orienting mechanisms which we have discussed so far can be understood best in line with these kinds of theoretical considerations (see also the excellent discussions of this subject by B. Hassenstein [1966] and N. Bischof [1966a, 1966b]).

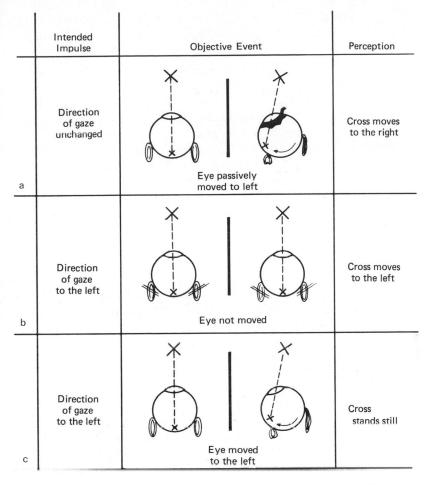

Intended Impulse	Objective Event	Perception
a Direction of gaze unchanged	Eye passively moved to left	Cross moves to the right
b Direction of gaze to the left	Eye not moved	Cross moves to the left
c Direction of gaze to the left	Eye moved to the left	Cross stands still

Figure 237. Eye movements and perception. Experiments for the determination of the functional schema (functional organization) of spatial constancy. (Explanation in the text.) (From E. v. Holst [1956].)

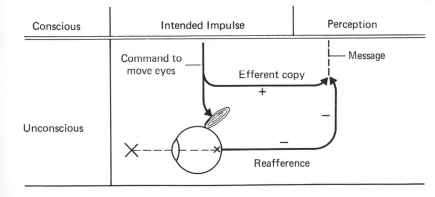

Conscious	Intended Impulse	Perception
	Command to move eyes	Message
	Efferent copy +	−
Unconscious		−
	Reafference	

Figure 238. Functional schema of spatial constancy. (From E. v. Holst [1956].)

Orientation toward objects

If we reach for an object, the grasping movement is controlled automatically by the eyes, which can detect each deviation to the left or right and which initiate the appropriate corrections of the hand movements by means of complex processes in the brain. Whereas here the grasping movement is under continuous corrective control of the eyes, in the very quick catching movements of the praying mantids the movement is not under the control of the eyes, because any corrective order given after the initial release of the action would be too late to have any effect, because of the speed with which the movement is carried out. The instance that regulates the orientation of the striking legs must be informed about the position of the fly in relation to the head as well as about the position of the head in relation to the body.

Before striking, the visual focusing on the target takes place in a very specific manner, as was investigated by H. Mittelstaedt (1953, 1954). The praying mantis first fixates the prey with the head and then strikes it by bringing forward its first pair of legs, which are folded under the prothorax and which are modified for catching prey. If necessary these forelegs can strike toward the side when the prey is not in the symmetrical plane of the prothorax.

The mantis has a cushion of sensory bristles ("neck organs") at the head joint (see Fig. 239) and the degree of bending of these bristles registers the degree by which the head is turned. If one cuts the nerve that comes from the left bristle pad, the animal will strike past the prey on the right for some time. Bilateral deafferentation does not produce a tendency to strike to the sides, but only prey is caught that is in a straight line with the prothorax. If the flies sit to the right of this line the strike misses on the left and vice

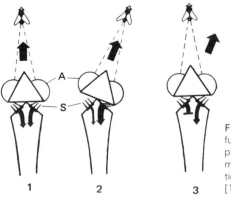

Figure 239. Schema to explain the functioning of the neck organs of a praying mantis. (1) and (2) intact animals; (3) left nerve cut; arrow, direction of strike. (From H. Mittelstaedt [1953].)

1   2   3

versa. Therefore the neck organs are involved in the aiming process.

It could be possible that the information which determines the direction of the strike consists of the additive components of optical and proprioceptive information something like this: If the mantis fixates upon a fly sitting directly in front of it, the image of the fly is equally represented in both eyes, and the report from the neck organs is symmetrical; the catching legs strike out straight forward. If the fly sits to the right of the mantis, the eyes again report symmetry, because the head is turned toward it, but in the neck organ the excitation coming from the right bristle pad is stronger. As a result the strike is directed to the right by the corresponding value

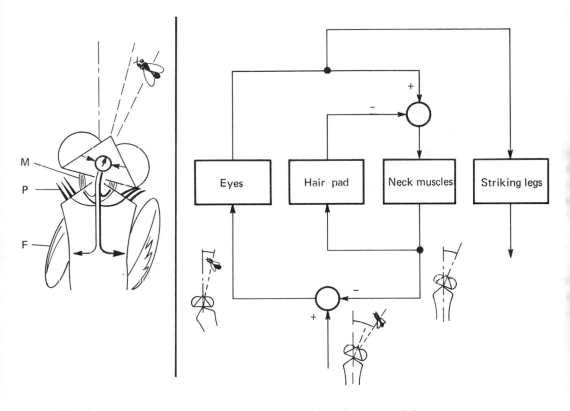

Figure 240. Functional organization of the aiming process in praying mantis. Left: overview of the organs involved: M, neck muscles; P, bristle pads; F, striking legs; right: abstracted schema. When aiming, the eyes give a turning command to the neck muscles which turn the head and the fixed eyes in the direction of the target. The bristle pads, which are stimulated by this action, seem to counteract this fixating reaction, because they work toward a return of the head to its original position, which results in the head aiming a short distance behind the fly. The influence of these bristle pads is considered within the system, however. The task of the bristle pad feedback system seems to be the nullification of external influences that might change the position of the head via an order to the neck muscles. (After H. Mittelstaedt from G. Wendler [1966].)

of this difference in excitation (Fig. 239). If this hypothesis is correct, the orienting mechanism should function if the head of the mantis is fixed asymmetrically to the prothorax by a drop of glue. This forced position would be reported by the neck organs and the animal that was so treated could fixate the prey by compensating the missing head movements by moving the body or the legs. Sixty to 80 percent of all strikes, however, miss in the opposite direction of the forced head position; it is as if the animal did not know that its head is at an angle to the body axis. Therefore, this proposed hypothesis cannot be correct. It appears as if information coming from the receptors in the orienting mechanism is processed only when the head is freely moveable. Mantids whose entire proprioceptive afference of the neck region was removed fixated and struck at their prey in a well-coordinated fashion, so one can assume that the excitatory pattern of the optical center, which directs the neck muscles and thus determines the position of the head, also determines the direction of the strike. The bristle pads of the neck organs signal the actual head position, but this information does not seem to be used when aiming the striking legs but only in the control of the head-positioning musculature. The task of this mechanism is to make the head position, which has been effected by the eyes, independent of further external disturbances. The neck organs then seem to control the neck muscles (Fig. 240).

To illustrate this with one example it means the following: A mantis which fixates upon a fly sitting to its right orients its striking legs to the right according to the amount of effort that was required to obtain this head position and which is measured by the innervation on the right. Quite anthropomorphically expressed: The praying mantis strikes in the direction in which it thinks it has turned its head. The knowledge about the actual head position, which is available through the neck organs, is not available to the localization apparatus but only to a lower motor center which has the task of making the normal position (zero position) of the head and the degree of its deviation independent of the mechanical strain on the neck muscles. This has been experimentally supported. Thus the head can be loaded with considerable rotational force (by attaching small weights) before the accuracy of the strikes is affected (H. Mittelstaedt 1953:106).

17 TEMPORAL FACTORS
      IN BEHAVIOR

Regularly recurring events, such as the changes from day to night, high and low tides, changes of the moon and the seasons, and so on, are of greatest importance to all organisms. Those who know animals are aware that different species are active at different times of the day or night. Some animals move about especially during the morning and evening hours, others are active during the day and sleep through the hours of darkness, and animals active at night rest during the day. This is as true for aquatic as well as for land animals (Fig. 241). Sometimes an animal changes from being active during the day to activity at night during the course of its development. The tortoises of the Galápagos Islands feed during the cooler evening and night hours when they are young. At that time they can even eat the dry grass because it is covered by dew. In contrast, young badgers play during the day in front of their den in the sun. Only gradually do they shift their activity toward dusk and night. This is accompanied by changes in their behavior. A young badger which until then has been quite trusting becomes quite shy during the day, but at night it is much less nervous (I. Eibl-Eibesfeldt 1950a). W. Kühme (1966) observed a similar confidence in lions during the night in the wild; during the day they exhibited a much larger flight distance.

During the time of rest and sleep the locomotive activity generally comes to an end, the electric activity pattern of the brain changes, and the thresholds of the sensory organs are raised. Often specific sleeping postures are adopted, and a number of activities continue during sleep, for example, ruminating. M. Holzapfel (1940) pointed to the fact that specific motivating mechanisms underlie sleep behavior, causing an appetitive behavior for sleeping. The sleepy

Figure 241. Fish fauna at a rocky shoreline in the southern Gulf of California (Baja California), Mexico. Top: at noon; bottom: around midnight without moonlight. The species shown: (1) *Eupomacentrus rectifraenum,* (2) *Epinephelus labriformis,* (3) *Holocentrus suborbitalis,* (4) *Thalassoma lucasanum,* (5) *Abudefduf troscheli,* (6) *Runula azalea,* (7) *Myripristis occidentalis,* (8) *Microlepidotus inornatus,* (9) *Bodianus diplotaenia,* (10) *Scarus californiensis,* (11) *Balistes verres,* (12) *Rypticus bicolor,* (13) *Chromis atrilobata,* (14) *Prionurus punctatus,* (15) *Heniochus nigrirostris,* (16) *Pareques viola,* (17) *Apogon retrosella,* (18) *Lutianus argentiventris,* (19) *Anisotremus interruptus,* (20) *Haemulon sexfasciatum,* (21) *Mycteroperca rosacea.* (From E. S. Hobson [1965].)

animal searches for a sleeping place and performs a number of preparatory acts, specific for the species. E. v. Holst and U. v. Saint Paul (1960) induced sleeping in the chicken through electrical stimulation of certain points in the hypothalamus. W. R. Hess (1954) in the same manner released sleeping in the cat. For further references about the physiology and ethology of sleep see J. Altmann (1966) and G. Tembrock (1964).

By specializing in activity at different times, day-active and night-active animals fill various ecological niches (for example, birds of prey). Many animals that live in tidal zones must seek shelter for several hours before and during the time when their home range is without water to protect themselves against dessication. The grunion of the California coast (*Leuresthes tenius*) must be ready to spawn at high tide, because they bury their eggs in the sand near the highest line reached by the water. Animals that are active during the day must go to their resting place for the night before it is too late; frogs must be ready to mate when the snows melt; in short, these changes must not find the animals unprepared. As we now know, many animal species are adapted in their endogenous activity cycles to the periodic changes in their environment. The 24-hour rhythms have been especially well studied (J. Aschoff 1962, 1964, 1965; J. Aschoff and R. Wever 1962a; H. Remmert 1965). If one registers the activity of animals in experimental cages one generally observes a distinct 24-hour periodicity. Usually it is accurately synchronized with the normal day–night cycle. If the animals are kept under constant conditions of continuous light or darkness at the same temperatures, they continue to show periodic activity, but the length of the periods deviate somewhat from the normal 24-hour periodicity. This proves that the periodicity is endogenous and is not induced by environmental factors. Species that are active during the day and night behave differently under these circumstances. Under continuous light conditions the length of the period increases in animals that are active at night, and it is shortened in those that are active during the day with an increase in the intensity of the light. To the contrary, during continuous darkness the length of the period of activity decreases for animals active at night and becomes longer for those active during the day (Aschoff's rule, K, Hoffmann 1965; see also Fig. 242). The internal clock of the animals runs faster or slower in a lawful manner approximating a 24-hour rhythm, which is why one speaks of "circadian" rhythms. They have also been demonstrated in man (J. Aschoff and R. Wever 1962b).

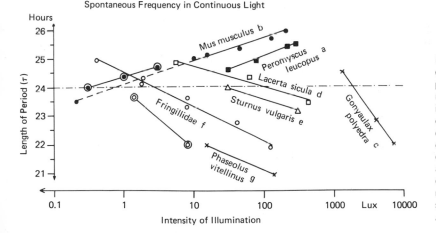

Figure 242. Dependence on the length of period in day- and night-active organisms in constant illumination of different luminescence. Filled symbols: dark-active animals; empty symbols: light-active animals; double circles, second experimental series of the same species. (From J. Aschoff [1960].)

An increase in the illumination determines not only the frequency but also the total amount of activity and the relationship of the activity time to the period of rest. This is achieved in such a way that the amount of activity and the relationship of the active and resting periods is positively correlated in light-active animals and negatively correlated in night-active animals—with the intensity of the illumination (circadian rule, J. Aschoff 1960; see also Fig. 243).

Figure 243. Activity periods of two chaffinches (*Fringilla coelebs*) under constant conditions during constant illuminations of 5, 0.2, and 5 lux. Period, activity time, and resting time are indicated by τ. (From J. Aschoff and R. Wever [1962a].)

The circadian rhythm is apparently inborn in many instances. Chicks that have been incubated and maintained under constant conditions show it in the same way as do lizards and mice that have been bred for several generations (J. Aschoff 1955b, J. Aschoff and J. Meyer-Lohmann 1954; K. Hoffmann 1959; see also Fig. 244). Lizards that had been hatched in an incubator under temperature and light periods that corresponded to a day length of 16 and 36 hours, respectively, exhibited the normal 24-hour rhythm when they were tested under constant conditions, just as did control animals that had been kept under a 24-hour rhythm after they had hatched (K. Hoffmann 1959).

If a human being is carefully kept isolated from all environmental influences in a subterranean bunker, one can also observe a spontaneous frequency (J. Aschoff and R. Wever 1962b; J. Aschoff 1966). Man's periodicity is also circadian, that is, it deviates slightly from the normal 24-hour periodicity, which proves its endogenous origin (Fig. 245). It was also found that various physiological processes

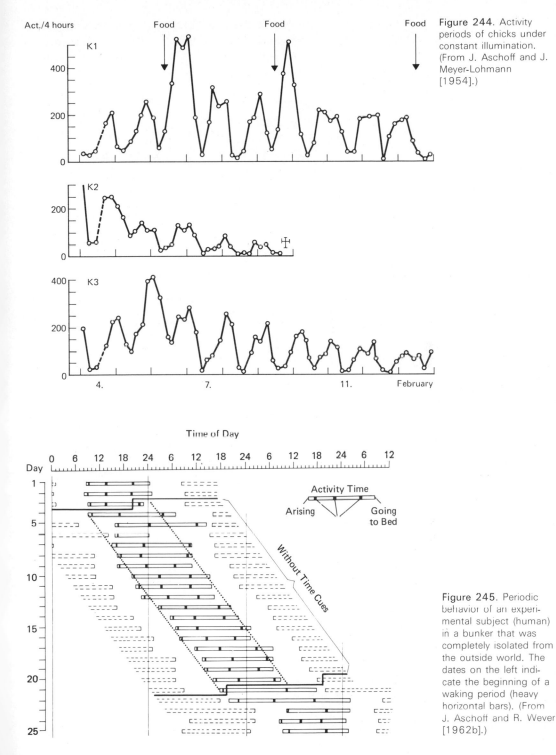

Figure 244. Activity periods of chicks under constant illumination. (From J. Aschoff and J. Meyer-Lohmann [1954].)

Figure 245. Periodic behavior of an experimental subject (human) in a bunker that was completely isolated from the outside world. The dates on the left indicate the beginning of a waking period (heavy horizontal bars). (From J. Aschoff and R. Wever [1962b].)

have their own distinct circadian rhythmicity, each with a slightly different frequency, which diverge from each other as time goes on (Fig. 246). Beyond this 24-hour rhythm a 7-day rhythm (circaseptan) was demonstrated (F. Halberg and others 1965).

Figure 246. Desynchronization of the circadian rhythms of a subject maintained isolated and without a clock. Horizontal bars and dotted lines, waking or sleeping periods; dashed line, the 37°C scale for body temperature. (From J. Aschoff [1966].)

The internal circadian rhythm is synchronized by external stimuli termed Zeitgeber, time setter, with the rhythm of the environment at large such as light, humidity, temperature, and sound (Fig. 247). Some lizards can still be synchronized by temperature cycles of small deviations. If the deviation is 1.6°C, then 75 percent of the animals are still fully synchronized; at 0.9°C about 25 percent are (K. Hoffmann 1968). In man even a weak electromagnetic field influences the circadian periodicity. When a field of 10 cycles per second was switched on, the period became shorter than previously. The "internal desynchronization phenomenon," in which the activity period becomes abnormally increased to 30 to 40 hours while the continuously

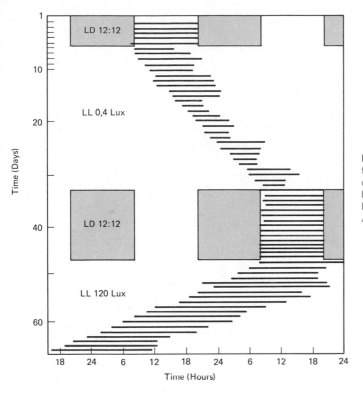

Figure 247. Activity periods of a chaffinch in constant illumination (L) of different intensities and under periodic light–dark alternation (LD) of 12 hours light and 12 hours dark. (From J. Aschoff [1965].)

monitored vegetative functions continue in a 25 to 26-hour periodicity, can only be observed when the magnetic field is switched off. Finally, it can be demonstrated that this field Zeitgeber is capable of maintaining the internal rhythm in nearly perfect synchrony as long as the timing of the Zeitgeber falls close to the time of spontaneous onset of the activity phase (R. Wever 1968). In man reciprocal social effects lead to synchronization of circadian rhythms (E. Pöppel 1968). The organisms are able to measure time by means of an internal clock whose mechanism is still unknown (E. Bünning 1963).

Upon removal of the pineal organ the activity of sparrows that are kept under constant conditions becomes arhythmic, but not if there is a change of day and night in illumination (S. Gaston and M. Menaker 1968).

In marine animals a distinct lunar periodicity has been demonstrated in several instances. Monthly and 14-day reproductive cycles are known (P. Korringa 1957). We already mentioned the grunion (*Leuresthes*), which spawns precisely at high tide along the California coast (B. W. Walker 1952). Another well-known example is given by the palolo worm (*Eunice viridis*) (H. Caspers 1961). In the closely

related polychaetes *Platynereis dumerilii* this monthly reproductive
cycle is maintained under constant conditions for at least two cycles
(C. Hauenschild 1960). More recently the creek planaria, a freshwater animal, was found to have lunar-periodic fluctuations in its
light preference (E. May and G. Birukow 1966).

Animals that live in the tidal zone show an activity rhythm which
corresponds to the daily rhythms of the tides. K. S. Rao (1954)
observed in mussels (*Mytilus edulis* and *M. californicus*) fluctuations
in the rhythmicity of rate of water propulsion corresponding to a
tidal rhythmicity. The rhythms were maintained for weeks in the
laboratory without deviating noticeably from the tidal rhythms.
According to F. A. Brown, Jr. (1965) this points to the involvement
of a still unknown Zeitgeber, because an endogenous rhythm of such

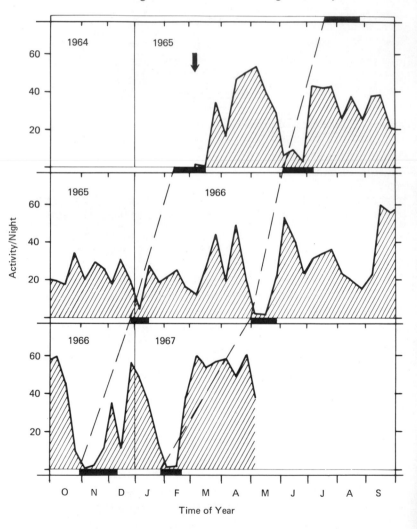

Figure 248. Molting time and nightly migratory restlessness of a willow warbler that lived for 27 months at 21°C (±2°C) in an artificial 12:12 hour day (200:0.2 lux). Ordinate: number of 10-minute intervals in which the bird was active per night, averaged over one third of a month. Black lines, molt. Since leaving the nest in June 1964 the bird lived first in a heated room with a natural light–dark cycle and then early in March was placed into the experimental conditions (arrow). (E. Gwinner [1967].)

precision is hard to imagine. J. T. Enright (1963) has expressed doubts about this interpretation. E. Naylor (1958) discovered a very precise tidal rhythm in the running activity of the crab *Carcinus maenas*. Under constant conditions this rhythm is lost by the sixth day. The sand beach hopper *Synchelidium* exhibits an activity rhythm of swimming and digging in at the sandy coasts of California that corresponds exactly to the local tidal rhythm. In captivity it is maintained for a few days with decreasing precision. The turbulence of the water is thought to be the Zeitgeber (J. T. Enright 1963).

The yearly rhythms in behavior are not only of an exogenous nature. E. T. Pengelley and K. Fisher (1963) kept golden mantled ground squirrels (*Citellus lateralis tescorum*) under constant conditions. The animals still exhibited a physiological cycle of approximately 1 year, which manifested itself in the uptake of food and periodic onsets of hibernation. Willow warblers (*Phylloscopus trochilus*) molt twice a year and exhibit migratory restlessness during the spring and fall. Willow warblers that were kept for 27 months on a 24-hour day (12 hours of light, 12 hours of darkness) retained this cycle; it was no longer synchronized exactly with the yearly seasons, but shifted from year to year as can be expected from a free-running, endogenous, circannual rhythm (E. Gwinner 1967; see also Fig. 248). We are indebted to J. Aschoff (1962) for a comprehensive review about the annual periodicity of activity. For a more general review of biological periodicity, see H. Remmert (1965). The comparison between animal and plant rhythms was discussed by E. Bünning (1963).

18   THE ETHOLOGY OF MAN

Since Darwin we know that one key to the understanding of human behavior lies in his phylogenetic development—in the process of how man came to be what he is. We have already made the observation that behavior mechanisms owing their adaptiveness to phylogenetic processes also determine human behavior sequences. This instinctive basis of human behavior has been recognized by many anthropologists, psychologists, and other scholars, but it has just as frequently been more or less radically denied by others (R. Bilz 1940, 1944; B. Berelson and G. A. Steiner 1964).

Thus M. F. A. Montagu (1962), for example, writes:

If there should be any instincts in man at all, they consist perhaps in the automatic reaction to a sudden noise or to the withdrawal of some support; otherwise man has no instincts.

A. Gehlen (1956, 1961) holds a view similar to Montagu's (see also p. 409). He also holds the position that human instincts are primarily chaotic and degenerative, in contrast to K. Lorenz, who sees degeneration as a secondary result caused by domestication and civilization.

In this book we have presented the facts that compel us to accept the existence of phylogenetic adaptations in the behavior of animals, and we have demonstrated that these adaptations are present in the form of fixed action patterns, internal motivating mechanisms, innate releasing mechanisms, releasers, and innate learning dispositions. We are now ready to examine to what extent similar adaptations also preprogram human behavior and to what degree they are still adaptive today, in the sense that they function in the service of the preservation of the species.

Fixed action patterns
and their release in infants

The newborn human being is equipped with a number of functional behavior patterns (A. Peiper 1951, 1953, 1963). In the main these are capacities located in the brain stem and the spinal cord; the cerebral cortex is at that time not functional in any real sense. Anencephalic children differ little in their behavior from that of healthy children during the first 2 months of their lives, although they lack a cerebral cortex (M. Monnier and H. Willi 1953; E. Gamper 1926).

Some behavior patterns, which serve the function of food intake, are phylogenetically quite old, and the human infant shares them with many other mammals. First we may list the rhythmic searching movements for the nipple, a turning of the head left and right, which may occur spontaneously or following a touch of the mouth region (H. F. R. Prechtl and W. M. Schleidt 1950). The seeking behavior ends when the infant gets the nipple into the mouth and when the lips close firmly around it. This rhythmic seeking of the breast is observed only during the first days after birth (Fig. 249).

Figure 249. Rhythmic search for the breast (searching automatism). (After H. F. R. Prechtl.)

It is soon replaced by an oriented search for the breast: When the mouth region is touched the infant turns toward the stimulus object, orienting in space so that he or she can get hold of it. This spatially oriented movement at first still has a rhythmic component which is soon lost (H. F. R. Prechtl 1958). The motor patterns of sucking also change within the first few weeks of life. Initially the lips close firmly around the nipple area (areola papillaris) and suction is produced by a partial vacuum in the mouth cavity (pump sucking); later the tongue alone does the work involved in sucking by pressing the nipple against the roof of the mouth. During this lick sucking the corners of the mouth remain open.

A characteristic reaction of the newborn infant is the grasping reflex with the hand. If one touches the palm of the infant's hand the fingers close firmly around the object and, as H. F. R. Prechtl (1955) has shown by motion picture analysis, in an ordered sequence of finger movements (Fig. 250). This reflectoric grasping is especially strong during sucking. Quantitative investigations show that children react especially to hair. The grasping reflex undoubtedly served originally the purpose of holding on to the mother's fur. This reflex is often considered a rudiment because man no longer possessess fur and therefore the reflex is thought to be no longer functional. The behavior does not seem to have completely lost its function. however;

Figure 250. Grasping reflex of the human infant. At first the middle finger closes; the others follow, and the thumb is last. (After H. F. R. Prechtl.)

one can observe how small infants sleep close to their mother's body and how they hold on to her clothing. The hand-grasp reflex is so strongly developed in premature babies that they are able to hang on to a stretched-out clothesline. This capacity is lost later, which is an indication of a beginning rudimentation (Fig. 251).

The climbing movements that can be seen in premature infants seem to be definite rudiments: Placed on their backs they perform alternating, well-coordinated arm and hand movements. One arm is moved downward with a closed hand, the other moves upward while the hand opens slowly.

Figure 251. Hanging by the hands and with the toes in a premature human child (7 months). (Photographs: A. Peiper [1963].)

Swimming movements can be released in infants that are a few weeks old by placing them into the water in a prone position and merely holding them up at their chin. They paddle in a coordinated fashion with their hands and legs. The behavior disappears at 3 to 4 months.

One can also release walking and crawling behavior in the newborn child. A newborn infant on its stomach will commence to perform crawling movements by moving the diagonally opposed limbs (Kreuzgang). If one supports the infant and places the feet onto a firm plane it will begin to walk and place one leg before the other (A. Peiper 1953; H. F. R. Prechtl 1955; see also Fig. 252). In addition, one can observe a number of movements in infants which serve to protect the body.

Figure 252. Walking newborn baby. (Photograph: A. Peiper.)

As examples of expressive behavior in newborn infants we can list crying and smiling. The first is a kind of "lost call"; a child can be easily quieted by picking it up or by imitating the presence of the mother by appropriate models (p. 206). The primary function of the smile seems to be to appease (p. 130). According to the legend, Cypselos, who later became the ruler of Corinth, was spared by those ordered to kill him while still a baby when he smiled at them. It is a fact that the smile releases delight in the mother, even those who initially were indifferent, and aids in the establishment of a strong emotional tie. The time of its first occurrence varies. Sometimes it can be observed in newborn or even prematurely born infants. The smile occurs spontaneously during sleep and also after drinking, diapering, and passing of wind (O. Koehler 1954a; J. A. Ambrose 1961). Laughing and cries of delight with a widely opened mouth mature around the fourth month of life.

Laughing as well as joyful shouts appear at a time where the laughing of adults does not facilitate the same behavior in the baby but startles it more than anything else, or can even cause the baby to cry when it has been laughing. The old imitation hypothesis does not hold up very well here (D. W. Ploog, 1964a:321).

By smiling and fixating the baby, unconsciously at first, strengthens the bond to the mother. K. S. Robson (1967) emphasized that the visual contact is at the base of human sociability. Mothers react very strongly when the child starts to look at them at approximately four to five weeks of age. It seems rewarding to them when their babies smile.[1] It is quite remarkable in this context that the visual fixating process occurs even in blind-born infants (p. 404).

During the first 3 months of life the children of deaf-mute parents are indistinguishable from those whose parents do not have this impediment (E. H. Lenneberg and others 1965). We have already mentioned that deaf-born children begin to babble (p. 114).

The initially spontaneous smile is later superseded by an answering smile. The former is frequently contrasted as a "grimace" to the later "genuine" smile. The latter is said to exist only when there exists a mutual relationship, that is, when the smile is a response to the smile of another person (A. Nitschke 1953). This type of distinction is a rather artificial break in a continuous maturational process. This can be clearly demonstrated in the answering smile, which matures quite independently of the mimic expressions of the

[1] "The human mother is subject to an extended, exceedingly trying and often unrewarding period of caring for her infant. Her neonate has a remarkably limited repertoire with which to sustain her. Indeed, his total helplessness, crying, elimination behavior and physical appearance, frequently elicit aversive reactions. Thus, in dealing with the human species, nature has been wise in making both eye-to-eye contact and the social smile, that often releases in these early months, behaviors that at this stage of development generally foster positive maternal feelings and a sense of payment for services rendered. . . . Hence, though a mother's response to these achievements may be an illusion, from an evolutionary point of view it is an illusion with survival value" (K. S. Robson 1967:15).

partner and becomes a personal greeting only very much later. R. A. Spitz and K. M. Wolf (1946) were able to release a smile in 3- to 6-month-old children by presenting them scarecrow faces and distorted grimaces as well as a normal human face. Within this wide spectrum everything was smiled at that was placed over the bed. R. Ahrens (1953) followed up the development of the recognition of mimic expressions. Until the onset of the second month eye-sized, well-defined, contrasting spots on a square or round two-dimensional place, representing a cardboard model of a head, release smiling better than a painted face or a rectangular bar on the same background. It makes no difference whether the pair of dots is presented in a parallel or vertical position or whether three pairs of dots are shown. One dot alone, on the other hand, is ineffective.

Around the second month of life dots presented in a horizontal plane in front of the infant's face are more effective than if they are presented vertically, and soon the child pays attention to the entire area around the eyes but not to the lower part of the face. This is included gradually toward the third month. At 4 months of age the child reacts to the movements of the mouth, without differentiating all details; it is not until the fifth month that the broadening of the mouth specifically releases smiling, and this is especially true for the 6-month-old child. The effectiveness of models then decreases. The child clearly distinguishes between models and faces of adults, but it does not understand the mimic expressions of smiling until it is 7 to 8 months old, when it reacts appropriately to a laughing person.

According to Ahrens the mimic expressions of the forehead were not adequately responded to until the children in the study were 14 months old. Vertical threat wrinkles frightened the institutionalized children; they turned away, ran off, and cried or screamed. Ahrens emphasized that the children had hardly ever seen threatening mimic expressions prior to these tests.

While the releasing mechanisms, expressive, and locomotor movements continue to mature and are increasingly integrated with individual experience, other behavior patterns become superimposed during the course of development and are exhibited only under special circumstances, for example, in brain-atrophy processes (p. 206). Some early-childhood behavior patterns are also taken over into the repertoire of expressive movements of adults (p. 122).

Behavior of children born blind or deaf-blind

With respect to the question of innate components in human behavior the behavior of children born blind or deaf-blind is most informative. We have here the accidental experiments of nature which can be assessed as deprivation (Kaspar-Hauser) experiments (p. 19). From this point of view J. Thompson (1941) studied the expressions of blind and blind-born children and compared them with those of seeing children. The results support and supplement the observations cited in the preceding section. Smiling, laughing, and crying, also the expressions of anger, pouting, fear, and sadness, looked the same in blind-born children, although they could not have imitated anyone. Blind-born children did, however, smile less as time went on in comparison with seeing children or those who had become blind later; no comparable decrease in crying was noted. In smiling a certain social feedback must play a role, which has yet to be investigated. When this feedback is missing the behavior atrophies somewhat. My own observations on a deaf-blind 7-year-old girl lead me to the conclusion that this stimulation is of a general nature. After the girl's teacher and sister intensively interacted with her and played with her, she laughed more frequently than before.

D. G. Freedman (1964) published a picture of a blind-born infant. She laughed when her mother spoke to him. The continuous nystagmus seen in the blind then ceased and the eyes fixated upon the source of the sound, although she could not see it (Fig. 253). This seems to be achieved with a central fixating process.

Figure 253. Two-month 20-day-old girl, blind from birth, smiling. Although the child cannot see, the eyes look, probably as the result of a central fixating process. Normally the eyes move about restlessly. (Photograph: D. G. Freedman [1964].)

Differences in the expressions of blind and seeing children are to be found not in the basic pattern, but in the additional occurrences of uncontrolled, superfluous "grimacing" movements in the blind, which was pointed out by G. Mackensen (1965) among others.

Of special interest is the behavior of deaf- and blind-born children, about which very little is known. These unfortunate children grow up in eternal night and silence. They have no means for imitation and education is very difficult. In spite of the lack of formal training these children nevertheless show a number of well-coordinated motor patterns. F. L. Goodenough (1932) reported that a 10-year-old deaf-blind-born girl, who grew up without instruction under conditions of poverty, was able to laugh heartily when she found her lost doll. She laughed also when she danced while standing on her toes, which she had learned by herself. When angry, she turned her head away, furrowed her brow, and pouted her lips. When she was very angry she threw back her head, shook it violently, and showed her clenched teeth.

As part of a still continuing investigation[1] I filmed the laughing and smiling of a 7-year-old deaf-blind-born girl and a 5-year-old boy who otherwise had no mental impairment. The motor patterns of laughing corresponded in all details to those of normal children. These two deaf-blind children threw back their heads during high-intensity laughing in a fashion typical for normal children and they also opened their mouths (Figs. 254A, B; 255). The rhythmic sounds are very clear, but their laughing is somewhat restrained, more like a giggle. The girl also showed a number of typical expressive movements, for example, crying. When angry she stomped with her feet. She rejects by shaking her head or by pushing away with her hand, when she also shakes the hand. If she stumbles she extends both hands forward. When taken on her caretaker's lap or shoulders, she liked to cuddle against him. This girl, who gave a very alert impression, especially when actively exploring her environment with her hands, is able to distinguish strange persons from familiar ones by sniffing briefly at the presented hand. Strangers are pushed away, a gesture that is often accompanied by turning the head away. This behavior is similar, with the exception of the sniffing, to that of healthy children. In short, a whole array of even quite complex behavior patterns, which are typical for human beings, have developed also in the deaf-blind and are therefore present as phylogenetic adaptations. Some characteristics of social behavior developed, even contrary to the educational efforts, such as, for example, the fear of strangers. Similarly, in a boy of the same institution who is approaching puberty now, certain aggressive inclinations developed and needed to be curbed by education.

The possible objection[2] that the deaf-blind child could have learned the complex movement coordinations of crying and laughing by a step-by-step reinforcement of these components can be answered as follows: If no phylogenetic adaptations existed,

[1] I wish to thank Karl Heinz Baaske, teacher for the deaf-blind at the State Institution for the Blind in Hannover, for his cooperation in making these observations possible.

[2] This objection was raised by R. L. Birdwhistell in a discussion in which I took part in Minneapolis.

Figure 254A. Deaf-blind 7-year-old girl, laughing. When fully laughing she throws her head back, opens her mouth, and laughs audibly, although restrained. (Photograph: I. Eibl-Eibesfeldt.)

Figure 254B. A deaf-blind girl (9 years old). Transition from smiling to weeping and finally to crying. This is the same girl as in Figure 254A, but she is older by now.

a large number of individual steps would be necessary to establish such complex movement coordinations. The mother would have to initially reward the child when she raised the corners of her mouth but not reward any other lip movement. She would have to shape the rhythmic vocalizations of laughing by step-by-step reinforcement of the appropriate inhaling and exhaling movements until finally the typical rhythm of typical laughing behavior were achieved. Again, different steps would have to be followed in the training of crying. The best Skinnerians would be confronted with great difficulties if they were asked to teach a higher mammal such movement coordinations. The deaf-blind children, on the other hand, are expected to have learned all this by the accidentally adequate behavior of the mother, although these children as a rule have great difficulty in learning such relatively simple skills as holding a spoon and bringing it to the mouth. A heavily brain damaged 12-year-old girl who was born deaf-blind in the same institution was able to laugh and cry. She clasps her caretaker and tries to climb up on him, whimpering, in the manner of a small child. But simple acts such as bringing a spoon to her mouth, she has been unable to learn despite the greatest efforts of teaching it. The hypothesis that the complex expressive movements of these children are learned thus lacks all foundation. If one still were to maintain this view, then one would have to assume special innate learning dispositions and would have to rely even more on the ethological concept of phylogenetic adaptation in behavior. The possibility that the deaf and blind-born children acquire the information about their mothers' facial expressions by touching her face and

Figure 255. Deaf-blind 5-year-3-month-old boy (Thalidomide child), laughing. (Photographs: I. Eibl-Eibesfeldt.)

consequently learning by imitation can also be excluded; I know a case of a deaf and blind-born boy who, in addition to this handicap, does not have arms to reach out but only very short stumps. Nevertheless he shows the basic facial expressions in a normal way.

A number of complex expressive behavior patterns, such as coquettish embarassment, cannot be seen in the deaf-blind. This may be due to lack of relevant experience or to the fact that the channels which usually receive such perceptions are closed in these children. That the latter is at least partially true can be deduced from the observation of blind-born children, whose mimic expressions are more highly differentiated than those of the deaf-blind. A 10-year-old blind-born girl showed embarrassed smiling with flushing, a lowering of the head, and incipient head rocking movements when I praised her for her performance at the piano. Those that are born blind have fewer expressive gestures but are otherwise most similar to seeing persons in their spontaneous expressions. They are, however, able to act out mimic expressions only in an incomplete form (F. Dumas 1932; M. N. Mistschenko 1933).

Some results
of the comparative method
in the study of human behavior

The observations on blind and deaf-blind people allow only limited statements to be made about human behavior. Such people lack the more complex behavior sequences which are normally released by visual and auditory signals. The question of if and how much in complex human behavior is inborn may be answered by the comparison of behavior in individuals of different cultures. If one can demonstrate communalities in expressions and gestures, then we may conclude that they derived from a common inherited root, especially the more specific the behavior patterns concerned are and the more widespread their occurrence in people of different ecology and cultural and racial background as well, and, again, because man is extremely inclined to culturally mold and change behavior in a relatively short time, as the evolution of language clearly demonstrates. This line of thought was expressed by C. Darwin (1872), and many present-day psychologists are aware of the basic agreement in mimic expression among different peoples:

As far as the accuracy of the data allow, the mimic expressions and pantomime which correspond to the described states, seem to appear in each people and race in a similar context or with the same meaning (perhaps one laughs without joy, but also because of it). It is true that important cultural differences exist in the expressive behavior, but this in no way detracts from the constancy of these primary expressions (N. H. Frijda, 1965:376).

S. Asch (1952) writes in a similar vein:

The findings of the ethnologists agree that there are basic expressions which occur without exception in all human societies. Cries of pain and of grief are universally distributed: When frightened one becomes pale and trembles; laughing and smiling are quite generally an expression of joy and happiness. It is probable that the agreements are even more encompassing and that reactions such as surprise, boredom, and astonishment are included. We may therefore speak of certain invariables in our emotional expressions, even though they have not been described in sufficient detail (p. 195, retranslated from the author's translation).

One is then quite surprised when one reads the statements of others, such as A. Gehlen (1956), who writes:

Inborn, instinctive behavior patterns are actually only demonstrable in very small children where they can hardly be distinguished from reflexes as sucking, grasping and holding-on movements. Otherwise and quite generally human motor patterns are bare of all instincts and this means according to the principle of complementarity: they are learned in their totality and concreteness in the way they are performed, they are built up individually through the integration of external stimuli and experiences.

R. L. Birdwhistell (1963, 1966) states that no expressive movement has any universal meaning, that they are all the product of culture and are not inborn. Such far-reaching generalizations are certainly unjustified. Even the comparison of the facial expressions of still photographs shows a widespread agreement, although a photograph captures only a short moment out of the entire behavior sequence. I will present a few pictures which show friendly smiles and laughing, rage and sadness in various peoples (Figs. 256–259).

a   b   c

Figure 256. Smiling people: (a) Schom-Pen man (Great-Nicobar Island); (b) Negro boy (near Bihamarulu, Tanzania); (c) Balinese (Sanur). (Photographs: I. Eibl-Eibesfeldt.)

a   b

Figure 257. (a) Kabuki actor showing rage (Tokyo); (b) expression of rage in a 4-year-old girl, whose motivation is probably jealousy. The father had photographed her sister repeatedly in her confirmation dress. Finally this little girl jumped forward and shouted "I want to have my picture taken too!" The child cries and is also very aggressive. Note the forward stance, which is an intention to attack, the clenched fists, and the expression of rage, especially around the corners of the mouth. (Photographs: (a) I. Eibl-Eibesfeldt; (b) from E. F. v. Eickstedt [1963].)

a

Figure 258. (a) Crying during great depression and bodily pain. A girl in Vietnam (near Don Xai), who lost her parents and was injured during the fighting in the city. (b) Indian baby, crying (Pisac, Peru). (Photographs: (a) Associated Press; (b) I. Eibl-Eibesfeldt.)

b

Figure 259. International language of facial expressions (Highland Indian woman from Pisac, Peru); (a) feeding her child; (b) smiling at the child, who replies in kind; (c) obviously in thought; (d) a little later smiling at her husband. (Photographs: I. Eibl-Eibesfeldt.)

The best proof, however, is motion picture film, but no relevant material is available. Insufficient documentation and hence a lack of knowledge are the causes for the statements cited above. N. H. Frijda says it very plainly:

Let us repeat, in a particular instance the explanation especially of facial expressions is lacking. Not only because of a failure to test hypotheses; but also because of insufficient or wrong hypotheses. Many of Darwin's explanations have been superseded or have always been somewhat far-fetched. But no one has since given his attention so conscientiously and in such detail to the facial expressions of man. Most people seem to be satisfied with global half-objective, half-interpretive descriptions, as Kirchhoff (1960) critically remarked. Especially: What work other than that of Darwin contains such detailed, theoretically unbiased descriptions of facial expressions as were actually observed in particular situations? We certainly take the easy way out (p. 399).

There are only a few isolated descriptive-analytic investigations of human facial expression. One example is a study by P. Lersch (1951), but there is hardly any comparative work. No one seems to have thought about documenting human behavior objectively, that is, through films taken without the subject's awareness. As unbelievable as it may seem, the ethogram of man has not yet been documented and registered in a way that would permit one scientist to examine data of another which are not already colored by the interpretation of the observer.

A search in the large film library of the Institute for Scientific Film in Göttingen, Germany, revealed numerous films about certain cultural activities, for example, the weaving of mats, making of pots, tilling of fields, building of boats and houses, dances, and so on. Almost always it is a staged activity (though this is frequently not even mentioned), which reduces the value of these films considerably. On the other hand, there are many film documents made in the field in which one can find material that is revelant to the kinds of questions we are asking, but they are always incidental to some other topic (S. R. Sorenson and D. C. Gajdusek 1966; R. Gardener and K. G. Heider 1968).

If one wants to know whether a Papuan, Bantu, Japanese, Italian stomps his foot when he is angry, one will search in vain in the film archives for unstaged films of people in rage. The same is true if one searches for comparative pictures of flirting, laughing, crying, and gestures of disdain and other expressive movements. This surprising discrepancy in the documentation of cultural activities, on the one hand, and of expressive behavior, on the other, is partially rooted in the historical development of psychology and ethnography. There are also certain methodological difficulties. Making of pots, weaving of mats, and cultural activities are readily performed for observers. The documents, then, do not actually reflect reality, but in general it is possible to capture the process. Facial expressions and gestures, on the other hand, must be recorded without the subject's awareness. Even a learned activity changes markedly when it is performed; this is even more true for emotional behavior. People are shy by nature and they do not like to be photographed. It is amazing over what great distances people perceive when a camera is pointing at them. Their behavior changes instantly. The facial expression becomes rigid, most people look restlessly toward the camera, smile in embarrassment or exaggerate, or overdo the behavior if they decide to continue it. Even in learned skills the smoothness of the performance is often lost.

A method that was developed by H. Hass and which we tested in various parts of the world in photographing people without their awareness overcomes all these difficulties (I. Eibl-Eibesfeldt and H. Hass 1959, 1967, 1968). An attachment that is mounted before the normal lens of the camera and contains a mirror prism makes it possible to film to the side (Fig. 260). With this technique it was possible to photograph people even from close distances without their awareness. They see, of course, that filming is going on, but the camera and the attention of the cameraman point into another direction, so they soon ignore it.

In these films we used the technique of time transformations (fast and slow motion) to make visible the lawfulness of the behavior sequences, which normally escape an observer. The method of time acceleration (1 to 7 frames per second) has so far not been used in the study of the behavior of higher vertebrates, and I would like to discuss its special advantages.

a   b

Figure 260. (a) Bolex camera with a mirror lens. A prism built into the attachment permits filming to the side. In the front the lens is covered by a mock lens. (b) H. Hass filming with the mirror lens at the marketplace in Arusha. (Photographs: I. Eibl-Eibesfeldt.)

First, by using the fast-motion technique one can obtain film protocols of behavior sequences of longer duration (for a small amount of film used per unit of time), which document the entire sequence. For example, if we are interested in the technique of making a pot we find the films taken by ethnologists unsatisfactory, in that they never show the entire sequence. Single episodes have always been selected: how the clay was brought, how it was kneaded, and how the bottom was shaped. Then we may see several stages of shaping, smoothing of the walls, and so on. The entire sequence is chopped up in this way and so includes the photographer's interpretation of what is important. What occurs between the cuts one learns, if at all, from the accompanying publication. This has been accepted as a necessary part of the methodology, as we can see from the writing of G. Spannaus (1961), who published guidelines for the preparation of ethnological films. He emphasized that an ethnographer has to depend upon a "representative" collection of complete movement sequences taken from the overall behavior because it is as a rule not possible to record the entire event, such as making a pot or a religious festival. "It is enough when all parts which occur once are recorded once, and those that are repeated are recorded once or twice. . . ." (G. Spannaus 1961:77). However, especially the example of making pots shows that we are dealing with a behavior sequence of a higher level of integration, which can be recorded in its entirety by means of the fast-motion technique.

If one selects the right number of frames per second, the movements run off quite rapidly, but each individual action remains clearly visible. One can see how the product grows under the shaping hands and one can count later how many individual movements were necessary to produce the particular pot or vase or whatever, that is, what amount of effort went into its making. This in turn makes it possible to compute the advantage of one technique over another with respect to their relative efficiency. With comparative fast-motion film records of, for example, tilling of a field, one can determine the amount of work involved in the different techniques. The comparison of fast-motion films of trained and untrained workers can also be very useful.

Of course the behavior patterns that belong to a lower level of integration, from which the more complex behavior sequence is made up, should also be recorded. These individual movements we film in slow motion and without the subjects' awareness.

The fast-motion technique opens up new paths for the documentation and analysis to the ethnographer. One may think of the investigation of religious ceremonies and other rites, which can now be recorded in their entirety. An event that takes half an hour can be recorded with 5 minutes of film at 4 frames per second. If later someone should become interested in the study of cultural ritualization, then fast-motion films, for example, of a Catholic mass, would be most useful. If this event were filmed at regular time intervals, one could see the changes directly from the films.

In addition to these uses, the fast-motion film technique can make visible certain regularities in behavior which normally escape direct observation. A newspaper seller whom H. Hass filmed in Vienna proved to be a most rewarding subject for demonstrating the value of this technique. When he was observed normally nothing unusual was detected about his behavior. The fast-motion technique revealed, however, that the man ran back and forth within a 1.5-meter-wide part of the wall, which was bordered on either side by a large display window. His behavior was so stereotyped that it appeared as if he were tied to this small spot before the walled portion of the building. It is possible that this is an indication of an inborn tendency to keep oneself covered from behind.

Pictures from an elevated position show that people approach conspicuous landmarks, for example, flagpoles, without any special reason, perhaps based on an innate orienting mechanism. This happens even if they must deviate from a more direct path to their goal to accomplish it.

During the analysis of fast-motion film of persons eating we noticed that individuals who ate alone looked up and around into the distance after each bite or two; their gaze often swept automatically to the sides as if scanning the horizon. Baboons and chimpanzees show the same behavior. This seems to be an alert behavior against enemies, a behavior that is also a phylogenetically inherited part of man, although today there is very little danger for man when he is eating.

The fast-motion technique is of special value in the investigation of people in smaller or larger groups. Normally even the observation of two persons is quite difficult, because the behavior of both cannot be recorded simultaneously. It is even more difficult when still more people are involved, as in a large family. On the other hand, if we have a fast-motion protocol we can view the film as often as is necessary and we can recognize how the behavior patterns of individuals are attuned to one another and how they are grouped. With this technique we have filmed playing children, mothers with children, couples, and larger congregations of people and the behavior of people in cultural ceremonies. Other experiments indicate that the fast-motion technique can be also used successfully for the investigation of human and animal behavior in other situations. I may mention also that for fast-motion filming one can mount a camera at an elevated place, such as the roof of a car, and let it run without the photographer nearby. To allay all suspicions we worked with the lens to the side even under these circumstances, although no one suspected that the camera standing by itself was operating.

When we want to analyze facial expressions and gestures we film the subject in slow motion (48 frames per second) without his awareness. For later analysis it is especially important to have a record of what the subject did just prior to and after the filming. We strive, therefore, to understand the behavior within the context of the situation and the sequence in which it takes place, in the same way as it is necessary in motivational analysis of animal behavior to avoid subjective interpretations afterward. In some cases it is possible to cause the releasing stimulus situation. When we were filming, curious onlookers would often gather around us, and we experimented with them without their being aware of it. By handing such a person a small box out of which popped a cloth snake when it was opened, we obtained the unrehearsed expression of fright. By casually looking at a person one can release greeting and sometimes even flirting behavior.

Until now we have primarily filmed in Europe, Kenya, Tanzania, Uganda, India, Siam, Bali, Hong Kong, New Guinea, Japan, Samoa, the United States, Mexico, Peru, and Brazil. We were especially interested in mother–child behavior, flirting, greeting, praying, and begging behavior, and we obtained many film records without the subjects' awareness. In countries with a highly developed theater culture (Japan, Thailand, Europe) we had the actors act out certain expressions according to a prepared list from us which we filmed with 48 frames per second. These scenes allow us to compare the natural expressions with the actor's portrayal of them.

Although the work is still in progress, we have filmed enough to say that some of the more complex human expressions can be traced back to the superposition of a few fixed action patterns which do not seem to be culturally determined. To give just one example, we found agreement in the smallest detail in the flirting behavior of girls from Samoa, Papua, France, Japan, Africa (Turcana and other Nilotohamite tribes) and South American Indians (Waika, Orinoko).

The flirting girl at first smiles at the person to whom it is directed and lifts her eyebrows with a quick, jerky movement upward so that the eye slit is briefly enlarged. This most probably inborn greeting with the eyes is quite typical (Figs. 261–264). Flirting men show the same movement of the eyebrow, which can also be observed during a friendly greeting between members of the same sex. After this initial, obvious turning toward the person, in the flirt there follows a turning away. The head is turned to the side, sometimes bent toward the ground, the gaze is lowered, and the eyelids are dropped (Fig.

Figure 261. Simultaneous ambivalence in flirting behavior. A 10-year-old flirting Samburu girl turning her eyes toward the observer while showing an intention to turn away. [Photograph: I. Eibl-Eibesfeldt (from 16-mm movie film, near Maralal, Kenya).]

a

b

c

d

Figure 262. Flirting Turkana woman (Lorukumu, Kenya) as an example of successive ambivalence. She makes contact with the eyes, laughs, lowers the head (in embarrassment?) and eyelids, and repeats full contact with the eyes. From a slow-motion sequence (48 frames per second). The entire sequence a–d includes 6.04 seconds (290 frames). (b) Frame 40 [0.83 second after (a)] and (c) is frame 177 [3.68 seconds after (a)]. (Photographs; H. Hass.)

a

b

c

d

Figure 263. Greeting with the eyes by a French woman: (a) neutral face, (b) and (c) raising the eyebrows; (d) a smile. Six frames after the first picture one notices a slight raising of the eyebrows. Slow motion at 48 frames per second. The sequence a–c includes 41 frames (0.87 second). Between frames 19 and 26 the eyes are maximally raised. Picture (b) shows frame 23 [0.047 after (a)]. Picture (c) is frame 41. The entire sequence of raising the brows and lowering them again includes 18 frames (0.37 second), and they are raised maximally only for 7 frames. [Photographs: H. Hass (16 mm movie film).]

a

b

c

d

Figure 264. Greeting with the eyes by a flirting girl from Samoa, from a slow motion film (48 frames per second). (a) Neutral face; (b) smiling at the partner (frame 41); (c) sudden raising of the eyebrows (frame 107). The entire sequence runs very fast. The eyebrows are clearly raised for only $\frac{1}{6}$ second. [Photographs: H. Hass (16 mm movie film).]

262). Frequently, but not always, the girl may cover her face with a hand and she may laugh or smile in embarrassment. She continues to look at the partner out of the corners of her eyes and sometimes vacillates between looking at and an embarrassed looking away. We were able to elicit this behavior when girls observed us during our filming. While one of us operated the camera the other would nod toward the girl and smile.

Here we already find that the superposition of a few invariable components (intention movements of turning toward someone, responsiveness, and turning away) yields a relatively complex and variable expression (see also p. 92). The assertion of R. L. Birdwhistell (1963, 1966) that there are no culturally independent expressions and that everything is learned is disproved by these results.

The comparative investigation of greeting behavior in people from different cultures showed additional agreement, even in the smallest details. During a friendly greeting over a distance the greeting person smiles, and if he is in an especially good mood will lift the eyebrows in the manner previously described, and then nod the head (Fig. 265). I filmed this behavior even in stone age Papuans, who had only recently come into contact with government patrols (I. Eibl-Eibesfeldt 1968).

Wide agreement is also found in many other expressions. Thus arrogance and disdain are expressed by an upright posture, raising of the head, moving back, looking down, closed lips, exhaling through the nose—in other words through ritualized movements of turning away and rejection. When enraged, people bare their teeth at the corners of the mouth.

With respect to gestures one also finds many agreements among peoples of different cultures. Bowing everywhere seems to be a gesture of submission, for example, during greeting or if one approaches a high-ranking person or in praying (Fig. 267) (T. Ohm 1948). Differences apply only to the extent; we may nod, while a Japanese bows very low. In triumph and when we are enthusiastic we throw up the arms (Fig. 268). Members of the most varied cultures greet by raising the open hand (Fig. 269). If one man wants to impress another—to display—it is again done quite similarly in different peoples by an erect posture, mean facial expression, and frequently with an artificial enhancement of the body size and width of the shoulders (p. 434). The only difference is in the means to achieve this expression in the various cultures. Some men place feathered crowns on their heads, others fur caps made of bear hide, another displays with weapons and colorful dress—the principle remains the same. When we are angry we become indignant, that is, we jump up into an intention movement for attack, make fists, and may even bang the table, which is a redirected attack behavior (p. 180). When angry

Figure 265. Eyebrow-flash during greeting. Upper row: Balinese of the Island Nusa Penida. Sequence a–d lasts 19 frames; (b) shows frame 6, (c) frame 11. With frame 6 raising of the eyebrows set in and they were maximally raised in frame 11. The lowering movement began with frame 17. Middle row: Papua, Huri tribe near Tari (New Guinea). Sequence a–d lasts 45 frames; (b) shows frame 30, (c) frame 36. Twenty-six frames after he had started to smile he began to lift the eyebrows and 4 frames later they were maximally raised and held so over 7 frames. Lower row: Papua, Woitapmin tribe near Bimin (New Guinea). Sequence a–d lasts 85 frames; (b) shows frame 75, (c) frame 79. At frame 16 he starts to smile and at frame 75 raises the eyebrows. They are maximally raised during frames 78 and 84. All photographs are reproduced from 16-mm taken at 48 frames per second. (Photographs: I. Eibl-Eibesfeldt.)

Figure 266. Greeting with the eyes by a Waika (upper Orinoco) man. The pictures were taken from a slow motion (16-mm, 48 fps) film and show the 1st, 15th, 33rd and 76th frame of a sequence. (Photograph: I. Eibl-Eibesfeldt.)

a

b

Figure 267. (a) A praying Japanese woman in the temple; (b) Portuguese before the King of the Congo. (b) from Ordoado Lopez [1597].)

Figure 268. Throwing up the hands in triumph: (a) An enthusiastic German sports fan during the world championship (1966) soccer games in England, when Germany was playing Spain, after the first German goal; (b) exuberant joy expressed at the carnival in Rio de Janeiro. (Photographs: (a) Associated Press; (b) I. Eibl-Eibesfeldt.)

Figure 269. Left: A Schom Pen (Great Nicobar Island) greets by raising the open hand; right: a woman from Karamojo (Africa) greeting in the same manner. (Photographs: I. Eibl-Eibesfeldt.)

we may stomp with a foot, an intention of attack which among Europeans is especially found in small, uncontrolled children; adults usually suppress it. I saw the same gesture in any angry Bantu boy. It remains to be investigated to what degree the gestures of approval or disapproval have an innate basis. Many races indicate a general "no" by shaking the head, closing the mouth, some by showing the tongue (ritualized spitting out; see Fig. 270), and they say "yes" by nodding their heads. Darwin points out that the first act of saying "no" (disapproving) in children is the rejection of food, by turning the head to the side from the breast or a spoon. One could think of a shaking-off movement.

a   b

Figure 270. Showing the tongue, a widespread gesture of disdainful rejection: (a) a Negro boy from the Bihamarulu region (Tanzania); (b) Warriors from New Guinea ridicule one another. (Photographs: (a) I. Eibl-Eibesfeldt; (b) from J. Cook [1784.])

The blind and deaf girl who was discussed earlier (p. 405) shook her head when she did not want to eat, and also when she refused something, for example, an invitation to play. These facts—that people also say "no" with different gestures, for example, a Sicilian by laying back his head—does not argue against Darwin's interpretation. We know that innate behavior patterns can be suppressed by training. One would like to know whether a rejecting shaking of the head is also used, for example, in Sicilian children.

It is possible that several primary forms of saying "no" exist, such as rejection or disapproval, and that people in different cultures accept one or the other by convention. One movement of rejection can be traced from the intention of turning away. In saying "no" a Greek, for example, lifts his head with a jerk backward, at the same time lowering his eyelids and often raising one or both hands with the open palms showing to the opponent. This behavior can be observed in northern Europe as a gesture of emphatic refusal ("for heaven's sake"). It is also very similar to the posture of arrogance. Sometimes instead of lifting the head backward we can observe a turning-to-the-side movement. Another widespread gesture of refusal or no is head shaking, and sometimes one can observe a rejecting form of shaking the hand, which may be a ritualized shaking off.

Nodding was derived according to Darwin from an intention movement to eat. Another possible interpretation is given by H. Hass (1968), who says that nodding could be taken as an intention movement to bow, as a ritualized gesture of submission, so to speak. When expressing agreement one does submit to the will of another. Much is to be said for this interpretation. Nodding is a widespread gesture of approval. Papuans nod and so do Waika Indians or Bantu. Like others, many Indians and Ceylonese also nod when stating, "Yes, this is so." However, when expressing their agreement to do something they were asked to do, they sway their head in a peculiar sideways movement. If one asks a Ceylonese, "Do you drink coffee?" he will nod upon confirmation. If we address him, however, saying, "Let us drink coffee," he sways his head in agreement. I have seen no nodding in the deaf and blind-born so far.

If the accounts are correct, the kiss is not found everywhere. In spite of this, however, one might think of it as a kind of ritualized feeding derived from the care-of-young behavior system which has been taken over as one of the expressions of tenderness (p. 106). In this connection the accounts of L. v. Hörmann (1912) are of special interest, in reference to the behavior of the inhabitants of the Hinterzillertal (mountain valley in Austria). It is the custom there to chew pine resin, which gradually changes into a viscous mass that is no longer sticky and is changed from one cheek into the other and sometimes is visible from the corners of the mouth (Flenken). "When chewing pitch the same custom, that of mutual exchanging of the wad prevails, as is also done with chewing tobacco. Among lovers this exchange plays an important role" (L. v. Hörmann 1912:99).[3] The boy exposes a piece of pitch from between his teeth and invites the girl to pull it out with her teeth, an attempt that the boys try to prolong as a kind of love play. When the dancing partner responds to this invitation of the boy, it is a sign of her interest and affection and even more.

Passing of food as a gesture of contact readiness can also be observed in small children. I recently observed this in a 3-year-old girl who was a guest in our house for the first time. The child observed her parents, who were engaged in friendly conversation with us, but she was at first shy. After lunch while we were drinking coffee the

---

[3] Hörmann previously reported that in the Zillertal, Pustertal, and Pinzgau (regions of Austria) the exchange of chewing tobacco is an expression of friendship between men. Acceptance of a chewed piece of tobacco by a girl is proof of love returned.

girl suddenly came up, took a cookie from a plate, and gave it to me smiling somewhat embarrassedly. She repeated this with an obvious flirtatious behavior and was happy when I accepted and ate the cookie. From then on she felt completely at ease.

This gesture appeases even those who are enemies, as an acquaintance of mine experienced during the war. He had been ordered to capture a prisoner from an enemy trench to obtain information from him, an act that he had carried out successfully on previous occasions. When he jumped into the trench with a drawn pistol and pointed it at the enemy soldier, the soldier, scared as he was, held out his hand with a piece of bread in it. This gesture so changed the mood of my acquaintance that he was unable to carry out his task and withdrew. After that he was unable to carry out similar missions. The food industry uses the function of forming bonds by means of food and drink in its advertising (Fig. 271).

Figure 271. Appeal to the group-uniting function of feeding in advertisements for food products.

In the cases described last, the agreement lies in the principle, not in the formal pattern of movements. The motor patterns are not innate but certain inclinations are. It remains to be ascertained whether these are caused by innate releasing mechanisms or by specific drives.

Not only the comparison of people from different cultures (p. 416), but also comparison with animals can be very revealing. In addition to true homologies, there are many analogies (p. 187), which were discussed in earlier chapters. Chimpanzees show a smile that is quite similar to the human smile (N. Kohts 1935; see also Fig. 272). Many similarities can be found in greeting behavior: We mentioned the eyebrow flash, nodding, kissing, clasping, and giving hands (p. 420). The form of greeting that is found among many peoples, the rubbing of noses, is probably not derived from the kiss but has

Figure 272. Smiling and laughing in chimpanzee. (After pictures by N. Kohts [1935].)

another origin. When Bali lovers greet each other in this way and they breathe in deeply, it is a kind of friendly sniffing. The sense of smell does play a larger role in the social relations between people than is generally realized. In the German language one speaks of not being able to stand another's odor (mann kann jemanden nicht riechen) when one cannot stand another person. T. Schultze-Westrum (1968) discovered a quote from K. Nevermann which says that among the Kanum-irebe tribe of southern New Guinea it is an expression of close friendship when one takes something of the odor of the person who leaves. The person who remains reaches under the arm pits of the one who is leaving, smells the hand, and rubs the odor over himself (Fig. 273).

It has frequently been stated that in cultures in which tenderness is expressed by rubbing noses, no kissing exists, for example, the Papuans, Polynesians, Indonesians, and Eskimos. This statement is based, however, on incomplete observation. In the first three cultural regions I observed that mothers hugged and kissed their children, even among stone age Papuans of a remote Kukukuku village who had only 7 months prior to my visit had their first brief contact

Figure 273. The author is greeted by a Waika (upper Orinoco) man with embracing and kissing. The Waika man took the initiative.

with a government patrol (I. Eibl-Eibesfeldt 1968). It is very unlikely that these mothers learned this behavior from the patrol members. In the same Papuans I also saw a father kiss his son on the cheek when he greeted him.

Of old inheritance is also our threat posture, which is expressed by rolling our arms inward in the shoulders, and during which the hair erectors on the shoulders and back contract, although we no longer have any fur. We experience this contraction only as a shudder. In chimpanzees, which assume the same posture, the hair becomes erect and their outline is enlarged (K. Lorenz 1943).

In response to strong acoustic stimuli people raise their shoulders, bend their head slightly foreward, and close their eyes. Reactions homologous to this "neck-shoulder reaction" are known from other mammals (P. Spindler 1958).

A very curious display behavior of many primates, including man, was pointed out by D. W. Ploog and others (1963) and W. Wickler (1966c). Squirrel monkeys display against conspecifics by presenting the erect penis when they meet. Even young animals show this. In the common marmoset (*Callithrix jacchus*) males display in defense of their families by raising their tail and exposing the rear to the opponent. The testes are pressed into the scrotum, an erection takes place, and the males urinate. After this display, they retreat to a marking place and mark it with urine. During the threat display they look back to the opponent. The posture of showing the rear to the viewer is probably explained by the flight motivation of the animal. Females display in a similar posture, and not knowing the behavior of the males one could be misled to assume that the posture derived from a female sexual presentation. This is not the case, however. The females imitate the male posturing.

Vervet monkeys, baboons, and many other monkeys have been observed where several males sit at the periphery of their group "on guard." It was believed that they were watching out for predators. But this precisely they do not do. Instead they slink away as inconspicuously as possible in such cases. W. Wickler has discovered that this behavior is directed against neighboring troops. The "guards" always sit with their backs to their own group and display their male genitals prominently, which in these animals are very conspicuously colored (Plate VIII and Fig. 274). When a strange conspecific approaches the penis becomes erect and in some species it is moved rhythmically. This behavior is a display that serves to mark the territory. Interestingly enough, the same behavior could be demonstrated in man. Some Papuan tribes emphasize their masculinity by artificial means (Fig. 275). In some male dresses of Europe this region is still emphasized today by decorative embroideries.

On the Nicobar Islands and on Bali I saw fetishes with an erect

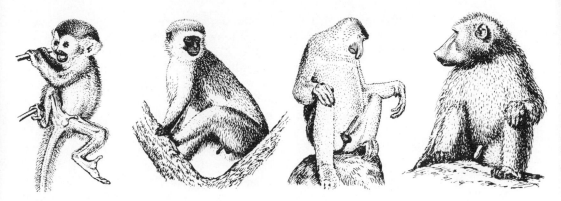

Figure 274. Genital displays of male primates. Left to right: squirrel monkey (*Saimiri*), vervet monkey (*Cercopithecus*), proboscis monkey (*Nasalis*), and baboon (*Papio*). (From W. Wickler [1966c].)

Figure 275. Genital displays in man. Left: two Papuans from Kogume on the Konca River to the right: Herme of Siphnos (490 B.C.), 66 cm high, Athens, National Museum; right: house guardian (Siraha) of the natives on the island of Nias. The man-high figures are still in use. In the Greek statue the beard is emphasized as a male symbol; in beardless peoples the male head ornaments are emphasized. (From W. Wickler [1966c].)

penis which are used to ward off ghosts (I. Eibl-Eibesfeldt and W. Wickler 1968). W. Wickler called attention to stone columns in ancient Greece with a man's head and a penis that were used as property markers. Phallic "guardians" carved in wood or stone can be discovered in Romanesque churches (in Lorch, West Germany, and St. Remy, France). In modern Japan phallic amulets are still used, for example, to protect against car accidents (Fig. 276). In the

Museum of Linz (Austria) one finds amulets that depict male sexual organs. It is possible that pathological exhibitionism can be traced back to a drive to display (Fig. 276). This hypothesis is supported by the observations of J. H. Schultz (1966). The sitting position of men differs clearly from that of women and is reminiscent of that of the monkeys we discussed (G. H. Hewes 1957).

Figure 276. Two amulets which the author acquired in Japan (Tagata Temple) that are intended to protect the wearer. One amulet shows a threatening face on its front. By removing a cover on the back a golden penis becomes visible. The cover is inscribed with the words: "To protect against traffic accidents." The threatening face and phallus are repeatedly found elements in figures which are meant to offer protection against demons. (See also I. Eibl-Eibesfeldt and W. Wickler [1968].)

W. Wickler derives the genital displays of primates from urine ceremonies which contain elements of copulation behavior (p. 97). In many mammals males mount conspecifics of the same sex during an aggressive assertion of rank. R. Schenkel (1947) described it in wolves. I. Eibl-Eibesfeldt (1950) in house mice, S. Zuckermann (1932) in baboons, and C. B. Koford (1963) in rhesus monkeys. In rhesus monkeys "rage copulations" were also observed during the course of aggressive conflicts, where the aggressively aroused individuals often mount a third one who is not involved (C. B. Koford 1963). It should be examined to what degree this occurs in humans. In a recent Polish novel, *The Painted Bird*,[4] I found the reference that young herdsmen rape strangers who enter their territory. A. P. Wilson (personal communication) told me that in some prisons in the United States new prisoners are occasionally beaten up by the other inmates. If the person does not fight back like a man, he is treated like a girl and is raped. In short, mounting, in many primates, possibly including man, is a demonstration of rank of an aggressive nature. It appears warranted to me to interpret the

[4] J. Kosinski, 1966, New York (Pocket Books).

"sitting on guard" of many primates as a further ritualization of this behavior (the threat to mount).

Releasing mechanisms, key stimuli, and releasers in man

The experiments carried out on a large scale by industry and the arts using various models show, just as certain miscarriages of our esthetic and ethical value judgments, that we react almost automatically to certain releasing stimulus situations in a predictable manner. This is likely to be on an innate basis, although a definite proof cannot be obtained because persons who have grown up without experience are not known. We know, however, from infants that they react innately by smiling to certain stimulus configurations (p. 403). R. L. Fantz (1967) furthermore showed that children already preferred a schematic representation of a human face over an array of other stimuli during their first week of life.

K. Lorenz (1943) stated that the behavior patterns of caring for young and the affective responses which a person experiences when confronted with a human child are probably released on an innate basis by a number of cues that characterize infants. Specifically the following characteristics are involved:
1. Head large in proportion to the body.
2. Protruding forehead large in proportion to the size of the rest of the face.
3. Large eyes below the midline of the total head.
4. Short, thick extremities.
5. Rounded body shape.
6. Soft-elastic body surfaces.
7. Round, protruding cheeks, which are probably genuine differentiations with a signal function. Sometimes it is said that in the corpus adiposum buccae we have a mechanical reinforcement of the sides of the mouth to aid in sucking, but this is not proved. Such an additional function is feasible, of course, but we notice that monkeys and other mammals can get along without this formation. This argues for a specifically human organ that evolved in the service of signaling.

These physical attributes are further enhanced by behavioral ones such as clumsiness. When an object possesses some of these characteristics it releases in children some typical affects and behavior patterns. We find these objects "cute" and may want to pick them up—to cuddle them. B. Hückstedt (1965) demonstrated experimentally that the rounded forehead and the relatively large brain case are important characteristics of "cuteness" which can be exaggerated in an experiment. The doll and film industry utilizes this possibility and constructs "supernormal" (p. 82) models to elicit behavior of caring for young. Animals are also considered cute if they have some of the child characteristics (Figs. 277 and 278). To be considered cute it is enough that the parakeet has a round head and that a young dog is clumsy and has feet much too big for his body. In Pekinese dogs breeders seem to have produced a perfect substitute object for the unfulfilled mothering reaction of older ladies. By offering cats to experimental persons of different ages P. Spindler (1961) released the typical patterns of caring for. The reactions (affection, euphory, patting, bending down the head, talking pet names in a high-pitched voice) mature at an age of 3.

It is also possible that the understanding of expressions is given a priori by innate releasing mechanisms, because we are easily de-

Figure 277. "Baby schema" of man. Left: head proportions that are generally considered to be "cute"; right: adult forms, which do not activate the drive to care for the young (brood care). (From K. Lorenz [1943].)

Figure 278. Left: overemphasizing the baby characteristics, from the December 1966 issue of *Ladies' Home Journal;* above: Disney dog as an example of the exaggeration of baby characteristics. Note the rounded shape and the head—body relation.

ceived by simple models. A crying or laughing face can be depicted with a few strokes.[5] When we see such expressions in animals we consider them friendly (Mandarin ducks), arrogant (camel), or daring (eagle), although this has nothing to do with the actual mood of the particular animal (Fig. 279). Finally, the automatic reactions to the

Figure 279. Many people misunderstand a camel's expression. Man has an innate releasing mechanism which responds to the relative position of the camel's eyes to his nose; only in man does this mean an "arrogant turning away." We therefore consider the camel to be an aloof animal. In the eagle the bony ridge above the eyes is seen as a wrinkling of the forehead. Together with the pulled-back corners of the mouth the expression is one of "proud decisiveness." (From K. Lorenz [1965a].)

[5] When we recognize the expression anyway it could be the result of a secondary process of abstraction, just as we recognized the caricature of a person. This needs to be tested in very small children. If it could be shown that at a certain age the sketchy representation is more effective than the natural object, one could suspect an innate releasing mechanism, because then the abstraction is available before any learning. Experiments are also needed on the blind-born who later had their vision restored.

expressions of another person argue for innate releasing mechanisms that determine a response to an expression. We already mentioned the disarming smile.

That we can respond innately to human expressions has been doubted, primarily on the basis of experiment. Subjects were presented with enlargements made from a film and other photographs of facial expressions for their evaluation (for example, B. M. Turhan 1960). The subjects judged the pictures quite differently, something that should not be surprising. Expressions are sequential structures. If one wants to examine their releasing effects, one must present the film sequences to the subjects and not present only single frames. When a biologist wants to test the releasing function of a certain bird song he does not present only a single tone. Undoubtedly there are static expressions that can be recognized even on a still photograph, but usually the releasing effect comes about as a result of the entire sequence.

P. Ekman and others (1969) found that observers in both literate and preliterate cultures (New Guinea, Borneo, United States, Brazil, and Japan) chose the predicted emotion for photographs of the face. The association between facial muscular movements and discrete primary emotions is evidently the same crossculturally.

Eye spots primarily excite the attention of an observer. R. G. Coss (1965, 1968) measured the pupil reactions of persons who viewed eye spots that were presented singly, pairs, or a group of three, each consisting of concentric circles imitating pupil and iris. The strongest responses were obtained to the paired spots, and the responses were stronger when the inside of the eye spot was dark. The response to double pairs of eye spots varied, depending upon their spacing. Horizontal spacing caused a stronger dilation of the pupils than a vertical one.

We judge our fellowmen on the basis of information which is most likely also inborn. The wide agreement of certain male and female ideals of beauty among people of different cultures points in this direction, as well as the fact that exaggeration of individual characteristics on models is so effective. In men broad shoulders are desirable, and rarely will we find a hero in art or literature who has narrow shoulders. The width of the shoulders in relation to the narrow hips is very effective, although it may be tremendously overdone, as is the case with Greek vases and statues (Fig. 280). The shoulders are also frequently emphasized by men through clothing. Furthermore, we value long limbs as well as slenderness as noble and against all reason consider gazelles and other animals with such characteristics as noble, whereas the plump hippopotamus is considered the opposite, although gazelle and hippopotamus each are perfect adaptations to a particular ecological niche. The female

Figure 280. Exaggerations of the width of the shoulders in artistic male representations. (a) early Greek bronze figure: Apollo, 20 cm high, formerly in the collection of Tyskiewicz, Boston Museum of Fine Arts; (b) Dionysos of A. Breeker; (c) Kabuki actor. (Photograph: I. Eibl-Eibesfeldt.)

ideal of beauty seems to consist of characteristics that may be illustrated, according to K. Lorenz (1943), by the shape of the classical Venus and the prehistoric Venus of Willendorf (Fig. 281). To this day people exist whose ideal of beauty corresponds to that of the prehistoric Venus in the main outlines, which is characterized, among other things, by a conspicuous steatopygie (fat deposit), a character displayed to this day by Hottentot women.

The secondary sexual characteristics of the woman are immediate indicators of the hormonal sexual functions, and Schopenhauer referred to this fact in his *Metaphysik der Geschlechtsliebe* (*Metaphysic of Sexual Love*):

A full bosom in a woman has a powerful attraction to the male sex, because it promises, in correspondence with the ability to propagate in the woman, abundant food for the newborn. On the other hand, excessively fat women release a feeling of disgust, the reason being that these attributes indicate an atrophy of the uterus, that is, infertility, which is not known by the head but by instinct.

Figure 281. (a) Venus of Willendorf, limestone carving from Aurignacia, Museum of Natural History of Vienna, Prehistoric Section; (b) Aphrodite of Cyrene, Therme Museum, Rome.

a   b

Some of the important components that are indicators of normal sexual function in a woman are a slender waistline (lacking heavy adipose deposits), red cheeks and lips, as well as perhaps the shape of the pubic-hair region. This is not a complete list nor does it apply to members of all races necessarily. We know of some of these characteristics that they are exaggerated in art and are emphasized by fashion. Fashion designers improve the releasing effect of the female breast by the use of padding, and in the previous century the buttock region was especially enlarged (cul de Paris).

D. Morris (1968) interprets the breasts and lips of women as sexual signals that are projected to the front. Our apelike ancestors, so argues Morris, mated by mounting from behind and reacted to releasers visible from behind (fleshy buttocks, red labia). Walking erect led to a redirection of the copulatory position and it became necessary to develop sexual releasers on the front of the body. According to Morris this came about by the evolution of copies of the labia as lips and buttocks as bosom on the front. The pulled-up breast of a movie star may evoke such associations, but a normal breast is just as dissimilar from buttocks as lips are from the labia. Morris also overlooks the fact that men also have red lips. His thesis is hard to support, especially because more likely interpretations exist for the development of these releasers. In higher vertebrates behavior patterns of care for young have a calming effect, for example, cuddling, feeding, clasping, and social grooming. Frightened young mammals run toward their mother to nurse, and children can be calmed with a pacifier. In many higher mammals behavior patterns of caring for young have been taken over as precopulatory behavior, in man among other things sucking behavior (not to be confused with kissing, which is a form of ritualized feeding). In connection with this the offering of the breast as a female contact gesture could have been taken over into the sexual domain and in this way the breasts acquired their specific releasing function (see also W. Wickler 1968). The lips in turn acquired their signal function as a result of their role in kissing. Kissing is a mutual activity, so both sexes evolved lips with signal functions.

The advertising industry uses our readiness to respond to sexual releasers to attract our attention and to direct it to the actual message (Fig. 282).

Figure 282. "The legs of your car." Tire advertisement of Pirelli, which utilizes the attention-getting quality of sexual releasers.

There is much evidence that even our esthetic judgments about the external environment beyond members of our own species are influenced by innate releasing mechanisms. The paintings of chimpanzees (p. 248) indicate that certain basic esthetic perceptions are already present in animals. This has been demonstrated by B. Rensch (1957, 1958, 1961) for other apes or monkeys and for some birds (carrion crows and jackdaws), who preferred regular geometric patterns over irregular ones.

We know only very little about releasing stimuli of other senses. With respect to odors we cited the investigations of J. LeMagnens (p. 69), who found that girls and women can smell musk substances when they are sexually mature which men cannot perceive, unless they have received estrogen injections. R. v. Krafft-Ebing (1924) reported the case of a young man who sexually aroused peasant girls by wiping their perspiring brows after a dance with a handkerchief that he had carried in his armpit. In Mediterranean countries forms of dancing exist where men dance around their female partners while waving a kerchief. It is said that in some areas it has also been carried previously in their armpits. It seems likely that certain pleasant as well as disgusting odors are reacted to on a primary basis; the same seems true for certain taste perceptions, although occasionally the key stimuli can be falsified, witness our reaction to saccharin. We seem to prefer sweet-tasting food. Normally such substances are rich in carbohydrates and hence in calories.

Certain perceptions of tactile stimuli release specific defensive reactions, "creeping things" on the back of the hand release a shaking movement of the hand, which K. Lorenz interprets as a defense against insects. A protection against damage of the teeth is prevented by a reaction to acoustical stimuli. Sharp, screeching noises release this reaction whether we bite on a hard object or scrape with a knife on a plate, which in some persons is felt as pain and projected into the teeth. The reaction consists of pulling the cheeks between the lips and performing cleaning movements with the tongue.

Acoustical releasing stimuli have been little investigated. The crying of babies, the sobbing of another person, the desperate cry for help of a child or a woman move and alarm us so strongly that one may suspect an innate basis. A pilot study by E. H. Hess and B. Beck (1967, personal communication) supports this hypothesis. Using the pupillometric technique which Hess developed, they presented male and female adult subjects with tape-recorded sounds of a baby (1) crying in pain, (2) crying in hunger, and (3) babbling. The greatest positive pupil reaction was recorded to the babbling. Negative pupil responses were recorded for the two types of crying. The reactions were not uniform in all subjects; that is, some showed clearer distinctions than others. Sex differences were not clear cut in the small sample of the study, and further work is being done. Undoubtedly the internal state of the subjects at the time of the study and their marital and parental status could have an effect on the responses. Soft or harsh conversation can also be recognized without specific knowledge of the particular language.

A basis of inborn releasing mechanisms may account for the effectiveness and appeal of the highest artistic expressions. In music rhythm undoubtedly plays a large role and we mentioned that various physiological rhythms can be brought into phase with a metronome even in animals. We also pointed out that all of our close primate relatives display by hitting resounding objects. Drums are among the oldest musical instruments of man, and as war drums they still serve the functions of threat display. This is true also of some other typical noise-producing instruments (horns) which are used to intimidate in place of shouting. There are strong indications favoring the hypothesis that we innately recognize particular melodies as touching, charming, soothing, and so on. We describe melodies by likening them to typically inborn vocalizations such as sobbing or joyous shouting. Tender or rumbling sounds stand for the linguistic expression of tender or angry words. Tender words resemble higher notes and we know that even little girls raise their voices when they talk to a little baby. The shrill vocalizations of an angry person are universally understood and we find them uncomfortable. J. Kneutgen (in preparation) studied the lullaby songs of various peoples and came to the conclusion that this is the most uniform musical expression throughout the world. A Chinese lullaby is just as soothing to a child as a German song or any other. When listening to lullabies, breathing becomes shallow and regular like that of a sleeping person. The characteristics of this form of breathing are also in the structure of the lullaby. The regularity of breathing is reflected in the regular components of the song. When a lullaby is played on tape, the breathing rhythm of the listener adjusts to the melody; that is, the breaths become as long as a phrase of the song. The inhaling phase of subjects generally coincides with the gradual rise of the melody; the exhaling accompanies the gradually falling melody at the end of the phrase. The breathing seems to accompany the music. The shallowness of breathing is matched by the simplicity of the melody. There are no large intervals, which gives the impression of a gentle gliding. The listeners feel relaxed, the frequency of heart beats decreases, and the psychogalvanic skin response shows little change. In another experiment, when jazz was presented under identical conditions, the subjects became excited. Breathing became irregular and the psychogalvanic skin response showed irregular changes. After the subject did kneebends, heart and breathing became normal within 3 minutes when they listened to a lullaby, and within 6 minutes without it. When jazz was played it took 8–9 minutes until breathing and pulse returned to the base line.

When we examine our music we discover that composers use these key stimuli intuitively to evoke various emotions in the listener—think for a moment of the rumbling drums of Beethoven's Fifth Symphony. The releasing stimuli are artfully encoded and lose much of their flashy obtrusiveness, which is a characteristic of popular music that is largely produced for commercial purposes. Because of this coding of key stimuli it also takes awhile before one is able "to listen one's self into the music," so to speak. By the artistic manipulation of the releasing stimuli the composer can create and dissolve tensions in the listener. The highs and lows of emotional experiences are touched in an ever-changing pattern that cannot be experienced in everyday life. This heightening of experiences is perhaps one of the most important effects of music. It is most certainly not the only component of artistic, musical creation, but it seems to be a substantial one. Added is also the pleasure in the playful experimentation and in the construction of new and different structures and sequences.

Inborn releasing mechanisms also seem to determine our need for cover and unobstructed vision into the distance. Persons who have had no fateful experiences with others or with predatory animals occupy corner and wall tables first in a restaurant, the tables in the center last. Children feel comfortable in niches and like to build such cover when they play.

That the biological processing of data relating to the perception of space is inborn has been shown by experiments with cats and human babies at the visual cliff (R. D. Walk 1966). Form constancy with various retinal representations of objects is not the result of long experience, as is widely believed. Babies less than 2 months old are capable of recognizing form constancies under various transformations (T. G. Bower 1966). By turning their heads the babies were able to operate switches in their cushion. The head-turning was rewarded by the appearance of a person smiling at the baby. The babies were conditioned to operate the switch when a 30 cm cube was shown at a distance of 1 meter. Once the children had learned this, they recognized the cube also when it was shown at a distance of 3 meters. They distinguished it from a 90 cm cube shown at a 3 meter distance, although the latter projects at the retina an image identical in size to the image that the 90 cm cube projects at a 1 meter distance. This supports earlier hypotheses by K. Lorenz (1943). Innate mechanisms are probably not only underlying our perceptions but also our forms of thought.

Quite remarkable finally is K. Lorenz's view (1943) that a number of releasing stimulus situations which affect our ethical value judgments are outlined by innate releasing mechanisms. In the art and literature of all peoples there are recurring themes, situational clichés: loyalty of friends, manly courage, love of homeland, love of wife or husband, love of children and parents—all are the noble basic motives of human actions that we follow from an inner disposition. They are the basic themes (Leitmotive) of literature and the theater from the ancient world to this day. We are gripped by the account of the friend who sacrifices himself for his fellow, and we identify with the hero of the legend or the western movie who liberates and protects the innocent girl or helpless child. If there were no innate predispositions for ethical behavior in man, if there were no binding norms for what is basically good or bad based on phylogenetically acquired adaptations, we would be in a dangerous position indeed. A cultural relativism is the logical consequence, and any cultural norm agreed upon by a society could exist rightfully. Environmentalism would provide the excuse, and I sometimes wonder why this danger is not seen.

I do not imply, however, that we have to follow every "inner" value judgment. We mentioned that animals carry along historic burdens—structure that evolved during phylogeny—which by the changing environmental situations have become maladaptive. The appendix in man, which caused many deaths in former centuries, is one example. In the same way we can consider some of our preprogrammed ethical "values" as being outdated. We mentioned, for example, that there is a strong conformity pressure in groups of men, and outsiders are reacted to strongly, a pattern found all over the world—even in recent times. Throughout history people have reacted emotionally against deviants and minorities, and demagogues justified this as a "sound popular instinct." Needless to say, this is maladaptive. We have furthermore reached a level of consciousness that makes us realize that those different from us are nonetheless basically the same and that the diversity constitutes the particular beauty of mankind. We have therefore to curb our archaic intolerance by encouraging this level of consciousness and taking advantage of our deeply rooted drive to bond.[6]

The reactions to these situational clichés, on the one hand, assures that in extreme situations we will behave correctly—in line with requirements for the preservation of the species. At the same time, however, these reactions also harbor great danger in the present age, which is why insight into the basic underlying mechanisms is needed. This is especially true for the social fighting reaction, whose function is to defend the group. The affective correlate of

[6] A monograph on the "bond" in man by the author is in preparation.

fighting, which is experienced as enthusiasm, is not released only by an actual threat. With the great aggressive readiness that is a part of man, even a reference to a possible threat is sufficient to release it—an observation that demagogues of all times have expertly used.

Primary and secondary means of stabilizing human social behavior

It is a part of the inborn and not to be removed inequality of human beings that they break down into leaders and followers. The latter are in the overwhelming majority; they have need of authority which makes decisions for them, to which they usually submit without question (S. Freud 1932 in a letter to A. Einstein, [1950, vol. 16:20]).

Are striving for rank and the readiness to submit actually characteristic of our species, as S. Freud, among others, assumes, or are they the products of education? To what degree are we by nature social and in the final analysis political beings? Or are we, as Hobbes asserted, forced together against our will by authority? In this section we want to examine and trace the biological bases of our social behavior and ask ourselves to what extent it may be determined by phylogenetic adaptations and what has been since acquired.

It is easiest to disprove Hobbes' statement when we take the family as our point of departure. Mother and child are in a natural, well-adjusted relationship to one another on a partnership basis. J. Bowlby (1958) has enumerated in detail how the bond between mother and child is at first brought about by a number of inborn reactions such as sucking, clasping, crying, smiling, following, and by the appropriate responses of the mother who loves her baby (p. 432), in which she responds to certain releasing stimuli. Theories which are based on the proposition that the child is only secondarily tied to the mother because she fills its need for food and warmth have to date been no more supported by fact than the assertion that the child resents being born and attempts to return to his mother's womb (M. Klein, cited by J. Bowlby 1958). The bond of the child to its mother is a primary one and develops not via the "self-love" of the child because it is fed by the mother (K. M. Banham 1950). Even the individualized relationships between mother and child, which develop gradually through learning processes, are, as we already discussed, programmed by innate learning dispositions.

However, not only the mother–child bond but also the permanent bond of the parents to each other seems to depend in man on something more than just tradition. Tradition does determine whether a man may have one or many wives. A permanent, long-lasting association of the partners, however, is generally the rule, and is necessitated by the slow development of the human child. Among the many mechanisms that tie people together, which we discussed earlier, we also find a form of sexual bonding that we do not know of in other mammals; the latter usually mate only during the short estrous periods of the females. Only in the chimpanzee have occasional copulations outside of this time been observed. R. M. Yerkes (1948) writes that chimpanzee females sometimes presented themselves successfully outside the estrous period and gained certain advantages from a particular male such as being first at the feeding place. In humans the limitation of the sexual drive and desire to specific cycles or seasons has largely been eliminated. A woman is physiologically ready to respond most of the time to the sexual desires of the man, although she is ready to conceive only during a fraction of that time. This enables her also to maintain a tie with the man on the basis of a sexual reward, and this is probably the function of this unique physiological adaptation. Also in the service of maintaining a bond between partners is the ability of the woman to experience an orgasm comparable to that of the man. This increases her readiness to submit and, in addition, strengthens her emotional bond to the partner.

Because of this, the sexual act of humans has acquired a significance in the social life of man which goes beyond the need for reproduction. One argument of the church against birth control by preventive, artificial means is the supposed unnaturalness of such a measure, which is based on the widespread assumption that the sexual act is only in the service of reproduction. This is so in animals. In man, in addition to this function, there is also the important one of maintaining the bond between partners. The sexual act enhances the relationship between people in a way not present in animals. The erroneous interpretation of this process often resulted in calling immoral the specifically human aspect of this behavior by admitting only the animal aspects of reproduction, which in the long run, results in a superficiality and disruption of the relationships between partners. We have already seen other comparable extentions of functions and want to refer here only to the ritualized feeding which is in the service of maintaining a pair bond (p. 106). W. Wickler (1965b) showed that the presenting movement of female primates has often become ritualized into a greeting and has been used as such by males. In a similar manner courtship movements in fish were ritualized into appeasement gestures. In all these instances sexual behavior became invested with a new meaning (W. Wickler 1966e). The specific sexual act is limited in animals to the function of reproduction, as far as we know today. Recently W. Wickler (1968c) discussed in more detail the bonding function of human sexuality.

We humans do not only live in family units but find ourselves in village communities, circles of friends, and so on, and in addition to all those whom we know individually, we associate with many people in an anonymous group (p. 351). We are also predestined to this type of group formation. This was emphasized by C. Darwin: "Since man is a social animal it is quite certain that he has inherited a tendency to be loyal towards his fellows and obedient to the leader of his tribe; since these characteristics are common to almost all social animals" (*The Descent of Man*).

Man is characterized as a gregarious being by a number of behavior patterns which serve in forming groups (pp. 116ff., 342ff., 425). The groupings or associations of men are exclusive to varied extents. His gregarious tendencies are opposed by nongregarious ones. As in most vertebrates, we observe in man distinct territorial behavior. Individuals maintain distinct distances between themselves and others. Specifically how close we are permitted to approach another person is determined by the various cultural patterns, but some generally valid basic outlines can readily be discerned (E. T. Hall 1966). One can experimentally overstep the individual distance by casually sitting close to a person in the library. The behavior of persons subjected to this experiment has been described by N. J. Felipe and R. Sommer (1966). The "victims" at first try to move away from the intruder, and failing that, they erect artificial barriers against him with books, rulers, and so on. If all these efforts to withdraw fail, they leave the table. The various forms of bodily contact such as shaking hands, putting arms around another during greeting, or sitting in close contact are restricted to certain situations and social circles. Children develop individual distances at the time they develop a feeling for property (D. W. Ploog 1964a). The expression of both tendencies seems based on a common mechanism.

We must expect also that human beings have certain needs for space which are based on an innate disposition and whose fulfillment is necessary for our well-being. It is true that man largely creates his own environment, but its structure is surely in line with his biological constitution (R. Sommer 1966). Even within a family each person has his own individual domain. The areas owned by each family are more clearly marked. Apartments and gardens are areas in which we assert territorial rights, and this "natural right" is almost everywhere recognized by the makers of laws. No one may enter another's dwelling without special permission; this is considered illegal entrance, breaking and entering (Hausfriedensbruch). Fences and signs designate our rightful ownership. The surprising interpretation of the function of the hermes and other artifacts by Wickler has been discussed (p. 429). Each trespass across territorial boundaries is accompanied by special ceremonies if it is to remain unpunished. Even when we visit friends we obey certain rituals that appease aggression, for example, giving presents, which have their parallels in the appeasing greeting ceremonies of animals (p. 125).

In everyday life we can observe examples of territorial behavior on many occasions. If one wants to sit down at an already occupied table in a restaurant, it is proper to ask politely if one may sit down. If one fails to observe this rule one releases anger in the other person. The same is true when one enters a partially occupied compartment in a train. If one does not greet in a friendly way one may experience an air of rejection. Mentally ill persons are especially territorial and aware of rank order (B. Staehelin 1953, 1954; A. H. Esser 1968). Only patients on a very regressed stage show no interest in establishing and defending a territory. A degree of aggressive behavior toward intruders is, therefore, considered a healthy sign (A. H. Esser and others 1965; E. Hackett 1966). It goes without saying that cultural variations exist on this basic theme.

In addition to family possessions humans also defend group territories. Basically, human territorial behavior has the same function as in other territorial animals. Anthropological investigations support this view. In the New Guinea edition of the *American Anthropologist* (1965) it is reported that before the intervention of the Australian government some highland tribes lived in areas in which survival was just barely possible. After the tribal wars had been outlawed the people moved into already settled and more suitable regions. Overpopulation and starvation were the result. Among these people who cultivate tuberiferous plants, food production soon reached an upper limit and tribal wars served here as the spacing-out mechanism that prevented a too close spacing of the various groups.

Occasionally it is asserted that the statements of biologists are apologetic, as if the assertion that a behavioral characteristic is inborn would imply that nothing, therefore, could be done about it. This is a false attribution. Certainly many of our behavior patterns and motivations evolved as adaptations in the service of specific functions. It is as true, however, that changes in the environment can convert the adaptive value of an adaptation into the contrary. This is certainly the case with territorial aggression in our overpopulated world bristling with arms. As has been emphasized, effective control of our aggressive impulses is needed for our survival and is certainly possible by environmental control or other measures, provided we explore the causes of this phenomenon without any bias. Those that simply deny man's inclination to act aggressively take the easy road. H. Helmuth (1967) and M. F. A. Montagu (1968) argued along that line, pointing at the existence of non-aggressive people, for example, the Eskimos, the Zunis, and the Arapesh (R. Benedict 1934; M. Mead 1935; K. Birket-Smith 1948). According to most reports these people lack territorial aggression, except some of the Eskimo tribes. But what has escaped their attention is the fact that they nonetheless show quite a number of aggressive acts within the group. The Eskimos engage in singing duels; they beat each other within the family (K. Rasmussen 1908). Of the Zunis R. Benedict reports quite aggressive initiation rites (see also P. Weidkuhn [1968]). There are many facts in favor of the dynamic instinct concept of aggression (pp. 326–332). This hypothesis demands consideration, therefore, this being the only responsible way of approach.

Within the group the aggressive behavior of individuals merely leads to the establishment of rank orders, which provide society with a certain kind of stability. The high-ranking persons usually assume some kind of leadership function. The development of such a ranking presupposes not only that some members of the group succeed in establishing their authority, either by fighting or special achievements, but also that the subordinates accept this rank order (p. 352).

This readiness to accept subordination, the opposite of the quest for rank, is very striking and poses particular problems for us. Obedience to the father or some other "recognized personage of public life" is generally considered of some ethical value. In all forms of government there is a tendency to the cult of personality. If necessary people will create models to be honored, and they seem to have a need to follow them. Human beings fight against the rule of brute force to be sure, but they also seem to have a distinct disposition to follow those whose authority they voluntarily accept. When one has voluntarily submitted one's self to authority, one is also at its mercy up to a certain point, as recent experiments by S. Milgram (1963, 1965a, 1966) have shown in a surprising manner. Milgram invited his subjects (men between 20 and 50 years of age) from various backgrounds (40 percent laborers, 40 percent white-collar workers, 20 percent professional men) to participate in a supposed learning experiment for a small honorarium. They were given the task to administer increasingly stronger electrical shocks for each error another person made. This person was supposed to be learning something but in actuality was an accomplice of the experimenter. In one sequence of trials the "learner" was strapped to a chair in a separate room from the teacher (subject) and electrodes were fastened to his body. The teacher was helped in this by the experimenter. Then the experimenter explained to the teacher that he was to administer a shock for each erroneous answer, and that he should begin with a low voltage which should be increased as more errors were made. In this way one would be able to study the effect of punishing stimuli on the learning process. The punishing stimuli were administered by the teacher by means of an apparatus that contained levers for 30 steps ranging from 30 to 450 volts. In addition to the voltage designation there were labels ranging from "low shock" to "danger: heavy shock." To test the role of the effect of immediate or delayed feedback from the victim various conditions of feedback were investigated.[7] In the first group of trials the "learner" protested against the treatment by pounding against the wall when the 300-volt lever had been pressed, and he ceased to respond at 315 volts and above. In the second series a tape recorder played back the protesting voice. From 75 volts on, each step had a specific response linked to the lever: at first mere mumbling, at 120 volts the report that the shocks were painful, finally protests with the demand to discontinue the test and to release the "learner." At 180 volts the victim cried that he could no longer stand it; from 315 volts on, he refused to answer but groaned in pain when he supposedly received a shock.

The third series was like the second, but here the teacher was in the room with the learner about 2 feet from him.

The fourth series of trials was like the third with the difference that the victim received a shock only when his hand rested on an electrified piece of metal. From 150 volts on, he refused to put his hand back on the shock device, and the experimenter then ordered the teacher, who was, in fact, unknowingly a subject, to force the hand of the victim onto the grid. Forty persons were tested in each group.

Under the condition of weak feedback 34 percent of the subjects ("teachers") opposed the experimenter's instructions; when voice feedback was heard 37.5 percent did. When they were next to the subject 60 percent refused and when they were touching him 70 percent refused. The less abstract and distant the suffering of the victims was, the more the pleading reactions were perceived by the subjects and the greater was the inhibition in the subject to administer suffering even under the authoritarian pressure of another person (the experimenter). However, 30 percent of the subjects still obeyed the instructions of the experimenter when they had to press down the victim's hands.

Frequently the subject would be unsure and ask the experimenter if they were to continue in view of the expressions of pain of the

---

[7] Pilot studies had shown that almost all experimental subjects went through the entire scale of punishing stimuli when they received no feedback from the subject.

learner. He received the stereotyped answer: "You have no choice; you must continue." In these instances a divergence between what the subjects said and what they did appeared. Although they protested that they could not do such things to the poor fellow, they nevertheless continued to administer the shocks in obedience to the authority of the experimenter.

In another set of experiments in which the degree of supervision by the experimenter was varied it was found thàt the subjects more readily disobeyed when the experimenter was absent. The number of agreeable subjects was three times as large when the experimenter was present than when he gave his instructions by telephone. In addition, many subjects did not increase the shock intensity when the experimenter was absent, although they claimed to have done so. When some of the subjects were allowed to observe a staged experiment in progress before they took part in it as "teachers,"· and where they could see the other subject refuse to obey the experimenter's order to apply further shocks, in 90 percent of the cases they also refused to obey his instructions (S. Milgram 1965b).

The results of these experiments prove that a large number of persons had difficulty in opposing the authority of the person in charge of the experiment. Even under conditions of vocal feedback subjects administered shocks to the victims in 62.5 percent of the cases which in reality would have killed or severely injured these people. And this took place in the United States, a culture that educates its children against blind obedience. This result, in effect, contradicts all expectations one would have based on the cultural ideal. Of 40 leading psychiatrists who were asked to make predictions about the outcome of the experiments, most expected that the subjects would not go beyond 150 volts and that only 0.1 percent would follow the instructions through all the steps. Between the expectation and reality there is a remarkable discrepancy. This points to innate tendencies which assert themselves against the cultural ideal.

The postscript that S. Milgram (1966) attached to his paper leaves one pondering (retranslated from the German):

With a numbing regularity we saw good people submit to the demands of authority and commit actions that were without feeling and cruel. Persons who in their daily lives were responsible and decent were led to commit cruel acts by the pretension of authority and the uncritical acceptance of the experimenter's definition of the situation.

Where is the boundary for such obedience? In many instances we tried to provide such limits. Screams of the victims were used; they were not enough. The victim complained about heart trouble; still some subjects shocked them when they were ordered to do so. The victim pleaded to be released and his answers were not registered by the signal apparatus; the subjects continued to shock them. Initially we had not expected that such drastic measures would be needed to obtain a refusal to cooperate, and each step in the experimental procedure was only added to the degree to which the ineffectiveness of the previous condition became revealed. The last attempt to erect a barrier was the condition in which the subject had to touch the victim. However, the very first subject already used force against the victim and proceeded to the highest shock intensity. One quarter of the subjects behaved similarly under these conditions.

The results—as they were observed and felt in the laboratory—cause great concern to this author. They allow for the possibility that human nature or—more specifically—the kind of person produced by American society would not afford much protection to its citizens against brutal and inhuman treatment at the behest of an evil authority. These people did to a large percentage what they were told, irrespective of the nature of their activity and without conscientious objections, as long as they saw that the order came from a legitimate authority. When, as in this study, an anonymous experimenter could successfully order adults to force a 50-year-old man into submission and administer painful electric shocks to him despite his protests, then one can only be apprehensive about what a government—with much more authority—could order its subjects to do (pp. 460ff.).

All of us know the example of the God-fearing Abraham, frequently glorified by artists, who was willing upon the command of God to sacrifice his own son (Fig. 283).

Figure 283. Abraham's sacrifice, etching by Rembrandt.

In the symbolism of Abraham's sacrifice is contained undoubtedly one of the greatest human problems. Obedience is an ethical value, just as love thy neighbor, but when does it cease to be that? When both are in conflict with one another, then obedience is often stronger, apparently based on our inborn disposition whose roots reach back probably into the rank-order structure of our primate ancestors. Generally it is of advantage for a primate group when they follow the stronger and usually more intelligent alpha males.

From this insight it follows, however, that love of neighbor and the morality of the individual are often not enough to resist the contrary orders of strong authorities.[8] Mankind during peace accepts certain humanitarian norms. If it were possible to establish them by law on an international scale and spell them out in detail, this would indeed be a decided advance in humanitarian development. The individual could then call upon the protection of the abstract authority of a law against the orders of an evil authority. He would no longer stand alone with his moral decision against one authority but would have another as his ally.

It is also important to educate people to be critical in their attitude toward authority, to avoid blind obedience. Any subordination should be based on reason. In this context the slogan "antiauthoritarian education" has recently been heard. But this seems to be but a phrase, because antiauthoritarian educators still take advantage of authorities. On the walls of antiauthoritatian kindergartens one may see pictures of politicians. Probably there is no other way. After all, society can only exist if individual egoism is, to a certain extent, curbed by society.

High rank in man is dependent up to a certain point on age, and respect of the old seems to have some biological roots. In baboons we find the relatively high-rank-order position of older males, who then also aid the group by their experience even after their physical strength has waned (p. 355). In the eyes of the group their physical weakness is somewhat compensated for by the development of a conspicuous coloration of the fur (a long, silvery mantle). In man, too, old people play an important social role (council of elders in a senate) and in many peoples the old impress the group members by special attributes (white hair, bushy eyebrows, white beard).

Membership in a particular group is learned. Some arguments can be made for suspecting imprinting-like fixations as factors in

---

[8] But of the cities of these people, which the Lord thy God doth give thee for an inheritance, thou shalt save alive nothing that breatheth: But thou shalt utterly destroy them; namely, the Hittites, and the Amorites, the Canaanites, and the Perizzites, the Hivites, and the Jebusites; as the Lord thy God has commanded thee. Deut. 20:16–17.

And they utterly destroyed all that was in the city, both man and woman, young and old, and ox, and sheep, and ass, with the edge of the sword. Josh. 6:21.

this process. The religious and political ideals of their youth are usually tenaciously adhered to by human beings. The same holds for the identification with an ethnic group. Once a young man has committed himself to a particular group, his attitude is decidedly determined for the rest of his life. Despite a similar genetic basis people can be something very different because of this imprinting-like process: Germans or Russians, Frenchmen or Americans. We owe to this tenacious clinging of a once-acquired attitude the colorful multiplicity of human cultures. However, because groups always close themselves off *against* others, we find here also the root of all ethnocentrism, whose consequences are often a destructive intolerance which remains to be overcome.

In these religious, political, and ethnic groupings we are already dealing with anonymous groups. The members of one religious body or of one nation do not know each other individually. However, they are united by common ideas, common representatives (head of nation, head of church), and frequently by very simple common symbols. These symbols may be badges or flags or forms of dress. How important such symbol identification is for the cohesion of a group can be seen from the fact that new nations and political groups see it as one of their first tasks to build tremendous and expensive memorials and to exhibit their insignia of state and the pictures of their presidents everywhere. An important group-binding function is also found in festivals and national holidays.

Within the anonymous group, aggression is largely well buffered, although not as well as in an individualized group. We are on the whole less altruistic toward an unknown person when he is a member of our anonymous group than toward our personal friends. The necessity of a morality based on reason (Kant) seems to be a necessary consequence of this fact. We will serve a friend without question because we are fond of him, but there are also situations in which people serve anonymous members of their group with a high emotional investment, especially when they direct their attention against members of another anonymous group whose members are adherents of other ideals. This aggressive characteristic of groups could possibly be overcome by symbol identification if it were possible to create ideals and symbols that could unite the whole of mankind. In this connection an idea of F. Fremont-Smith (1962) deserves special attention. He asked himself on the occasion of attending a conference in Russia which common interest possibly could unite people of the various political orientations, and he found as an answer: the protection of the child. In an address before his Russian hosts, he said, among other things:

We have now reached that point in our history when no nation can any longer protect its own children. No government, may it be ever so powerful, can any longer guarantee the safety of its most precious possession, the safety of its children. If a nuclear holocaust should come, New York, London and Moscow will perish and with it all its children. If the nations could, however, agree to protect the children of the others, then the children could be saved. If the USSR would guarantee the safety of all American children, then all children could be saved.

Out of this common desire a common symbol could be formed: the helpless child that needs protection, which among others, was used by the Christian religions as a uniting symbol. It is interesting that it is again the care-of-young behavior which even on this high plane is probably the most effective "cement" to keep a group together (p. 104).

In all anonymous groups there exists a strong tendency toward conformism, an assimilation to the other members in appearance and behavior. Outsiders who resist this assimilation release aggression (p. 333). This conformity in behavior becomes especially obvious in large-scale activities such as sports festivals and holiday parades. This phenomenon is based on certain mood-facilitating mechanisms that are similar to those we already know exist in animals. A significant means of inducing a similar mood among people to bring them into harmony is music. Marching music virtually forces everyone into step. In this connection it is interesting to note that, in animals, simple behavior patterns such as breathing of fishes, song of birds, and simple movement stereotypies (squirrels) can be forced into the rhythm of the monotonous sound of a metronome (J. Kneutgen 1964).

It has sometimes been stated that basically man was made only for life in an individualized group such as the family and a circle of friends. What we have said so far should have shown that man also possesses all the prerequisites for the formation of anonymous groupings that contain millions of members. He is adapted to both, and the step toward the anonymous group seems to be of a more recent date, and the processes of adaptation are still in progress.

We see, for example, that men, in an adaptation to the large society of millions, have shed their individualistic displays with respect to clothing and behavior. A Papuan may still develop a full masculine display within his small community. The members of such a small group are so closely tied together by personal familiarity that this basically aggressive display is not disruptive. However, where this bond of personal friendship is lacking such display behavior would be a cause for conflict. By giving up individualistic challenging displays in dress and behavior, the members of large societies are adapted to the new conditions. Shaved and in the gray flannel suit we live with less friction in the crowded conditions of modern times. Women, on the other hand, are allowed to enhance their attractiveness and to appear colorful, because their "display" activates binding mechanisms.

In our feelings toward our fellow men we are in the process of developing new attitudes of social responsibility. Emotionally we are still better adapted to a life in individualized communities, the emotional ties with members of the anonymous group being less strong than those with our family and friends. The need is urgent, however, to develop a new social responsibility also toward members of the anonymous group, and ways to achieve this goal are offered by different ideologies. Christianity, as well as Marxism, propagate the identification of the individual with the anonymous group, fighting at the same time egoism and individualism. Because the source of individualism is the family, attempts have been made repeatedly in theory and practice to dissolve this core group as well as to fight the establishment of individual bonds. Thus, it was hoped, man would become one with the collective. These attempts have failed so far, owing to man's inborn inclination to form families and seek personal relationships. One will thus be compelled to look for methods that achieve social responsibility of individuals toward the anonymous community but also allow the formation of individualistic bonds. The one does not necessarily exclude the other.

In examining the possible innate dispositions of man, as far as his social behavior is concerned, special attention needs to be given to the question of whether there are any significant differences in the social dispositions of men and women. K. Lorenz (1956) and more recently L. Tiger and R. Fox (1966) point out the group formation of males which excludes women, which is found in all cultures and which may be a direct adaptation to hunting and fighting behavior. Men also seem more ready to form anonymous groups than women, who are more oriented toward the family. This leads us to ask if there is a predisposition for the various roles in a society which are played by men and women. Assertions that differences between boys and girls are the direct result of planned education by the parents are not lacking (M. Mead 1965). The proof for these statements remain to be provided, and the fact that in almost all cultures, as far as is known, men are more aggressive and less passive than women should give us something to think about. Furthermore, male characteristics, such as a desire for higher rank and increased aggressiveness, are characteristics we share with other primates. For these reasons one now discusses again the probability that constitutional differences may exist (J. Kagan and H. A. Moss 1962; D. G. Freedmann 1967).[9]

We have sketched a rough framework into which human social behavior has been placed by phylogenetic adaptation. These adaptations consist less of rigid behavior patterns and more of innate mo-

---

[9] That women are inherently less curious than men must be doubted. More likely they are interested in different things than men. The possible sex-specific shift of interests should be studied in various cultures in a comparative way.

tivations and learning dispositions (p. 217). We mentioned drumming and the phallic display as examples. Both dispositions are probably basic drives and innate releasing mechanisms—adaptations on the receptor side that allow the recognition of the biologically adequate signal and thus shape the activity of people in principle. The detachment of these territorial displays from rigid motor patterns allows, however, a greater range of expression. Man does not need to sit guard the way other primates do (p. 429) but may carve statues instead and thus create symbols. Learning dispositions allow a wider range of freedom. Despite a basic similarity, this leads to a multiplicity of cultural modifications of human social behavior, where each culture and subculture developed their rites in diverging ways. Once formed, they are as rigid as phylogenetically developed rites. Just as the phylogenetically evolved rites of animals control the inborn motivations, so cultural rites do this in man, and for this reason they are just as important for an orderly life together in groups (K. Lorenz 1966). To gloss over them as just so much "cultural whitewash"—as a sort of superficial varnish—is basically wrong. Our inborn mechanisms are insufficient to control our drives. They became secondarily reduced during the course of phylogenesis and were replaced by cultural control patterns. This is a gain in adaptive modifiability, because various patterns of culture could be developed that made possible the exploitation of various habitats. An Eskimo does require different patterns for the control of his sexual or aggressive impulses than a modern city dweller of central Europe or the United States. Cultural control patterns can also be changed more quickly along with a change in living conditions, but in all cases they are indispensable for social communal life and man is, as A. Gehlen aptly remarked in this case, a cultural creature by nature.

The cultural rites are probably often developed upon the basis of innate learning dispositions. To examine these questions along ethological lines is a most attractive task for us in the future. We recall the greeting ceremonies, which contain basic components despite the multiplicity of expressions, which all serve the function of appeasement. This holds for the various forms of making presents when entering a strange house, making an inaugural visit, and so on, and for the various forms of symbolic submission such as bowing, concealment of aggression-releasing characteristics, removal of weapons and armor, and similar behavior. However, in a greeting there is also a distinct display component. One shows the partner who one is, and a firm handshake is already a mutual taking stock of one another, one could say a ritualized tournament, and it is distinctly unpleasant when the other presents his hand in such a way that one cannot properly return his grip when shaking hands. Greetings with displays are seen during state visits. The gun salute is a form of ritualized aggression; it is a demonstration just like the parade of troops in honor of the guest. However, that these forms of greeting are also intended to be friendly is shown by presentation of arms. The response of the greeted person also follows according to rules that have not been investigated further. If he is of high rank and the greeter of lower rank, the former may place his arm protectively around the shoulder or symbolically on the head of the greeter. Persons of equal rank frequently put their arms around each other. Comparable behavior is also shown by men when dealing with revered or feared supernatural powers. They submit to them by giving presents of various kinds, including ritualized feeding, to gain their favor. On Bali there are daily offerings of some rice and flowers at designated places and during large festivals abundant food sacrifices are made. Such remnants of old customs are also found in the Enns Valley in Austria. A special food made of milk and bread (Peschtmilch) is eaten on Epiphany, and the remainder of the food is left on the table with the spoon in it. The Pescht (fairytale character) will come and eat it (V. R. v. Geramb 1918). If man fears the ghosts, he tries to chase them away with fetishes that exhibit human threat gestures and expressions.

The human dance forms show culturally determined differences but also remarkable agreement. In exhibition dances the men boast of the physical prowess by high jumps, clapping their hands, stomping their feet, showing their weapons, and so on. This is true in the Cossack dances, the Scottish Highlander's dances, or those of the Nilotohamites (Fig. 284).

The human desire to impress leads to parallel symptoms in the various cultures. Ethnologists describe them as "prestige economy" (J. Faublée 1968). This includes expenses for official robes as well as the annual change of automobiles, unnecessary from a technical point of view; the herds of the Massai, which are often impractically large; and the huge expensive stone covers on the graves of some shepherd tribes in Southern Madagascar, made to demonstrate distinction and power. The savings for these power demonstrations are in all cases drawn from daily life.

This prestige displays take the most fantastic form in the Kwakiutl of the Vancouver Islands (R. Benedict 1934, 1955). There rival chiefs compete in the destruction of property. The rivals are invited and in front of them valuables are broken in pieces and large amounts of highly prized oil poured into the fire. The guests have to answer by inviting their hosts in turn and stepping in this destructive competition by destroying even more. The higher in value the destroyed property, the higher is the prestige achieved. Certain aspects of the race to the moon, both in the East and West, seem motivated by prestige thinking.

a   b

Figure 284. Dancing Samburu warriors as an example of human display activity (near Maralal, Kenya). (a) The young warriors show off their strength by jumping high up into the air, one after another, to the rhythm of songs and the clapping of hands. We find parallels in many exhibition dances of men (Cossack dances, Scottish Highlander's dances, and so on). Girls sitting nearby seemingly ignore the proceedings. (b) Somewhat later the girls joined the warriors in the dance. In some phases of the dance the warriors also display their weapons by throwing their spears high and turning them around the long axis so that the blade glitters in the sun. The dance is not arranged and has been filmed without the awareness of the dancers. (Photographs: I. Eibl-Eibesfeldt.)

Along these lines many cultural activities have biological bases that have hardly begun to be investigated. One may think of man's various festivals. On the one hand, they serve the strengthening of a bond between group members, whether within a family, a clan, or a larger group. Sports events unite by their very competition, and for "the fights of the chariot and songs" even the various quarreling Greek tribes used to come together. Festivals also serve to separate groups. Certain rituals are performed within a group, and being unique is emphasized. The separating function of rites can be so perfect that such groups behave as if they were ethospecies, an aspect that was pointed out by K. Lorenz (1966). Festivals also make possible the formation of new relationships and are also often the vents for unfulfilled appetitive behavior (Fig. 285). Just as during recess the dammed-up drive to run around and play is discharged in children, so the various unfulfilled drives (aggression) are discharged in people who have no occasion to do so during their routine daily lives. The carnival has a uniting function—in Rio it has the character of the folk festival—but it is also a custom that functions as a safety valve. The comparative study of customs along ethological lines has hardly begun. Here human ethology as a "biology of cultural achievements" opens up a wide and interesting area of research.

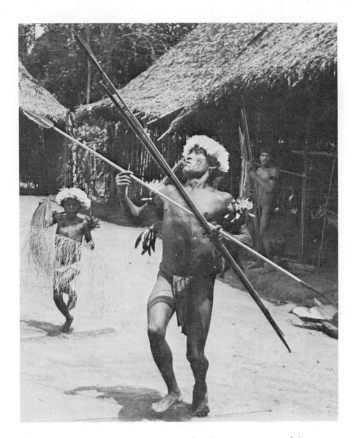

Figure 285. Display and appeasement, two basic components of human expressive behavior: During the palm-fruit festivities different Waika villages (upper Orinoko) extend invitations to one another in order to reaffirm alliances which unite them. They exchange, among other things, gifts and promises. The visitors greet their hosts by a dance through the village, with the men displaying themselves in full regalia and war-like postures, waving of weapons, and stomping on the ground with fierce expressions on their faces. As with all threat-greeting (salute, handshake p. 454) the individual thus demonstrates his worth as a warrior and at the same time has value as an ally. The aggressive and as a result possible fight-releasing demeanor is counteracted by the presence of the dancing child (see also p. 330) who follows behind with two ragged palm fronds in his hands. The rituals of man are not arbitrarily invented; they are also shaped by phylogenetically acquired dispositions.* (Photograph: I. Eibl-Eibesfeldt.)

* An ethological interpretation of the palm-fruit festivities of the Waika can be found in a book that is in preparation by this author: *Liebe und Hass: Zur Naturgeschichte elementarer Verhaltensweisen* (Piper—Munich, Germany) or roughly translated: *Love and Hatred: A Natural History of Basic Behavior Patterns.*

Intelligence to use tools and language

Along the lines of J. G. Herder, A. Gehlen (1940) spoke of man as a "deficient creature" (Mängelwesen), referring to the nakedness of man and the lack of inherited weapons. He is more helpless than any animal in this world and would hardly have a chance to survive if he did not have the aid of technology. This view of man as a deficient being continues to be held in the anthropological literature and has more recently found its way into the popular literature. This view is, however, very one-sided. It overlooks the fact that there is no such thing as a perfect organism. Each specialization in one area means a loss in another. A mammal such as a seal, which has been adapted to life in the sea, can move only with difficulty on land. In addition, each presently living organism is the result of countless restructuring processes. The fact that all land vertebrates have descended from fishes is evident as an historical burden in the blood circulatory system of these vertebrates. The still incompletely separated circulatory system of amphibians and many reptiles can be considered a deficiency from the point of view of an engineer. The step-by-step transformation of the fish-blood circulatory system to include the circulation through the lungs as an adaptation to life on land at first leads to an incomplete separation of venous and arterial blood. Frogs, salamanders, and even the lively lizards, as "vertebrates with mixed blood," do not have as much endurance to reach the peak performance during, say an escape, as fish, birds, or mammals can easily reach (G. Kramer 1949). That whalebone whales form tooth rudiments as embryos only to resorb them again, and that we in an earlier developmental stage develop branchial arches, can only be interpreted as an historical burden. Whenever a way of life changes, the morphological and physiological adaptations limp after them. The tendency of man to develop fallen arches and varicose veins in the legs shows that these systems still lack some of the necessary adaptations to upright walking (K. Saller 1963). These things are not, however, specific deficiencies of man, but they are the expression of evolution in progress.

Finally, many characteristics of man considered "faults" prove upon close observation true adaptive characteristics. This is true for hairlessness, which, along with the numerous sweat glands, makes it possible to chase prey animals with perseverance in warm climates. Bushmen pursue antelopes until they collapse from fatigue. Mammals that have fur often suffer from overheating.

Furthermore, it can hardly be considered a deficiency that man is not specialized in a particular way. He is, in K. Lorenz' (1959) words "a specialist in not being specialized." To this fact man owes his worldwide distribution. His sense organs are excellent and his physical capacities were illustrated by K. Lorenz (1959) by comparing them with those of other mammals.

If one were to set the three tasks: to march 35 km in one day, to climb up a 5-meter-long rope, and to swim a distance of 15 meters in 4 meters depth and pick up a number of objects in a certain order from the bottom, all activities which a highly nonathletic person who sits much of his time behind a desk—like myself—can do without difficulty, then there is no single other mammal which can duplicate this feat (p. 154).

This universality in the physical realm is matched by a surprising individual adaptability. Man is, as K. Lorenz said, a "curiosity being open toward the world." Whereas most mammals are curious only in their youth, this characteristic of youth is retained by man for the rest of his life. He is forever ready to actively explore new things and to experiment with things in his environment (see play, p. 238). From his tree-dwelling apelike ancestors he inherited a few adaptations which were originally developed for climbing, the hand for grasping, binocular vision, and spatial intelligence—the ability to comprehend spatial relationships "centrally." He who climbs about in trees by the use of hands must be able to judge distances well and to integrate the observed relationships centrally. Even to this day our thinking is organized in terms of space; we translate all obscure relationships into concepts that we can centrally "grasp."

We gain insight into an intertwined relationship—like an ape in a confusion of branches—but we have really comprehended the object only when we have completely "grasped" it. In this last expression the age-old haptic primate reveals himself as a predecessor to the optical one (K. Lorenz 1959:153).

All these adaptations to life in the trees—the central representation, binocular vision, the grasping hand—became useful in a new way when our ancestors were forced, with an increasing change from jungle to grassland, to move on the ground. At first they probably moved from one group of trees to another, but even this requires erection of the body to look out for enemies above the grass. The hands could become free for carrying objects. Chimpanzees carry fruit in their hands while walking erect.

With the continued progression of the grasslands, the necessity to hunt prey probably arose and this produced a strong selection pressure toward the use of tools, which eventually led to the making of them. The Australopithecines of South Africa and the Olduvai Gorge made tools (R. A. Dart 1957; L. S. B. Leakey 1963; G. Heberer 1965). With these instruments they killed their prey.

Simple chipped-stone tools for a long time determined the state of human culture, and the explosive development of tool-producing cultures came about only in relatively recent times. Why the development stagnated for over hundreds of thousands of years after the invention of the first stone implements, and then changed to the sudden evolution of a tool-producing culture, we do not know. Perhaps this was directly tied up with the development of language. Language makes it possible to pass on the experiences of individuals to a greater extent. In a brilliant investigation H. Hass (1968) discussed the various selective advantages of tool using. Tools serve the function of organs; they are "artificial organs," so to speak. They do not need to be nourished, however, and one does not need to carry them along all day. It is also possible to exchange them against others and thus change one's specialization. Different people may use the same artifacts and cooperate in their production. Not only such instruments as a fork or knife have to be considered artificial, but also trucks, planes, or bridges.

The human hand shows a number of adaptations which enable it to be used especially in making and using tools, adaptations which are already present in rudimentary form in other primates. Thus all primates can grasp an object with a hand (adaptation for climbing). The thumb becomes more and more specialized in the primates and finally can be independently moved and opposed to the other fingers (Fig. 286). This development is furthest advanced in man; we can hold an object with one or several fingers and the opposing thumb (precision grip). This firm thumb–finger grip is further improved by the fact that the terminal phalanges are broadened. The thumb is long in relation to the index finger and it is moved by strong muscles. The joints connecting the thumb to the metacarpal and the trapezoid bone enable man to rotate it 45° about the long axis so that it can be opposed to all other fingers (J. Napier 1962).

Higher mammals have the capacity to solve problems without actually trying out all possibilities. A chimpanzee, confronted with the task of getting a banana fastened at the roof of the cage out of reach, may sit down quietly and, by looking around, spot a box and finally, without moving, discover the solution of placing the box beneath the banana. In man the capacity to experiment in imagination is advanced so far that we may rightly call him the "phantasy being" (A. Gehlen 1940). We combine our engrams and thoughts in ever-new ways, and not only when a concrete task demands a solution. We play with these contents of our mind, build castles in the air, devise plans for our actions, and thus dissolve old habits. This capacity enables us to stay open to adaptive modifications. Sometimes, however, we create images in our phantasy which determine as imaginary forces our future behavior (H. Hass 1967). This is apt to happen when our phantasy constructs originate under the influence of strong motivations (aggression, sex). To a degree man lives a second life in phantasy, and this may have cathartic effects sometimes, when real life denies the role a man would like to play.

Figure 286. Hands of living primates. (a) The hand of the shrew (*Tupaia*) shows a beginning of the specialization of the thumb which is typical for primates. (b) In the tarsiers (*Tarsius*) the thumb has already become separated from the other fingers and is able to rotate about the joint it forms with the carpal bone. (c) In the capuchin monkey (*Cebus*) the angle between the thumb and other fingers is still wider and movement can occur independently at the base of the first carpal joint. (d) In the Old World gorilla the first carpal is connected by a saddle joint with the first trapezium. This makes possible a rotating motion of the first carpal bone. (e) The hand of modern man; the development shown in (a)–(d) has continued. The thumb is quite long in relation to the index finger and curves inward considerably. Strong muscles move the thumb to and away from the palm of the hand. Saddle joints between thumb and carpal bones and carpal bone and trapezium make it possible to rotate the thumb 45° about its long axis so that it can be opposed to all other fingers. The flattening of the distal phalanges makes possible the firm grip of the thumb with fingers. (From J. Napier [1962].)

Remarkable is the capacity of man to learn new motor patterns purely centrally without exercise. We are not only capable of reproducing an oral or written word in an instant, but we can also invent new movement patterns in our phantasy and reproduce them in behavior according to our phantasy.

The necessity of communal hunting enforced the evolution of a highly differentiated communication system. Our repertoire of expressive movements is very rich, we can signal even with our eyes (I. Eibl-Eibesfeldt 1968, M. v. Cranach (in press)). It is a reasonable guess that the white eyeballs in man developed to allow the perception of minute eye movements.

Typical for humans is language with words. We have already emphasized that almost all vocalizations of animals are interjections; only rarely can we observe anything like naming or other language-like communications (p. 114). Whereas animals merely reflect their internal state with their primarily innate repertoire of vocalizations, man gives names to objects in his environment and hence he can make statements about it (J. B. Lancaster 1966).

To communicate emotions we do not necessarily require language even today, because our innate expressive behavior repertoire is quite sufficient. It is possible that the function of language originally was only to communicate certain environmental contingencies, such as collaboration for the hunt. Children also use their first sentences for communicating about the environment and only much later express their emotions by the use of language. R. J. Andrew (1963a) calls the emancipation of vocalizations from emotions an important prerequisite for human language. This fits in with the observation that the most sensitive parts of our auditory range, around 3000 cycles per second, is not utilized in speech. In this range lies the frequency of the distress calls of a child or a woman, to which we probably react innately; in other words, this range is emotionally occupied. For speech we use the available frequencies around 1000 cycles per second. Here we are also aware of the freedom from instincts, and only because of this can language become the basis for objective communication. This objectivity is a distinctly human characteristic, but the independence from instincts has its limits. It is true that, in man, learned behavior patterns predominate over innate ones. In absolute terms, however, we most likely do not have fewer fixed action patterns than other primates; more likely we have more. In addition, we have inborn motivations such as play, hunting, and gathering drives, striving for rank and status, which have no definite correlated motor patterns and which are in part recent acquisitions in our phylogeny; we may think here of our drive to speak (p. 114). Such primary motivations are probably the cause of the principal analogies in various cultures, because they have a determining influence as learning dispositions. In the course of this learning, man learns by innate mechanisms when his behavior is drive reducing or when a terminal, drive-rewarding situation has been met. This liberation of the drives from the bonds of strictly programmed courses of motor behavior allows for the wide range of adaptive modifiability of our behavior and is one more of man's distinct behavioral characteristics.

A look into the future

The claim of ethology to be able to contribute to a better understanding of human behavior has often been rejected by scholars of a more philosophical orientation. The methods of ethology, it is said, are not adequate for human psychology precisely because we are not animals (Rosenkötter 1966). Animal behavior is said to run its course like a programmed computer, and this model is not applicable to human behavior.

First we must reply that an ethologist does not uncritically extrapolate from insights gained in the study of one animal species to another. Instead he develops working hypotheses whose applicability must be verified in each individual case. However, the hypotheses that were developed during the study of one species are of great help in studying a new species. Physiologists have for a long time successfully used this approach, for example, when they use a drug that was tested in rabbits and apply it to the cure of human ailments. In so doing they in no way imply that human beings are rodents. The closer we are to an animal species, the greater is the probability that insights gained from studying one species are applicable to another. In this connection let me point out that a sharp separation of animal and man in a scientific sense has been indefensible since Darwin, especially in view of the more recent primate studies in Japan and Africa. When we apply objective criteria, an earthworm has less in common with a chimpanzee than the latter with us. Whoever continues to contrast earthworm and chimpanzees as "animals" with man, states a personal preference based on his philosophy but does not classify them according to scientific criteria. When we speak of man, then, we do so in the sense of one particular species. We also contrast the human species frequently to others, but not by excluding it from the animal kingdom. We are united with all animals by a natural relationship. We have shown that ethological working hypotheses are also useful in the study of human behavior. In human behavior we were able to demonstrate the existence of phylogenetic adaptations in the form of fixed action patterns, innate releasing mechanisms, releasers, internal motivating mechanisms, and inborn learning dispositions. Our knowledge is still filled with gaps, but we are encouraged to continue our research in line with biological viewpoints, especially as it has been shown that, especially with respect to our social behavior, our actions are to an important extent preprogrammed by phylogenetic adaptations. Our social behavior, particularly, is clearly disrupted today by certain changes in our environment, and only insight into the causal relationships can lead to successful therapy. Only the exact knowledge of the determinants of our behavior will lead to its eventual mastery. If we know, for example, which releasing stimulus situations arouse certain impulses to act, then we can either seek them out or avoid them. Insight into the workings of our innate motivating mechanisms will make their control easier, as is true for all insights into relationships which allow us to extricate ourselves from a rigid stimulus–response chain by virtue of our mental capacities. We have already discussed the capacity to detach ourselves from a problem and view it from a distance, an ability that is highly developed in man (p. 238). This ability to gain perspective allows us to contemplate the consequences of our actions and to choose among several alternative courses of action. Yes, up to a certain point man can, with the help of this capacity, act against his drives. It is in a way the basis of our specifically human freedom. The prerequisite for a responsible decision is, however, the causal understanding of those behavior mechanisms which underlie our behavior. The less we know about them, the more blindly will we follow their dictates.

BIBLIOGRAPHY

ABEL, E. (1960a) Liaison facultative d'un poisson (*Gobius bucchichii* Stein-dachner) et d'une Anémone (*Anemonia sulcata*) en Mediterranée. Vie et Milieu, 11:518–531.

———— (1960b) Fische zwischen Seeigel-Stacheln. Natur und Volk, 90:33–37.

ADAMSON, J. (1960) Born Free. London (Collins).

ADLER, J., LINN, G., and MOORE, A. V. (1958) Pushing in Cattle: Its Relations to Instinctive Grasping in Humans. Animal Behavior, 6:85–86.

ADRIAN, E. D., and BUYTENDIJK, F. J. J. (1931) Potential Changes in the Isolated Brainstem of the Goldfish. J. Physiol. (London), 71:121–135.

AGRANOFF, B. W. (1967) Memory and Protein Synthesis. Sci. Am., 216(6): 115–123.

AHRENS, R. (1953) Beitrag zur Entwicklung des Physiognomie- und Mimik-erkennens. Z. Exptl. Angew. Psychol., 2:412–454, 599–633.

ALBRECHT, H. (1966a) *Tilapia mariae (Cichlidae)*, Kampf zweier Männchen. Encycl. cinem. E 603. Göttingen (Inst. wiss. Film).

———— (1966b) Zur Stammesgeschichte einiger Bewegungsweisen bei Fischen: untersucht am Verhalten von *Haplochromis (Pisces, Cichlidae)*. Z. Tierpsy-chol., 23:270–302.

ALBRECHT, H., and WICKLER, W. (1968) Freilandbeobachtungen zur "Begrü-ßungszeremonie" des Schmuckbartvogels *Trachyphonus d'arnaudii* (Prevost u. Des Murs). J. Ornith., 109:255–263.

ALEXANDER, B. K. (in preparation) Peer Deprivation in Mother-Reared Rhesus Monkeys. Child Develop. Monogr. Summary in Am. Zool. 6:560.

ALLEE, W. C. (1926) Studies in Animal Aggregations: Causes and Effects of Bunching in Land Isopods. J. Exptl. Zool., 45:255–277.

———— (1938) The Social Life of Animals. London.

ALLEMANN, C. (1951) Die Spieltheorien, Menschenspiel und Tierspiel. Zürich.

ALTEVOGT, R. (1955) Beobachtungen und Untersuchungen an indischen Winkerkrabben. Z. Morphol. Ökol. Tiere, 43:501–522.

———— (1957) Untersuchungen zur Biologie and Ökologie indischer Winker-krabben. Z. Morphol. Ökol. Tiere, 46:1–110.

ALTMAN, J. (1966) Organic Foundations of Animal Behavior. New York (Holt, Rinehart and Winston).

ALTMANN, M. (1952) Social Behavior of Elk, *Cervus canadensis nelsoni* in the Jackson-Hole Area of Wyoming. Behaviour, 4:116–143.

ALTUM, B. (1868) Der Vogel und sein Leben. Münster.

AMBROSE, J. A. (1960) The Smiling and Related Responses in Early Human Infancy. An Experimental and Theoretical Study of Their Course and Significance. Univ. London, Ph.D. Dissertation.

——— (1961) The Development of the Smiling Response in Early Infancy. In Foss, B. M. (ed.), Determinants of Infant Behavior. London (Methuen).

——— (1963) The Age of Onset of Ambivalence in Early Infancy: Indications from the Study of Laughing. J. Child Psychol. Psychiat., 4:167–181.

ANDREW, R. J. (1963a) The Origin and Evolution of the Calls and Facial Expressions of the Primates. Behaviour, 20:1–109.

——— (1963b) Evolution of Facial Expression. Science, 142:1034–1041.

ANTHONEY, T. R. (1968) The Ontogeny of Greeting, Grooming and Sexual Motor Patterns in Captive Baboons (Superspecies *Papio cynocephalus*). Behaviour, 31:358–372.

ANTONIUS, O. (1939) Über Symbolhandlungen und Verwandtes bei Säugetieren. Z. Tierpsychol., 3:263–278.

——— (1947) Beobachtungen an einem Onagerhengst. Umwelt, 1:299–300.

APFELBACH, R. (1967a) Kampfverhalten und Brutpflegeform bei *Tilapia*. Naturwiss., 54:72.

——— (1967b) *Tilapia macrochir (Cichlidae)* Laichablage. Wiss. Film E1019, Göttingen, Publikationen zu Wiss. Filmen 1A, pp. 63–67.

ARDREY, R. (1962) African Genesis. London (Collins).

——— (1966) The Territorial Imperative. New York (Atheneum).

ARMSTRONG, E. A. (1947) Bird Display and Behavior. 2nd ed., London (Lindsay and Drummond Ltd.).

ARONSON, L. R. (1949) An Analysis of the Reproductive Behavior of the Mouth-Breeding Cichlid Fish *Tilapia macrocephala* Bleeker. Zoologica, 34:133–157.

——— (1951) Orientation and Jumping Behavior in the Gobiid Fish *Bathygobius soporator*. Am. Museum Novitates, 1486.

——— (1956) Further Studies on Orientation and Jumping Behavior in the Gobyfish *Bathygobius soporator*. Anat. Rec., 125:606.

ASCH, S. (1952) Social Psychology. Englewood Cliffs, N.J. (Prentice-Hall).

ASCHOFF, J. (1955a) Jahresperiodik der Fortpflanzung bei Warmblütlern. Studium Gen., 8:742–776.

——— (1955b) Tagesperiodik bei Mäusestämmen unter konstanten Umgebungsbedingungen. Arch. Ges. Physiol., 262:51–59.

——— (1960) Exogenous and Endogenous Components in Circadian Rhythms. Cold Spring Harbor Symp. Quant. Biol., 25:11–26.

——— (1962) Spontane lokomotorische Aktivität. In KÜKENTHAL, Handb. d. Zool., 8, 11. (4), 1–74.

——— (1964) Die Tagesperiodik licht- und dunkelaktiver Tiere. Rev. Suisse Zool., 71:528–558.

——— (1965) Circadian Clocks. Proc. Feldafing Summer School, Sept. 1964. Amsterdam (North-Holland).

——— (1966) Tagesrhythmus des Menschen bei völliger Isolation. Umschau, 12:378–383.

——— (1967) Human Circadian Rhythms in Activity, Body Temperature and Other Functions. Life Sciences and Space Research. Amsterdam (North-Holland), 159–173.

Aschoff, J., Gerecke, U., and Wever, R. (1967) Phasenbeziehungen zwischen den circadianen Perioden der Aktivität und der Kerntemperatur beim Menschen. Arch. Ges. Physiol., 295:173–183.

Aschoff, J., and Meyer-Lohmann, J. (1954) Angeborene 24-Stunden-Periodik bei Küken. Arch. Ges. Physiol., 260:170–176.

Aschoff, J., and Wever, R. (1962a) Aktivitätsmenge und a:α-Verhältnis als Meßgröße der Tagesperiodik. Z. Vergleich. Physiol., 46:88–101.

———— (1962b) Spontanperiodik des Menschen bei Ausschluß aller Zeitgeber. Naturwiss., 49:337–342.

Autrum, H. (1943) Über kleinste Reize bei Sinnesorganen. Biol. Zbl. 63:209–236.

———— (1948) Über Energie- und Zeitgrenzen der Sinnesempfindungen. Naturwiss. 35:361–369.

———— (1952) Nerven- und Sinnesphysiologie. Fortschr. Zool., 9:537–604.

———— (1958) Electrophysiological Analysis of the Visual Systems in Insects. Exptl. Cell. Res. Suppl., 5:436–439.

———— (1962) Die Sinnesorgane und ihre Arbeitsweise. Der große Herder, Ergeb., 1:519–534. Freiburg (Herder).

———— (1966) Tier und Mensch in der Masse. Festrede. Munich (Verl. Bayr. Akad. Wiss.).

Autrum, H., and Holst, D. v. (1968) Sozialer "Stress" bei Tupajas (*Tupaia glis*) und seine Wirkung auf Wachstum, Körpergewicht und Fortpflanzung. Z. Vergleich. Physiol., 58:347–355.

Azrin, N. H., Hutchinson, R. R., and McLaughlin, R. (1965) The Opportunity for Aggression as an Operant Reinforcer during Aversive Stimulation. J. Exptl. Anal. Behavior, 8:171–180.

Babich, F. R., Jacobson, A. L., Bubash, S., and Jacobson, A. (1965) Transfer of a Response to Naive Rats by Injection of Ribonucleid Acid Extracted from Trained Rats. Science 149:656–657.

Backhaus, D. (1960) Über das Kampfverhalten beim Steppenzebra. Z. Tierpsychol., 17:345–350.

———— (1961) Giraffen in zoologischen Gärten und freier Wildbahn. Inst. des Parcs Nat. du Congo et du Ruanda-Urundi, Brussels.

Baerends, G. P. (1941) Fortpflanzungsverhalten und Orientierung der Grabwespe, *Ammophila campestris*. Tijdschr. Entomol., 84:68–275.

———— (1950) Les Sociétés et les Familles de Poissons. Colloq. Intern. Centre Natl. Rech. Sci., 34:207–219 (Paris).

———— (1956) Aufbau tierischen Verhaltens. In Kükenthal, Handb. d. Zool., 8, 10 (3), 1–32.

———— (1957) Behavior: The Ethological Analysis of Fish Behavior. In Brown, M. (ed.), Physiology of Fishes II. New York (Academic Press), 229–269.

———— (1958) Comparative Methods and the Concept of Homology in the Study of Behaviour. Arch. Neerl. Zool., 13:401–417.

Baerends, G. P., and Baerends-Van Roon, J. M. (1950) An Introduction to the Study of the Ethology of Cichlid Fishes. Behaviour, Suppl., 1:1–243.

Baerends, G. P., Bril, K. A., and Bult, P. (1965) Versuche zur Analyse einer erlernten Reizsituation bei einem Schweinsaffen. Z. Tierpsychol., 22:394–411.

Baerends, G. P., Brower, R., and Waterbolk, H. T. (1955) Ethological Studies on *Lebistes reticulatus* Peter. I. Analysis of the Male Courtship Pattern. Behaviour, 8:249–334.

Baeumer, E. (1955) Lebensart des Haushuhns. Z. Tierpsychol., 12:387–401.

BAK, I. J. (1965) Electron Microscopic Observations in the Substantia nigra of Mouse during Reserpine Administration. Experientia, 21:568.

BALLY, G. (1945) Vom Ursprung und von den Grenzen der Freiheit, eine Deutung des Spieles bei Tier und Mensch. Basel (Birkhäuser).

BANHAM, K. M. (1950) Development of Affectionate Behavior in Infancy. J. Gen. Psychol., 76:283–289.

BANKS, E. M. (1962) A Time and Motion Study of Prefighting Behavior in Mice. J. Gen. Psychol., 101:165–183.

BARDACH, J. E., WINN, H. E., and MENZEL, D. W. (1959) The Role of the Senses in the Feeding of Nocturnal Reef Predators Gymnothorax moringa and G. vicinus. Copeia, 2:133–139.

BARLOW, H. B., HILL, R. M., and LEVICK, W. R. (1964) Retinal Ganglion Cells Responding Selectively to Direction and Speed of Image Motion in the Rabbit. J. Physiol. (Cambridge), 173:377–407.

BARTHOLOMEW, G. A. (1953) Behavioral Factors Affecting Social Structures in the Alaska Fur Seal. Trans. 18th North Am. Wildlife Conf., Wildlife Management Inst., Washington.

BASEDOW, H. (1906) Anthropological Notes on the Western Coastal Tribes of the Northern Territory of South Australia. Trans. Roy. Soc. South Australia, 31:1–62.

BASTOCK, M. (1956) A Gene Mutation which Changes a Behavior Pattern. Evolution, 10:421–439.

BASTOCK, M., MORRIS, D., and MOYNIHAN, M. (1953) Some Comments on Conflict and Thwarting in Animals. Behaviour, 6:66–84.

BATESON, G., and MEAD, M. (1942) Balinese Character: A Photographic Analysis. Special Publ. N. Y. Acad. Sci., 2.

BATESON, P. P. B. (1966) The Characteristics and Context of Imprinting. Biol. Rev., 41:177–220.

BATHAM, E. J., and PANTIN, C. F. A. (1950) Inherent Activity in the Sea-Anemone Metridium senile. J. Exptl. Biol., 27:290–301.

BEACH, F. A. (1937) The Neural Basis of Innate Behavior. I. Effects of Cortical Lesions upon the Maternal Behavior Pattern in the Rat. J. Comp. Physiol. Psychol., 24:393–439.

——— (1938) The Neural Basis of Innate Behavior. II. Relative Effects of Partial Decortication in Adulthood and Infancy upon the Maternal Behavior of the Primiparous Rat. J. Gen. Psychol., 53:108–148.

——— (1940) Effects of Cortical Lesions upon the Copulatory Behavior of Male Rats. J. Comp. Physiol. Psychol., 29:193–245.

——— (1942) Comparison of Copulatory Behavior of Male Rats Raised in Isolation, Cohabitation and Segregation. J. Gen. Psychol., 60:121–136.

——— (1947) Evolutionary Changes in the Physiological Control of Mating Behavior in Mammals. Psychol. Rev., 54:297–315.

——— (1948) Hormones and Behavior. New York (Hoeber).

——— (1958) Normal Sexual Behavior in Male Rats Isolated at Fourteen Days of Age. J. Comp. Physiol. Psychol., 51:37–42.

BECHTEREW, W. (1913) Objektive Psychologie oder Psychoreflexologie. Leipzig.

——— (1926) Reflexologie des Menschen. Leipzig.

BECKER, G. (1965) Zur Magnetfeldorientierung von Dipteren. Z. vergleich. Physiol., 51:135–150.

BEEBE, W. (1953) A Contribution to the Life History of the Euchromid Moth Aethria carnicauda Butler. Zoologica, 38:155–160.

BELLOWS, R. T. (1939) Time Factors in Water Drinking in Dogs. Am. J. Physiol., 125:87–97.

BENEDICT, R. (1934) Patterns of Culture. Boston (Houghton Mifflin).
——— (1955) Urformen der Kultur. Rowohlts Deut. Enzykl., 7.
BENSON, C. W. (1948) Geographical Voice Variation in African Birds. Ibis, 90:48–71.
BERELSON, B., and STEINER, G. A. (1964) Human Behavior. New York (Harcourt).
BERKOWITZ, L. (1962) Aggression. A Social Psychological Analysis. New York (McGraw-Hill).
BEST, J. B. (1963) Protopsychology. Sci. Am., 208(2):55–62.
BETHE, A. (1898) Dürfen wir den Ameisen und Bienen psychische Qualitäten zuschreiben? Arch. Ges. Physiol. 70:15–100.
BIERENS DE HAAN, J. A. (1940) Die tierischen Instinkte und ihr Umbau durch Erfahrung. Leiden.
BILZ, R. (1940) Pars pro toto. Ein Beitrag zur Pathologie menschlicher Affekte. Leipzig (Thieme).
——— (1944) Zur Grundlegung einer Paläopsychologie. I. Paläophysiologie. II. Paläopsychologie. Schweiz. Z. Psychol., 3:202–212, 272–280.
BIRCH, H. G. (1945) The Relation of Previous Experience to Insightful Problem-Solving. J. Comp. Physiol. Psychol., 38:367–383.
BIRDWHISTELL, R. L. (1963) The Kinesis Level in the Investigation of the Emotions. In KNAPP, P. H. (ed.), Expressions of the Emotions in Man. New York (International Univ. Press).
——— (1966) Communication Without Words. In ALEXANDRE, P. (ed.), L'Aventure Humaine (Paris).
BIRKET-SMITH, K. (1948) Die Eskimos. Zürich (Orell Füssli).
BIRUKOW, G. (1953) Photogeomenotaxische Transpositionen bei *Geotrupes sylvaticus*. Rev. Suisse Zool., 60:534–540.
——— (1956) Angeborene und erworbene Anteile relativ einfacher Verhaltenseinheiten. Zool. Anz., Suppl., 19:32–48.
BIRUKOW, G., and BUSCH, E. (1957) Lichtkompaßorientierung beim Wasserläufer *Velia currens* am Tage und zur Nachtzeit. Z. Tierpsychol., 14:184–203.
BISCHOF, N. (1966a) Psychophysik der Raumwahrnehmung. Handb. d. Psychol., 1(1), Göttingen, 307–408.
——— (1966b) Stellungs-, Spannungs- und Lagewahrnehmung. Handb. d. Psychol., 1 (1). Göttingen, 409–497.
BISCHOFF, H. (1927) Biologie der Hymenopteren. Biol. Stud.bücher. Berlin (Springer).
BITTERMANN, M. E. (1965) The Evolution of Intelligence. Sci. Am., 212:92–100.
BLAIR, W. F. (1957a) Mating Call and Relationship of *Bufo hemiophrys*. Texas J. Sci., 9:99–108.
——— (1957b) Structure of the Call and Relationship of *Bufo microscaphus*. Copeia., 208–212.
——— (1958) Structure and Species Groups in US Tree Frogs (*Hyla*). Southwestern Naturalist, 3:77–89.
BLAUVELT, H. (1964) Dynamics of the Mother-Newborn Relationship in Goats. Group Proc. Josiah Macy Jr. Found. New York, 221–258.
BLEST, A. D. (1957) The Function of Eye-Spot Patterns in the Lepidoptera. Behaviour, 11:209–255.
——— (1960) The Evolution Ontogeny and Quantitative Control of Settling Movements of Some New World Saturnid Moths, with Some Comments on Distance Communication by Honey-Bees. Behaviour, 16:188–253.
——— (1966) Learning, Instinct and Evolution. Nature, 212:564.
BLODGETT, H. C. (1929) The Effect of the Introduction of Reward upon Maze Performance of Rats. Univ. Calif. Publ. Psychol., 4:113–134.

BLÖSCH, M. (1965) Untersuchungen über das Zusammenleben von Korallen-fischen *(Amphiprion)* mit Seeanemonen. Dissertation, Univ. Erlangen.

BOGERT, C. M. (1961) The Influence of Sound on the Behavior of Amphibians and Reptiles. In LANYON, W. E., and TAVOLGA, W. N. (eds.), Animal Sounds and Communication. Washington.

BOHANNAN, P. (1966) Law and Warfare. New York (Natural History Press).

BOLWIG, N. (1964) Facial Expression in Primates with Remarks on Parallel Development in Certain Carnivores. Behaviour, 22:167–192.

BONNER, J. (1964) The Molecular Biology of Memory. Summary of the Sym-posium on the Role of Macromolecules in Complex Behavior, Kansas State University, Manhattan, 89–95.

BOPP, P. (1954) Schwanzfunktionen bei Wirbeltieren. Rev. Suisse Zool., 61: 83–151.

BORELL DU VERNAY, W. v. (1942) Assoziationsbildung und Sensibilisierung bei *Tenebrio.* Z. Vergleich. Physiol., 30:84–116.

BOURLIERE, F. (1950) Classification et Caractéristiques des Principaux types de groupements sociaux chez les Vertébres sauvages. Colloq. Intern. Centre Natl. Rech. Sci. (Paris), 34:71–79.

——— (1955) The Natural History of Mammals. London (Harrap).

BOWER, T. G. (1966) Slant Perception and Shape Constancy in Infants. Science, 151:832–834.

BOWLBY, J. (1952) Maternal Care and Mental Health. World Health Organ-ization Monogr. Ser., 2.

——— (1958) The Nature of the Child's Tie to His Mother. Intern. J. Psy-choanal., 39:350–373.

——— (1969) Attachment and Loss. Vol. 1. Attachment. The Int. Psycho-Analytical Library No. 79, London 428 S (Hogarth Press).

BOYCOTT, B. (1965) Learning in the Octopus. Sci. Am., 212 (3):42–50.

BRAEMER, W. (1960) A Critical Review of the Sun-Azimuth Hypothesis. Cold Spring Harbor Symp. Quant. Biol., 25:413–427.

BRAUM, E. (1963) Die ersten Beutefanghandlungen junger Blaufelchen *(Coregonus wartmanni* Bloch) und Hechte *(Esox lucius* L.). Z. Tierpsychol., 20: 257–266.

BRELAND, K. and BRELAND, M. (1966) Animal Behavior. New York (Mac-millan).

BROCK, F. (1927) Das Verhalten des Einsiedlerkrebses *Pagurus arrosor* wäh-rend des Aufsuchens, Ablösens und Aufpflanzens seiner Seerose *Sagartia parasitica.* Roux Arch. Entwicklungsmech., 112:204–238.

BROWER, L. P., and ZANDT-BROWER, J. v. (1962) Investigations into Mimicry. Natural Hist., 71:8–19.

BROWN, A. W. A., and WEBB, H. M. (1960) A "Compass-Direction Effect" for Snails in Constant Conditions and Its Lunar Modulation. Biol. Bull., 119: 307.

BROWN, F. A., Jr. (1965) A Unified Theory for Biological Rhythms: Rhythmic Duplicity and Genesis of "Circa" Periodisms. In ASCHOFF, J. (ed.), Cir-cadian Clocks. Amsterdam (North-Holland), 231–261.

BRUCE, H. M. (1961) Time Relations in the Pregnancy-Block Induced in Mice by Strange Males. J. Reprod. Fertility, 2:138.

BRUHIN, H. (1953) Zur Biologie der Stirnaufsätze bei Huftieren. Physiol. Comp. Oecol., 3:63–92, 93–127.

BUBENIK, A. B. (1968) The Significance of the Antlers in the Social Life of the Cervidae. Deer, 1:208–214

BUCHHOLZ, K. F. (1957) Das Sitzverhalten einiger *Orthretum*-Arten (Odonata). Bonner Zool. Beitr., 9:296–301.

BUDDENBROCK, W. v. (1952) Vergleichende Physiologie, 1. Basel (Birkhäuser).

BUECHNER, H. K. (1961) Territorial Behavior in the Uganda Kob. Science, 133:698–699.

BULLOCK, T. H. (1953) Predator Recognition and Escape Responses of Some Intertidal Gastropods in the Presence of Starfish. Behaviour, 5:130–140.

——— (1961) The Origins of Patterned Nervous Discharge. Behaviour, 17: 48–59.

——— (1962) Integration and Rhythmicity in Neural Systems. Am. Zoologist, 2:97–104.

BULLOCK, T. H., and HORRIDGE, G. A. (1965) Structure and Function in the Nervous System of Invertebrates. I. and II. San Francisco (Freeman).

BÜNNING, E. (1963) Die physiologische Uhr. 2nd ed. Heidelberg (Springer).

BÜRGER, M. (1959) Eine vergleichende Untersuchung über Putzbewegungen bei Lagomorpha und Rodentia. Zool. Garten, 24:434–506.

BURGHARDT, G. M. (1966) Stimulus Control of the Prey Attack Response in Naive Garter Snakes. Psychol. Sci., 4:37–38.

BURGHARDT, G. M., and HESS, E. H. (1966) Food Imprinting in the Snapping Turtle Chelydra serpentina. Science, 151:108–109.

BURKHARDT, D. (1958) Kindliches Verhalten als Ausdrucksbewegungen im Fortpflanzungszeremoniell einiger Wiederkäuer. Rev. Suisse Zool., 65: 311–316.

——— (1960) Die Eigenschaften und Funktionstypen der Sinnesorgane. Ergeb. Biol., 22:226–267.

——— (1961) Allgemeine Sinnesphysiologie und Elektrophysiologie der Rezeptoren. Fortschr. Zool., 13:146–189.

BURKHARDT, D., SCHLEIDT, W. M., and ALTNER, H. (1966) Signale in der Tierwelt. Vom Vorsprung der Natur. Munich (Moos).

BUSNEL, R. G. (1964) Acoustic Behaviour of Animals. Amsterdam (Elsevier).

BUTENANDT, A. (1955) Über Wirkstoffe des Insektenreiches. II. Zur Kenntnis der Sexual-Lockstoffe. Naturw. Rundschau, 12:457–464.

BUTENANDT, A., BECKMANN, R., STAMM, D., and HECKER, E. (1959) Über den Sexuallockstoff des Seidenspinners Bombyx mori. Reindarstellung und Konstitution. Z. Naturforsch., 143:283–284.

BUTLER, R. A. (1953) Discrimination Learning by Rhesus Monkeys to Visual Exploration Motivation. J. Comp. Physiol. Psychol., 46:95–98.

BUYTENDIJK, F. J. J. (1933) Wesen und Sinn des Spieles. Berlin.

——— (1940) Wege zum Verständnis der Tiere. Zurich.

BYRNE, W. L., et al. (1966) Memory Transfer. Science, 153:658–659.

CALHOUN, J. B. (1962) Population Density and Social Pathology. Sci. Am., 206(2):139–148.

CARMICHAEL, L. (1926) The Development of Behavior in Vertebrates Experimentally Removed from the Influence of External Stimulation. Psychol. Rev., 33:51–58.

——— (1927) A Further Study of the Development of the Behavior of Vertebrates Experimentally Removed from the Influence of External Stimulation. Psychol., Rev., 34:34–47.

——— (1928) A Further Experimental Study of the Development of Behavior. Psychol. Rev., 35:253–260.

CARPENTER, C. R. (1942) Societies of Monkeys and Apes. Biol. Symp., 8:177–204.

——— (1965) The Howlers of Barro Colorado Island. In DE VORE, I. (ed.), Primate Behavior. New York (Holt, Rinehart and Winston, Inc.), 250–291.

CARR, A. (1965) The Navigation of the Green Turtle. Sci. Am., 212(5):78–86.

CARTHY, J. D. (1956) Animal Navigation. How Animals Find Their Way about. London (G. Allen).

CASPARI, E. (1964) Behavior Genetics. Am. Zoologist, 4:95–99.

CASPERS, H. (1961) Beobachtungen über Lebensraum und Schwärmperiodi-zität des Palolowurmes *Eunice viridis*. Intern. Rev. Ges. Hydrobiol., 46: 175–183.

CASTELL, R., and PLOOG, D. W. (1967) Zum Sozialverhalten der Totenkopf-Affen (*Saimiri sciureus*). Auseinandersetzung zwischen zwei Kolonien. Z. Tierpsychol., 24:625–641.

CAVILL, G. W. K., and ROBERTSON, P. L. (1965) Ant Venoms Attractants and Repellents. Science, 149:1337.

CHANCE, M. R. A. (1963) The Social Bond in Primates. Primates, 4:1–22.

CHANCE, M. R. A., and RUSSELL, W. M. S. (1959) Protean Displays: a Form of Allaesthetic Behaviour. Proc. Zool. Soc. London, 132:65–70.

CHAUVIN, R. (1952) L'effet de groupe. Structure et physiologie des sociétés animales. Colloq. Intern. Centre Natl. Rech. Sci. (Paris), 34:81–97.

CHMURZYNSKI, I. A. (1967) On the Role of Relations between Landmarks and the Nest-Hole in the Proximate Orientation of Female *Bembex rostrata* (L.). Acta Biol. Exptl. (Warsaw), 27:221–254.

CHRISTIAN, J. J. (1959) The Roles of Endocrine and Behavioral Factors in the Growth of Mammalian Populations. In GORBMAN, A. (ed), Comparative Endocrinology. New York (Wiley), 71–97.

——— (1960) Factors in Mass Mortality of a Herd of Sika Deer (*Cervus hippon*). Chesapeake Sci. 1:79–95.

——— (1963) Endocrine Adaptive Mechanisms and the Physiologic Regula-tion of Population Growth. In MEYER, M. V., and GELDER, R. van (eds.), Physiological Mammalogy. 1. New York (Academic Press), 189–353.

CLARK, E., ARONSON, L. R., and GORDON, M. (1954) Mating Behavior Pat-terns in Two Sympatric Species of Xiphophorin Fishes: Their Inheritance and Significance in Sexual Isolation. Bull. Am. Mus. Nat. Hist., 103:135–226.

COGHILL, G. E. (1929) Anatomy and the Problem of Behaviour. Cambridge (Univ. Press).

COOK, J. (1784) A Voyage to the Pacific Ocean (1776–1780): London (C. J. KING, ed.).

CORNING, W. C., and JOHN, E. R. (1961) Effect of Ribonuclease on Retention of Conditioned Response in Regenerating Planaria. Science, 134:1363–1365.

COSS, R. G. (1965) Mood Provoking Visual Stimuli, their Origins and Applica-tions. Industrial Design Graduate Program, Los Angeles, The Regents of the University of California.

——— (1968) The Ethological Command in Art. Leonardo I. Great Britain (Pergamon Press). 273–287.

COTT, H. B. (1957) Adaptive Coloration in Animals. London (Methuen).

COWLES, J. T. (1937) Food-Tokens as Incentives for Learning by Chimpan-zees. Comp. Psychol. Monogr., 14:1–96.

CRAIG, W. (1918) Appetites and Aversions as Constituents of Instincts. Biol. Bull. Woods Hole, 34:91–107.

——— (1928) Why do Animals fight? Intern. J. Ethics, 31:264–278.

CRANACH, M. v. (in press) Über die Signalfunktion des Blickes. Soziale Theorie und Soziale Praxis. Festschrift Baumgarten. Meisenheim (A. Hain).

CRANE, J. (1943) Display, Breeding and Relationship of the Fiddler Crabs (*Brachyura Genus Uca*). Zoologica, 28:217–223.

——— (1949) Comparative Biology of Salticid Spiders at Rancho Grande. Venezuela IV. An Analysis of Display. Zoologica, 34:159–214.

——— (1952) A Comparative Study of Innate Defensive Behavior of Trinidad Mantids. Zoologica, 37:259–293.

——— (1957) Basic Patterns of Display in Fiddler Crabs. Zoologica, 42:69–82.

———— (1966) Combat, Display and Ritualisation in Fiddler Crabs. Phil. Trans. Roy. Soc. London B 251:459–472.

CRAWFORD, M. P. (1937) The Cooperative Solving of Problems by Young Chimpanzees. Comp. Psychol. Monogr., 14(2).

CREUTZBERG, F. (1961) On the Orientation of Migrating Elvers (*Anguilla vulgaris*) in a Tidal Area. Neth. J. Sea Res., 1:257–338.

CRISLER, L. (1962) Wir heulten mit den Wölfen. Munich (dtv), 75.

CULLEN, E. (1957) Adaptations in the Kittiwake to Cliffnesting. Ibis. 99:275–302.

———— (1960) Experiments on the Effects of Social Isolation on Reproductive Behaviour in the Three-Spined Stickleback. Animal Behavior, 8:235.

CUNNING, D. C. (1968) Warning Sounds of Moths. Z. Tierpsychol., 25:129–138.

CURIO, E. (1960) Ontogenese und Phylogenese einiger Triebäußerungen von Fliegenschnäppern. J. Ornithol., 101:291–309.

———— (1961) Versuche zur Spezifität des Feinderkennens durch Trauerschnäpper. Experientia, 17:516.

———— (1963) Probleme des Feinderkennens bei Vögeln. Proc. 13th Intern. Ornithol. Congr., 206–239.

———— (1965a) Die Schlangenmimikry einer südamerikanischen Schwärmerraupe. Natur u. Mus., 95:207–211.

———— (1965b) Zur geographischen Variation des Feinderkennens einiger Darwinfinken (*Geospizidae*). Zool. Anz., Suppl., 28:466–492.

———— (1966a) Die Schutzanpassungen dreier Raupen eines Schwärmers (*Lepidopt., Sphingidae*) auf Galapagos. Zool. Jb. System, 91:1–29.

———— (1966b) Färbung und Ruheverhalten dreier Raupenformen eines Schwärmers. Umschau, 14:475.

———— (1966c) Wie Insekten ihre Feinde abwehren. Naturwiss. Med., (11):3–21.

CUSHING, J. E. (1941) An Experiment on Olfactory Conditioning in *Drosophila guttifera*. Proc. Natl. Acad. Sci. U.S., 27:296–299.

DAANJE, A., (1950) On the Locomotory Movements in Birds and the Intention Movements Derived from it. Behaviour, 3:48–98.

DANE, B., and van der KLOOT, W. G. (1964) An Analysis of the Display of the Goldeneye Duck (*Bucephala clangula* L.) Behaviour, 22:282–328.

DARLING, F. Fraser (1937) A Herd of Red Deer. Oxford (Univ. Press).

DART, R. A. (1957) The Osteodontokeric Culture of *Australopithecus prometheus*. Transvaal. Mus. Mem., 10.

DARWIN, C. (1872) The Expression of Emotions in Man and Animals. London.

———— (1881) Das Bewegungsvermögen der Pflanzen. J. v. CARUS. Stuttgart.

DATHE, H. (1964) Zur Körperpflege der Tiere in freier Wildbahn und Gefangenschaft. Wiss. kult. Mitt. Tierpark Berlin. 1:349–383.

DAVENPORT, D. (1955) Specificity and Behavior in Symbioses. Quart. Rev. Biol., 30:29–46.

———— (1966) The Experimental Analysis of Behavior in Symbiosis. In HENRY, S. M. (ed.), Symbiosis I., New York (Academic Press), 381–429.

DAVENPORT, D., and NORRIS, K. (1958) Observations on the Symbiosis of the Sea Anemone *Stoichactis* and the Pomacentrid Fish *Amphiprion percula*. Biol. Bull., 115:397–410.

DAVENPORT, D., ROSS, D. M., and SUTTON, L. (1961) The Remote Control of Nematocyst Discharge in the Attachment of *Calliactis parasitica* to Shells of Hermit Crabs. Vie et Milieu, 12:197:209.

DAVID, F. A., BRONFENBRENNER, U., HESS, E. H., MILLER, D. R. SCHNEIDER,

D. M. and SPUHLER, J. N. (1963) The Incest Taboo and the Mating Patterns of Animals. American Anthropologist 65:253–265.

DAVIS D. E. (1962) The Phylogeny of Gangs. In BLISS, E. (ed.), Roots of Behavior. New York (Harper & Row).

DAWKINS, R. (1968) The Ontogeny of a Pecking Preference in Domestic Chicks. Z. Tierpsychol., 25:170–186.

DEEGENER, P. (1918) Die Formen der Vergesellschaftung im Tierreiche. Leipzig.

DELGADO, J. M. R. (1967) Aggression and Defense under Cerebral Radio Control. UCLA Forum in Medical Science, 7:171–193. In CLEMENTE, C. D. and LINDSLEY, D. B. (eds.), Aggression and Defense. Berkeley (Univ. California Press).

DEMBOWSKI, J. (1955) Tierpsychologie. Berlin (Akademie-Verlag).

DENNIS, W. (1960) Causes of Retardation among Institutional Children. Iran. J. Genet. Psychol., 96:47–59.

DETHIER, V. G. (1957) Communication by Insects: Physiology of Dancing. Science, 125:331–336.

DETHIER, V. G., and BODENSTEIN, D. (1958) Hunger in the Butterfly. Z. Tierpsychol., 15:129–140.

DEVORE, I. (1965) Primate Behavior: Field Studies of Monkeys and Apes. New York (Holt, Rinehart and Winston, Inc.).

DIEBSCHLAG, E. (1940) Psychologische Beobachtungen über die Rangordnung bei der Haustaube. Z. Tierpsychol., 4:173–188.

DIESSELHORST, G. (1965) Klasse Aves, Vögel. Handb. d. Biol., 6, 2. Frankfurt (Athenaion), 745–866.

DIETERLEN, F. (1959) Das Verhalten des syrischen Goldhamsters (Mesocricetus auratus). Z. Tierpsychol., 16:47–103.

DILGER, W. C. (1960) The Comparative Ethology of the African Parrot Genus Agapornis. Z. Tierpsychol., 17:649–685.

———— (1962) The Behavior of Lovebirds. Sci. Am., 206(1):88–98.

DOFLEIN, F. (1905) Beobachtungen an Weberameisen. Biol. Zbl., 25:497–507.

DOLLARD, J., DOOB, L., MILLER, N., MOWRER, O., and SEARS, R. (1939) Frustration and Aggression. New Haven (Yale Univ. Press).

DOLLO, L. (1909) Poissons Voiliers. Zool. Jahrb. System, 27:419.

DREES, O. (1952) Untersuchungen über die angeborenen Verhaltensweisen bei Springspinnen (Salticidae). Z. Tierpsychol., 9:169–207.

DÜHRSSEN, A. (1960) Psychogene Erkrankungen bei Kindern und Jugendlichen. Göttingen (Verlag f. medizin. Psychologie).

DUMAS, F. (1932) La mimique des aveugles. Bull. Acad. Med., 107:607–610.

DUPEYRAT, A. (1963) Papua—Beast and Men. London (MacGibbon and Kee).

EBHARDT, H. (1958) Verhaltensweisen verschiedener Pferdeformen. Säugetierkdl. Mitt., 6:1–9.

ECCLES, J. C. (1953) The Neurophysiological Basis of Mind: The Principles of Neurophysiology. London (Oxford Univ. Press).

EDWARDS, D. (1968) Mice: Fighting by Neonatally Androgenized Females. Science, 161:127–128.

EIBL-EIBESFELDT, I. (1949) Über das Vorkommen von Schreckstoffen bei Erdkrötenquappen. Experientia, 5:236.

———— (1950a) Über die Jugendentwicklung des Verhaltens eines männlichen Dachses (Meles meles L.) unter besonderer Berücksichtigung des Spieles. Z. Tierpsychol., 7:327–355.

———— (1950b) Ein Beitrag zur Paarungsbiologie der Erdkröte (Bufo bufo L.). Behaviour, 2:217–236.

———— (1950c) Beiträge zur Biologie der Haus- und der Ährenmaus nebst einigen Beobachtungen an anderen Nagern. Z. Tierpsychol., 7:558–587.

——— (1951a) Nahrungserwerb und Beuteschema der Erdkröte (*Bufo bufo* L.). Behaviour, 4:1–35.

——— (1951b) Gefangenschaftsbeobachtungen an der persischen Wüstenmaus (*Meriones persicus* Blanford): Ein Beitrag zur vergleichenden Ethologie der Nager. Z. Tierpsychol., 8:400–423.

——— (1951c) Zur Fortpflanzungsbiologie und Jugendentwicklung des Eichhörnchens. Z. Tierpsychol., 8:370–400.

——— (1953a) Zur Ethologie des Hamsters (*Cricetus cricetus* L.). Z. Tierpsychol., 10:204–254.

——— (1953b) Vergleichende Studien an Ratten und Mäusen. Prakt. Desinfektor, 45:166–168.

——— (1953c) Eine besondere Form des Duftmarkierens beim Riesengalago, *Galago crassicaudatus*. Säugetierkdl. Mitt., 1:171–173.

——— (1953d) Ethologische Unterschiede zwischen Hausratte und Wanderratte. Zool. Anz., Suppl., 16:169–180.

——— (1954) Paarungsbiologie der Anuren. Wiss. Film, C628. Göttingen (Inst. wiss. Film).

——— (1955a) Über Symbiosen, Parasitismus und andere zwischenartliche Beziehungen bei tropischen Meeresfischen. Z. Tierpsychol., 12:203–219.

——— (1955b) Ethologische Studien am Galapagos-Seelöwen *Zalophus wollebaeki* Sivertsen. Z. Tierpsychol., 12:286–303.

——— (1955c) Sexualverhalten und Eiablage beim Alpenmolch. Wiss. Film, C698 und Beiheft. Göttingen (Inst. wiss. Film)

——— (1955d) Der Kommentkampf der Meerechse (*Amblyrhynchus cristatus* Bell) nebst einigen Notizen zur Biologie dieser Art. Z. Tierpsychol., 12:49–62 (siehe auch wiss. Film der Encycl. cinem., E591. Göttingen [Inst. wiss. Film] 1964).

——— (1955e) Biologie des Iltisses (*Putorius putorius*). Wiss. Film C 697, Göttingen (Inst. wiss. Film).

——— (1956a) Vergleichende Verhaltensstudien an Anuren. 2. Zur Paarungsbiologie der Gattungen *Bufo, Hyla, Rana* und *Pelobates*. Zool. Anz., Suppl., 19:315–323.

——— (1956b) Einige Bemerkungen über den Ursprung von Ausdrucksbewegungen bei Säugetieren. Z. Säugetierkde., 21:29–43.

——— (1956c) Zur Biologie des Iltis (*Putorius putorius* L.). Zool. Anz., Suppl., 19:304–314.

——— (1957a) Ausdrucksformen der Säugetiere. In KÜKENTHAL, Handb. d. Zool., 8(6), 1–26.

——— (1957b) *Rattus norvegicus*. Kampf I (erfahrener Männchen). Encycl. cinem., E131. Kampf II (unerfahrener Männchen). Encycl. cinem., E132. Göttingen (Inst. wiss. Film).

——— (1958a) Das Verhalten der Nagetiere. In KÜKENTHAL, Handb. d. Zool., 8(10), 13, 1–88.

——— (1958b) Versuche über den Nestbau erfahrungsloser Ratten. Wiss. Film B 757, Göttingen (Inst. wiss. Film).

——— (1959) Der Fisch *Aspidontus taeniatus* als Nachahmer des Putzers *Labroides dimidiatus*. Z. Tierpsychol., 16:19–25.

——— (1960a) Beobachtungen und Versuche an Anemonenfischen der Malediven und der Nicobaren. Z. Tierpsychol., 17:1–10.

——— (1960b) Naturschutzprobleme auf den Galapagos-Inseln. Acta Trop., 17:97–137.

——— (1961a) The Interactions of Unlearned Behavior Patterns and Learning in Mammals. In DELAFRESNAYE, J. F. (ed.). Brain Mechanisms and Learning. Oxford (Blackwell), 53–73.

——— (1961b) The fighting Behavior of Animals. Sci. Am., 205(6):112–121.

────── (1961c) Eine Symbiose von Fischen *(Siphamia versicolor)* und Seeigeln. Z. Tierpsychol., 18:56–59.

────── (1962a) Die Verhaltensentwicklung des Krallenfrosches *(Xenopus laevis)* und des Scheibenzünglers *(Discoglossus pictus)* unter besonderer Berücksichtigung der Beutefanghandlungen. Z. Tierpsychol., 19:385–393.

────── (1962b) Freiwasserbeobachtungen zur Deutung des Schwarmverhaltens verschiedener Fische. Z. Tierpsychol., 19:165–182.

────── (1963) Angeborenes und Erworbenes im Verhalten einiger Säuger. Z. Tierpsychol., 20:705–754.

────── (1964a) *Tropidurus delanonis* (Iguanidae). Kommentkampf der Männchen. Encycl. cinem., E609. Göttingen (Inst. wiss. Film).

────── (1964b) Galapagos, die Arche Noah im Pazifik. 3rd ed. Munich (Piper).

────── (1964c) Im Reich der tausend Atolle. Munich (Piper).

────── (1965a) *Amphiprion percula:* Kampfverhalten. Encycl. cinem., E752. Publ. zu wiss. Filmen, 2 A. Göttingen (Inst. wiss. Film).

────── (1965b) *Nannopterum harrisi (Phalacrocoracidae):* Brutablösung. Encycl. cinem., E596. Publ. zu wiss. Filmen, 1 A, 303–306 Göttingen (Inst. wiss. Film).

────── (1965c) Das Duftmarkieren des Igeltanreks (*Echinops telfairi* Martin). Z. Tierpsychol., 22:810–812.

────── (1966a) Ethologie, die Biologie des Verhaltens. In Gessner, F., and Bertalanffy, L. v. (eds.), Handbuch der Biologie 2. Frankfurt (Athenaion), 341–559.

────── (1966b) *Dicranura vinula*—Feindabwehr. Encycl. cinem., E973. Göttingen (Inst. wiss. Film).

────── (1967) Concepts of Ethology and Their Significance for the Study of Human Behavior. In Stevenson, H. W. (ed.), Early Behavior: Comparative and Developmental Approaches. New York (Wiley), 127–146.

────── (1968). Zur Ethologie des menschlichen Grußverhaltens. I. Beobachtungen an Balinesen, Papuas und Samoanern nebst vergleichenden Bemerkungen. Z. Tierpsychol., 25:727–744.

Eibl-Eibesfeldt, I., and Eibl-Eibesfeldt, E. (1967) Die Parasitenabwehr der Minima-Arbeiterinnen der Blattschneiderameise *Atta cephalotes.* Z. Tierpsychol., 24:279–281.

Eibl-Eibesfeldt, I., and Hass, H. (1959) Erfahrungen mit Haien. Z. Tierpsychol., 16:733–746.

────── (1966) Zum Projekt einer ethologisch orientierten Untersuchung menschlichen Verhaltens. Mitt. Max-Planck-Ges., 6:383–396.

────── (1967) Neue Wege der Humanethologie. Homo, 18:13–23.

Eibl-Eibesfeldt, I., and Klausewitz, W. (1961) *Gnathypops rosenbergi annulata* n. ssp. von den Nikobaren. Senckenbergiana Biol., 42:421–426.

Eibl-Eibesfeldt, I., and Sielmann, H. (1962) Beobachtungen am Spechtfinken *Cactospiza pallida.* J. Ornithol., 103:92–101.

────── (1965) *Cactospiza pallida* (Fringillidae): Werkzeuggebrauch beim Nahrungserwerb. Encycl. cinem., E597. Publ. zu wiss. Filmen, 1 A, 385–390. Göttingen (Inst. wiss. Film).

Eibl-Eibesfeldt, I., and Wickler, W. (1962) Ontogenese und Organisation von Verhaltensweisen. Fortschr. Zool., 15:354–377.

────── (1968) Die ethologische Deutung einiger Wächterfiguren auf Bali. Z. Tierpsychol., 25:719–726.

Eickstedt, E. Fr. v. (1963) Die Forschung am Menschen. Teil 3: Psychologische und philosophische Anthropologie. Stuttgart (Enke), 1513–2645.

Eisenberg, J. F. (1965) The Social Organisations of Mammals. In Kükenthal, Handb. d. Zool., 8(10), 7, 1–92.

EISNER, T., and MEINWALD, J. (1966) Defensive Secretion of Arthropods. Science, 153:1341–1350.

EKMAN, P., SORENSON, E. R., and FRIESEN, W. V. (1969) Pan-Cultural Elements in Facial Displays of Emotion. Science 164:86–88.

ELLEFSON, J. O. (1968) Territorial Behavior in the Common White-Handed Gibbon *Hylobates lar*. In Jay, P.C. (ed.): Primates. New York (Holt, Rinehart and Winston, Inc.).

ELLIOTT, M. H. (1930) Some Determining Factors in Maze Performance. J. Psychol., 42:315–317.

EMERSON, A. E. (1956) Ethospecies, Ethotypes, and the Evolution of *Apicotermes* and *Allognathotermes* (Isoptera, Termitidae). Am. Museum Novitates, 1771:1–31.

EMLEN, STR. T. (1969) Bird Migration: Influence of Physiological State upon Celestial Orientation. Science 165:664–672.

ENRIGHT, J. T. (1961) Lunar Orientation of *Orchestia corniculata*. Biol. Bull., 120:148–156.

――― (1963) The Tidal Rhythm of Activity of a Sand-Beach Amphipod. Z. Vergleich. Physiol., 46:276–313.

ERICKSON, C. J., and LEHRMAN, D. S. (1964) Effect of Castration of Male Ring Doves upon Ovarian Activity of Females. J. Comp. Physiol. Psychol., 58:164–166.

ERIKSON, E. H. (1953) Wachstum und Krisen der gesunden Persönlichkeit. Stuttgart (Klett).

――― (1966) Ontogeny of Ritualisation in Man. Phil. Trans. Roy. Soc. London, B 251:337–349.

ESCH, H. (1967) The Evolution of Bee Language. Sci. Am., 216(4):97–104.

ESCHERICH, K. (1906) Die Ameise. Braunschweig (Vieweg).

ESSER, A. H. (1968) Dominance Hierarchy and Clinical Course of Psychiatrically Hospitalized Boys. Child Develop., 39:147–157.

ESSER, A. H., CHAMBERLAIN, A. S., CHAPPLE, E. D. and KLINE, N. S. (1965) Territoriality of Patients on a Research Ward. In WORTIS, J. (ed.), Recent Advances in Biological Psychiatry, 7. New York (Plenum Press).

EVERETT, G. M. (1961) Some Electrophysiological and Biochemical Correlates of Motor Activity and Aggressive Behavior. Neuropharmacology, 2:479–484.

EVERETT, G. M., and WIEGAND, R. G. (1962) Central Amines and Behavioral States. A Critique and New Data. Proc. 1st Intern. Pharmacol. Meeting, 8, Pharmacol. Analysis of Central Nervous Action. New York (Macmillan), 85–92.

EWER, R. F. (1963) The Behaviour of the Meerkat *Suricata suricatta*. Z. Tierpsychol., 20:570–607.

――― (1968) Ethology of Mammals. London (Logos Press) and New York (Plenum Press).

FABER, A. (1953a) Ausdrucksbewegungen und besondere Lautäußerungen bei Insekten als Beispiel für eine vergleichend-morphologische Betrachtung der Zeitgestalten. Zool. Anz., Suppl., 16:106–115.

――― (1953b) Laut- und Gebärdensprache bei Insekten. Orthoptera I. 278. Mitt. Mus. Naturkde, Stuttgart.

FABRE, J. H. (1879–1910) Souvenier entomologique. Paris (Delagrave), 1–10.

FABRICIUS, E. (1951) Zur Ethologie junger Anatiden. Acta Zool. Fenn., 68:1–178.

FANTZ, R. L. (1967) Visual Perception and Experience in Infancy. In STEVENSON, H. W. (ed.), Early Behavior. New York (Wiley), 181–224.

FARRIS, H. E. (1967) Classical Conditioning of Courting Behavior in the

Japanese Quail *Coturnix coturnix japonica.* J. Exptl. Anal. Behavior, 10:213–217.

FAUBRÉE, J. (1968) Note sur l'economie ostentatoire. Rev. Ticrs-Monde, 9:17–23.

FEDER, H. M. (1966) Cleaning Symbiosis in the Marine Environment. In HENRY, S. M. (ed.), Symbiosis I., New York (Academic Press), 327–380.

FELIPE, N. J., and SOMMER, R. (1966) Invasions of Personal Space. Social Prob., 14:206–214.

FESHBACH, S. (1961) The Stimulating Versus Cathartic Effects of a Vicarious Aggressive Activity. J. Abnorm. Soc. Psychol., 63:381–385.

FESTETICS, A. (1959) Ökologische Untersuchungen an Brutvögeln des Saser. Vogelwelt, 80:1–21.

―――― (1961) Ährenmaushügel in Österreich. Z. Säugetierkde., 26:1–14.

FEUSTEL, H. (1966) Anatomische Untersuchungen zum Problem der *Aspido-siphon-Heterocyathus*-Symbiose. Zool. Anz., Suppl., 29:131–143.

FIEDLER, K. (1954) Vergleichende Verhaltensstudien an Seenadeln, Schlangen-nadeln und Seepferdchen. Z. Tierpsychol., 11:358–416.

―――― (1964) Verhaltensstudien an Lippfischen der Gattung *Crenilabrus*. Z. Tierpsychol., 21:521–591.

FISCHER, H. (1965) Das Triumphgeschrei der Graugans *(Anser anser).* Z. Tier-psychol., 22:247–304.

FISCHER, K. (1961) Untersuchungen über die Sonnenkompaßorientierung und Laufaktivität von Smaragdeidechsen (*Lacerta viridis* Laur). Z. Tierpsychol., 18:450–470.

―――― (1963) Spontanes Richtungsfinden nach dem Sonnenstand bei *Chelonia mydas* (Suppenschildkröte). Naturwiss., 51:203.

FISHER, J., and HINDE, R. A. (1949): The Opening of Milk Bottles by Birds, Brit. Birds, 42:347–358.

FLETCHER, R. (1948) Instinct in Man. Aberdeen (Univ. Press).

FOREL, A. (1892) Die Nester der Ameisen. Zürich.

FRAENKEL, G. S., and GUNN, D. S. (1961) The Orientation of Animals. Oxford (Clarendon Press).

FRANCK, D. (1966) Möglichkeiten zur vergleichenden Analyse auslösender und richtender Reize mit Hilfe des Attrappenversuches, ein Vergleich der Successiv- und Simultanmethode. Behaviour, 27:150–159.

FRANK, F. (1953) Über den Zusammenbruch von Feldmausplagen. Zool. Jahrb. Abt. System, 82:1–156.

FRANZISKET, L. (1955) Die Bildung einer bedingten Hemmung bei Rücken-marksfröschen. Z. Vergleich. Physiol., 37:161–168.

FREEDMAN, D. G. (1964) Smiling in Blind Infants and the Issue of Innate vs. Acquired. J. Child Psychol. Psychiat., 5:171–184.

―――― (1965a) Hereditary Control of Early Social Behavior. In FOSS, B. M. (ed.), Determinants of Infant Behavior. 3. London (Methuen).

―――― (1965b) An Ethological Approach to the Genetic Study of Human Behavior. In VANDENBERG, S. G., Methods and Goals in Human Behavior Genetics. New York (Academic Press), 141–161.

―――― (1967) A Biological View of Man's Social Behavior. In ETKIN, W., and FREEDMAN, D. G. (eds.), Social Behavior from Fish to Man. Chicago (Phoenix Books).

FREEDMAN, D. X., and GIARMAN, N. J. (1963) Brain Amines, Electrical Activity and Behavior. In GLASER, G. H. (ed.), EEG and Behavior. New York (Basic Books), 198–243.

FREMONT-SMITH, F. (1962) Saving the Children Can Save Us. Saturday Rev., Aug. 11.

FREUD, S. (1950) Gesammelte Werke. 18 vols., London (Imago Publ.).

FREUD, J., and UYLERT, J. E. (1948) Micturation and Copulation Behavior in Dogs. Acta Brevia, 16:49–53.

FREYE, H. A., and GEISSLER, H. (1966) Das Ohrenspiel der Ungulaten als Ausdrucksform. Wiss. Z. Univ. Halle, 15:893–915.

FRIJDA, N. H. (1965) Mimik und Pantomimik. In KIRCHHOFF, R. (ed.), Handb. d. Psychol., 5, Ausdruckspsychol., 351–421.

FRINGS, H., and FRINGS, M. (1959) Reactions of American and French Species of *Corvus* and *Larus* to Recorded Communication Signals Tested Reciprocally. Ecology, 39: 126–131.

FRISCH, K. v. (1914) Der Farben- und Formensinn der Biene. Zool. Jahrb. Allgem. Zool. Physiol., 40:1–186.

———— (1923) Ein Zwergwels, der kommt, wenn man pfeift. Biol. Zbl., 43:439–446.

———— (1941) Über einen Schreckstoff der Fischhaut und seine biologische Bedeutung. Z. Vergleich Physiol., 29:46–145.

———— (1950) Die Sonne als Kompaß im Leben der Bienen. Experientia, 6: 210–221.

———— (1959) Aus dem Leben der Bienen. Verständl. Wissenschaft, I, 6th ed.

———— (1965) Die Tanzsprache und Orientierung der Bienen. Berlin (Springer).

———— (1968) Honeybees: Do They Use Direction and Distance Information Provided by Their Dancers? Science, 158:1072–1076.

FRISCH, K. v., LINDAUER, M., and DAUMER, K. (1960) Über die Wahrnehmung polarisierten Lichtes durch das Bienenauge. Experientia, 16:289–302.

FRISCH, O. v. (1958) Die Bedeutung der elterlichen Warnrufe für Brachvogel- und Limicolenkücken. Z. Tierpsychol., 15:381–382.

———— (1962) Zur Biologie des Zwergchamäleons. Z. Tierpsychol., 19:276–289.

FROMME, A. (1941) An Experimental Study of the Factors of Maturation and Practise in Behavioral Development of the Embryo of the Frog *Rana pipiens*. Genet. Psychol. Monogr., 24:219–261.

FUCHS, P. (1967) Tatauierung in Afrika. Bild d. Wiss., 2:109–117.

FÜLLER, H. (1958) Symbiose im Tierreich. Wittenberg (Ziemsen).

FULLER, J. L., and THOMPSON, W. R. (1960) Behavior Genetics. New York (Wiley).

FUNKE, W. (1965) Untersuchungen zum Heimfindeverhalten und zur Ortstreue von *Patella* L. Zool. Anz., Suppl., 28:411–418.

GAITO, J. (1964) Nucleic Acids and Brain Function. Symp. on the Role of Macromolecules in Complex Behavior. Kansas State University, Manhattan, 68–75.

———— (1966) Macromolecules and Behavior. Amsterdam (North-Holland).

GAMPER, E. (1926) Bau- und Leistungen eines menschlichen Mittelhirnwesens (Arhinencephalie mit Encephalocelie). Z. ges. Neurol. Psychiat., 102:154–235, 104:49–120.

GARCIA, J., and ERVIN, F. R. (1968) Gustatory-visceral and Telereceptor-cutaneous Conditioning-Adaptation in Internal and External Milieus. Communications in Behavioral Biol., Part A, 1:389–415.

GARCIA, J., McGOWAN, B. K., ERVIN, F. R., and KOELLING, R. A. (1968) Cues: Their Relative Effectiveness as a Function of the Reinforcer. Science, 160, 794–795.

GARCIA, J., McGOWAN, B. K., and GREEN, K. F. (1969) Sensory Quality and Integration: Constraints on Conditioning. In Black and Prokasy (eds.) 1969,

Conference on Conditioning at Dept. Psychol. Ontario (McMaster University), (in print).

GARDENER, R. A. and GARDENER, B. T. (1969) Teaching Sign Language to a Chimpanzee. Science 165:664–672.

GARDENER, R. and HEIDER, K. G. (1968) Gardens of War, Life and Death in the New Guinea Stone age. New York, (Random House).

GASTON, S., and MENAKER, M. (1968) Pineal Function: The Biological Clock in the Sparrow. Science, 160:1125–1127.

GATENBY, J. B. (1960) The New Zealand Glow-Worm. Tuatera, 8:86–92.

GATENBY, J. B., and COTTON, S. (1960) Snare Building and Pupation in Bolitophila luminosa. Trans. Roy. Soc. New Zealand, 88:149–156.

GAUL, A. T. (1952) Audiomimicry: An Adjunct to Color Mimicry. Psyche, 59:82–83.

GAZZANIGA, M. S. (1967) The Split Brain in Man. Sci. Am., 217(2):24–29.

GEHLEN, A. (1940) Der Mensch, seine Natur und seine Stellung in der Welt. Berlin.

——— (1956) Urmensch und Spätkultur. Bonn.

——— (1961) Anthropologische Forschung. Rowohlts Deut. Enzycl., 138.

GEIST, V. (1966a) The Evolution of Horn-Like Organs. Behaviour, 27:175–214.

——— (1966b) The Evolutionary Significance of Mountain-Sheep Horns. Evolution, 20:558–566.

——— (1966c) Ethological Observations on Some West-African Cervids. Zool. Beitr., 12:219–250.

GELBER, B. (1965) Studies on the Behavior of Paramaecium aurelia. Animal Behavior, Suppl., 1:21–29.

GENTZ, K. (1935) Zur Brutpflege des Wespenbussards. J. Ornithol., 83:105–115.

GERAMB, V. R. v. (1918) Zur Volkskunde des Gesäusegebietes. Z. Deut. Oesterr. Alpenver., 49:33–66.

GERARD, R. W. (1961) The Fixation of Experience. In DELAFRESNAYE, J. F. (ed.), Brain Mechanisms and Learning. Oxford (Blackwell), 21–35.

GERLACH, S. (1965) Tierwanderungen. Handb. d. Biol., 5. Frankfurt (Athenaion), 413–472.

GESSNER, F. (1942) Die Leistungen des pflanzlichen Organismus. Handb. d. Biol., 4:34–187. Frankfurt (Athenaion).

GIARMAN, N. J., and FREEDMAN, D. X. (1965) Biochemical Aspects of the Actions of Psychomimetic Drugs. Pharmacol. Rev., 17:1–25.

GIBSON, E. J., and WALK, R. D. (1960) The Visual Cliff. Sci. Am., 202(4):64–71.

GILLIARD, E. T. (1963) The Evolution of Bowerbirds. Sci. Am., 209(2):38–46.

GINSBURGH, B. and ALLEE, W. (1942) Some Effects of Conditioning on Social Dominance and Subordination in Inbred Strains of Mice. Physiol. Zool., 15:485–506.

GLICKMANN, S. E., and SROGES, R. W. (1966) Curiosity in Zoo Animals. Behaviour, 24:151–188.

GNADENBERG, W. (1962) Erlebnisse mit Hunden. Z. Tierpsychol., 19:586–596.

GOETHE, F. (1938) Beobachtungen über das Absetzen von Witterungsmarken beim Baummarder. Deut. Jäger, 13.

——— (1939) Über das "Anstoß-Nehmen" bei Vögeln. Z. Tierpsychol., 3:371–374.

——— (1955) Beobachtungen bei der Aufzucht junger Silbermöwen. Z. Tierpsychol., 12:402–433.

GOETSCH, W. (1940) Vergleichende Biologie der Insektenstaaten. Leipzig.

GOLDFARB, W. (1943) The Effects of Early Institutional Care on Adolescent Personality. J. Exptl. Educ., 12:106–129.

GOODALL, J. (1963) My Life among Wild Chimpanzees. Natl. Geogr. Mag., 125(8):272–308.

——— (1965) Chimpanzees of the Gombe Stream Reserve. In DE VORE, I. (ed.), Primate Behavior. New York (Holt, Rinehart and Winston), 425–473.

GOODENOUGH, F. L. (1932) Expressions of the Emotions in a Blind-Deaf Child. J. Abnorm. Soc. Psychol., 27:328–333.

GOTTLIEB, G. (1965a) Imprinting in Relation to Parental and Species Identification by Avian Neonates. J. Comp. Physiol. Psychol., 59:345–356.

——— (1965b) Prenatal Auditory Sensitivity in Chickens and Ducks. Science, 147:1596–1598.

——— (1966) Species Identification by Avian Neonates: Contributory Effects of Perinatal Auditory Stimulation. Animal Behavior, 14:282–290.

GOTTLIEB, G., and KUO, Z. Y. (1965) Development of Behavior in the Duck Embryo. J. Comp. Physiol. Psychol., 59:183–188.

GOULD, E. (1957) Orientation in Box Turtles Terrapene c. carolina. Biol. Bull., 112:336–348.

GRABOWSKI, U. (1941) Prägung eines Jungschafes auf den Menschen. Z. Tierpsychol., 4:326–329.

GRAHAM BROWN, T. (1911) The Intrinsic Factors in the Act of Progression in the Mammal. Proc. Roy. Soc. London B84:308–320.

——— (1912) The Factors in Rhythmic Activity of the Nervous System. Proc. Roy. Soc. London B85:278–289.

GRANIT, R. (1955) Receptors and Sensory Perception. New Haven (Yale Univ. Press).

GRASSÉ, P. P. (1949) Ordre des Isopteres ou Termites. Traité de Zoologie 9: 408–544.

GRAY, J. (1950) The Role of Peripheral Sense Organs during the Locomotion of Vertebrates. Physiological Mechanisms in Animal Behavior. Symp. Soc. Exptl. Biol. (Cambridge Univ. Press), 112–126.

GRAY, J., and LISSMANN, H. W. (1946a) Further Observations on the Effect of Deafferentiation on the Locomotory Activity of Amphibian Limbs. J. Exptl. Biol., 23:121–132.

——— (1946b) The Coordination of Limb Movements in the Amphibia. J. Exptl. Biol., 23:133–142.

GRAY, P. H. (1958) Theory and Evidence of Imprinting in Human Infants. J. Psychol., 46:155–160.

GRETHER, W. F. (1939) Color Vision and Color Blindness in Monkeys. Comp. Psychol. Monogr., 15:1–38.

GRIFFIN, D. R. (1958) Listening in the Dark. New Haven (Yale Univ. Press).

——— (1962) Echo-Ortung der Fledermäuse, insbesondere beim Fangen fliegender Insekten. Naturwiss. Rundschau, 15:169–173.

GRIFFITH-SMITH, N. (1966) Evolution of Some Arctic Gulls (Larus). An Experimental Study of Isolating Mechanisms. Ornithol. Monogr., 4.

GROHMANN, J. (1939) Modifikation oder Funktionsreifung? Z. Tierpsychol., 2:132–144.

GROOS, K. (1933) Die Spiele der Tiere. 3rd ed. Jena.

GROOT, C. (1965) On the Orientation of Young Sockeye Salmon (Oncorhynchus nerka) during Their Seaward Migration Out of Lakes. Behaviour, Suppl., 14.

GROSSMANN, K. E. (1967) Behavioral Difference between Rabbits and Cats. J. Gen. Psychol., 3:171–182.

GRZIMEK, B. (1949a) Die "Radfahrer-Reaktion." Z. Tierpsychol., 6:41–44.
——— (1949b) Ein Fohlen, das kein Pferd kannte. Z. Tierpsychol., 6:391–405.
——— (1951) Affen im Haus. Stuttgart.
——— (1954) Beobachtungen an Schimpansen, *Pan tr. troglodytes* (Blumenbach, 1775) in den Nimbabergen. Säugetierkdl. Mitt., 1:1–5.
GUITON, P. (1960) On the Control of Behavior during the Reproductive Cycle of *Gasterosteus aculeatus.* Behaviour, 15:163–184.
GÜNTHER, K. (1956) Systematik und Stammesgeschichte der Tiere 1939–1953. Fortschr. d. Zool., 10:33–278.
GUTHRIE, E. R. (1952) The Psychology of Learning. New York (Harper).
GWINNER, E. (1961) Über die Entstachelungshandlung des Neuntöters *(Lanius collurio).* Vogelwarte, 21:36–47.
——— (1964) Untersuchungen über das Ausdrucks- und Sozialverhalten des Kolkraben *(Corvus corax).* Z. Tierpsychol., 21:657–748.
——— (1966) Über einige Bewegungsspiele des Kolkraben. Z. Tierpsychol., 23:28–36.
——— (1967) Circannuale Periodik der Mauser und Zugunruhe bei einigen Vögeln. Naturwiss., 54:447.
GWINNER, E., and KNEUTGEN, J. (1962) Über die biologische Bedeutung der "zweckdienlichen" Anwendung erlernter Laute bei Vögeln. Z. Tierpsychol., 19:692–696.
HAAS, A. (1962) Phylogenetisch bedeutungsvolle Verhaltensänderungen bei Hummeln. Z. Tierpsychol., 19:356–370.
——— (1965) Weitere Beobachtungen zum "generischen Verhalten" bei Hummeln. Z. Tierpsychol., 22:305–320.
HACKETT, E., ESSER, A. H., and KLINE, N. S. (1966) Heterogeneous Dimensions of Chronic Schizophrenic Behavior. Manuscript, Research Center, Rockland Hospital, Orangeburg, N.Y.
HALBERG, F., ENGELI, M., HAMBURGER, C., and HILLMAN, D. (1965) Spectral Resolution of Low-Frequency Small-Amplitude Rhythms in Excreted 17-Ketosteroids: Probable Androgen-Induced Circasepton Desynchronisation. Acta Endocrinol., Suppl., 103:1–54.
HALL, E. T. (1966) The Hidden Dimension. New York (Doubleday).
HALL, K. R. L. (1965) Social Learning in Monkeys. J. Zool. 148:15–87.
HALL, K. R. L., and SCHALLER, G. B. (1964) Tool Using Behavior of the California Sea Otter. J. Mammal., 45:287–298.
HALL, M. F. (1962) Evolutionary Aspects of Estrildid Song. Symp. Zool. Soc. London, 8:37–55.
HAMBURGER, V. (1963) Some Aspects of the Embryology of Behavior. Quart. Rev. Biol., 38:342–365.
HAMBURGER, V., and OPPENHEIM, R. (1967) Prehatching Motility and Hatching Behavior in the Chick. J. Exptl. Zool., 166:171–203.
HAMBURGER, V., WENGER, R. E., and OPPENHEIM, R. (1966) Motility in the Chick-Embryo in Absence of Sensory Input. J. Exptl. Zool., 162:133–160.
HARLOW, H. F. (1932) Social Facilitation of Feeding in the albino Rat. J. Gen. Psychol., 41:211–221.
——— (1953) Higher Functions of the Nervous System. Ann. Rev. Physiol., 15:493–514.
HARLOW, H. F., and HARLOW, M. K. (1962a) The Effect of Rearing Conditions on Behavior. Bull. Menninger Clin., 26:213–224.
——— (1962b) Social Deprivation in Monkeys. Sci. Am., 207:137–146.
HARLOW, H. F., HARLOW, M. K., and MEYER, D. R. (1950) Learning Motivated by a Manipulation Drive. J. Exptl. Psychol., 40:228–234.
HARRIS, G. W. (1964) Female Cycles of Gonadotrophic Secretion and Female

Sexual Behavior in Adult Male Rats Castrated at Birth. J. Physiol. (London), 175:75–76.

HARRIS, T. V. (1950) Habitat Selection of *Peromyscus*. Ph.D. Thesis Univ. Michigan, Ann Arbor.

HARRISON, C. J. O. (1965) Allopreening as Agonistic Behavior. Behaviour, 24:161–209.

HARTMANN, H., KRIS, E., and LOEWENSTEIN, R. M. (1949) Notes on the Theory of Aggression. Psychoanalytic Study of the Child 3–4. New York (International Univ. Press).

HARTMANN, M. (1956) Die Sexualität. Stuttgart (G. Fischer).

HARTRY, A. L., KEITH-LEE, P., and MORTON, W. D. (1964) Planaria, Memory Transfer through Cannibalism Reexamined. Science, 146:274–275.

HASKINS, C. P., and HASKINS, E. F. (1958) Note on the Inheritance of Behavior Patterns for Food Selection and Cocoon Spinning in F_1-hybrids of *Callosamia promethea* and *C. angulifera*. Behaviour, 13:89–95.

HASLER, A. D. (1954) Odour Perception and Orientation in Fishes. J. Fisheries Res. Board Can., 11:107–129.

—— (1956) Perception of Pathways by Fishes in Migration. Quart. Rev. Biol., 31:200–209.

—— (1960) Guideposts of Migrating Fishes. Science, 131:785–792.

HASLER, A. D., and SCHWASSMANN, H. O. (1960) Sun Orientation of Fish at Different Latitudes. Cold Spring Harbor Symp. Quant. Biol., 25:429–441.

HASS, H. (1951) Drei Jäger auf dem Meeresgrund. Zürich (Füssli).

—— (1957) Wir kommen aus dem Meer. Berlin (Ullstein).

—— (1968) Wir Menschen. Wien (Molden).

HASS, H., and EIBL-EIBESFELDT, I. (1964) *Dotilla sulcata (Brachyura):* Fressen und Graben. Encycl. cinem., E538. Publ. zu wiss. Filmen, 1 A, 165–168. Göttingen (Inst. wiss. Film).

HASSENSTEIN, B. (1955) Abbildende Begriffe. Zool. Anz., Suppl., 18:197–202.

—— (1965) Biologische Kybernetik. Heidelberg (Quelle u. Meyer).

—— (1966) Kybernetik und biologische Forschung. Handb. d. Biol., 1, 631–719. Frankfurt (Athenaion).

HASSLER, R., and BAK, I. J. (1966) Submikroskopische Catecholaminspeicher als Angriffspunkte der Psychopharmaka Reserpin und Mono-Amino-Oxydase-Hemmer. Nervenarzt, 37:493–498.

HAUENSCHILD, C. (1960) Lunar Periodicity. Cold Spring Harbor Symp. Quant. Biol., 25:73–86.

HAYES, C. (1951) The Ape in Our House. New York (Harper & Row).

HEBB, D. O. (1949) The Organisation of Behaviour. New York.

—— (1953) Heredity and Environment in Mammalian Behaviour. Brit. J. Animal Behavior, 1:43–47.

HEBERER, G. (1965) Über den systematischen Ort und den psychisch-physischen Status der Australopithecinen. In HEBERER, G. (ed.), Menschliche Abstammungslehre. Stuttgart (G. Fischer), 310–356.

HEDIGER, H. (1933) Beobachtungen an der marokkanischen Winkerkrabbe *(Uca tangeri)*. Verhandl. Schweiz. Naturforsch. Ges., 114:388–389.

—— (1934) Zur Biologie und Psychologie der Flucht bei Tieren. Biol. Zbl., 54:21–40.

—— (1942) Wildtiere in Gefangenschaft. Basel.

—— (1949) Säugetierterritorien und ihre Markierung. Bijdr. tot de Dierkde, 28:172–184.

—— (1954) Skizzen zu einer Tierpsychologie im Zoo und im Zirkus. Zürich (Gutenberg).

————— (1961) The Evolution of Territorial Behavior. In WASHBURN, S. L. (ed.), Social Life of Early Man. Chicago (Aldine), 34–57.

————— (1963) Weitere Dressurversuche mit Delphinen und anderen Walen. Z. Tierpsychol., 20:487–497.

HEGH, E. (1922) Les Termites. Brussels (L. Desmet).

HEILIGENBERG, W. (1963) Ursachen für das Auftreten von Instinktsbewegungen bei einem Fische (Pelmatochromis subocellatus kribensis). Z. Vergleich. Physiol., 47:339–380.

————— (1964) Ein Versuch zur ganzheitsbezogenen Analyse des Instinktverhaltens eines Fisches (Pelmatochromis subocellatus kribensis). Z. Tierpsychol., 21:1–52.

————— (1965) A Quantitative Analysis of Digging Movements and Their Relationship to Aggressive Behaviour in Cichlids. Animal Behavior, 13:163–170.

HEINROTH, O. (1910) Beiträge zur Biologie, insbesondere Psychologie und Ethologie der Anatiden. Verh. 5. Int. Ornith. Kongr., Berlin, 589–702.

HEINROTH, O., and HEINROTH, M. (1928) Die Vögel Mitteleuropas. Berlin-Lichterfelde (Bermühler).

HEINROTH-BERGER, K. (1965) Über Geburt und Aufzucht eines männlichen Schimpansen im Zool. Garten Berlin. Z. Tierpsychol., 22:15–35.

HEINZ, H. J. (1949) Vergleichende Beobachtungen über die Putzhandlungen bei Dipteren und bei Sarcophaga carnaria L. im besonderen. Z. Tierpsychol., 6:330–371.

HELD, R., and HEIN, A. (1963) Movement-Produced Stimulation in the Development of Visually Guided Behavior. J. Comp. Physiol. Psychol., 56:872–876.

HELMUTH, H. (1967) Zum Verhalten des Menschen: die Aggression. Z. Ethnol., 92(2):265–273.

HERAN, H. (1966) Sinnesphysiologie. Handb. d. Biol., 5, 473. Frankfurt (Athenaion).

HERING, H. E. (1896) Über Bewegungsstereotypien nach centripetaler Lähmung. Arch. Exptl. Pathol. Pharmakol., 38:266–283.

HERING, E. (1921) Über das Gedächtnis als eine allgemeine Funktion der organisierten Materie. 3rd ed. Leipzig (Akadem. Verl. Ges.).

HERNANDEZ-PEON, R., and BRUST-CARMONA, H. (1961) Functional Role of Subcortical Structure in Habituation and Conditioning. In DELAFRESNAYE, J. F. (ed.), Brain Mechanisms and Learning. Oxford (Blackwell), 393–412.

HERREBOUT, W. M., KUYTEN, P. J., and RUITER, L. de (1963) Observations on Color Patterns and Behavior of Caterpillars Feeding on Scots Pine. Arch. Neerl. Zool., 15(3):315–357.

HERTER, K. (1943) Beziehungen zwischen der Ökologie und der Thermotaxis der Tiere. Biol. Gen., 17:243–309.

————— (1952) Der Temperatursinn der Säugetiere. Beitr. Tierkde–Tierzucht, 3:1–171.

————— (1953) Der Temperatursinn der Insekten. Berlin (Dunker u. Humblot).

HERTER, K., and SGONINA, K. (1938) Vorzugstemperatur und Hautbeschaffenheit bei Mäusen. Z. Vergleich. Physiol., 26:366–415.

HESS, C. v. (1913) Experimentelle Untersuchungen über den angeblichen Farbensinn der Bienen. Zool. Jahrb. Allgem. Zool. Physiol., 34:81–106.

HESS, E. H. (1956) Space Perception in the Chick. Sci. Am., 195:71–80.

————— (1959) Imprinting: an Effect of Early Experience. Science, 130:133–141.

————— (1965) Attitude and Pupil Size. Sci. Am., 212(4):46–54.

HESS, E. H., SELTZER, A. L., and SHLIEN, J. M. (1965) Pupil Response of Hetero- and Homosexual Males to Pictures of Men and Women: A Pilot Study. J. Abnorm. Psychol., 70:165–168.

HESS, W. R. (1954) Das Zwischenhirn. 2nd ed. Basel (Schwabe).
———— (1957) Die Formatio reticularis des Hirnstammes im verhaltensphysiologischen Aspekt. Arch. Psychiat. Nervenkr., 196:329–336.
HESSE, R., and DOFLEIN, F. (1943) Tierbau und Tierleben. 2. Das Tier als Glied des Naturganzen. Jena (G. Fischer).
HEUSSER, H. (1960) Über die Beziehungen der Erdkröte (Bufo bufo L.) zu ihrem Laichplatz II. Behaviour, 16:93–109.
HEWES, G. H. (1957) The Anthropology of Posture. Sci. Am. 196(2):123–132.
HILGARD, E. R. (1956) Theories of Learning. New York (Appleton).
HINDE, R. A. (1953) Appetitive Behaviour, Consummatory Act and the Hierarchical Organization of Behaviour: with Special Reference to the Great Tit (Parus major). Behaviour, 5:189–224.
———— (1956) The Behaviour of Certain Cardueline F_1 Interspecies Hybrids. Behaviour, 9:202–213.
———— (1958) The Nestbuilding Behaviour of Domesticated Canaries. Proc. Zool. Soc. London, 131:1–48.
———— (1959) Some Recent Trends in Ethology. In KOCH, S. (ed.), Psychology: a Study of Science. Study, 1/2:561–610.
———— (1965) Interaction of Internal and External Factors in Integrations of Canary Reproduction. In BEACH, F. A. (ed.), Sex and Behavior. New York (Wiley), 381–415.
———— (1966) Animal Behaviour: A Synthesis of Ethology and Comparative Psychology. New York (McGraw-Hill).
HINZE, G. (1950) Der Biber. Berlin (Akademie-Verlag).
HOBSON, E. S. (1965) Diurnal-Nocturnal Activity of Some Inshore Fishes in the Gulf of California. Copeia, 3:291–302.
HOEBEL, B. G., and TEITELBAUM, P. (1962) Hypothalamic Control of Feedings and Self-Stimulation. Science, 135:375–377.
HOFFMANN, K. (1954) Versuche zu der im Richtungsfinden der Vögel enthaltenen Zeitschätzung. Z. Tierpsychol., 11:453–475.
———— (1955) Aktivitätsregistrierungen bei frisch geschlüpften Eidechsen. Z. Vergleich. Physiol., 37:253–262.
———— (1959) Die Aktivitätsperiodik von im 18- und 36-Stundentag erbrüteten Eidechsen. Z. Vergleich. Physiol., 42:422–432.
———— (1960) Experimental Manipulation of the Orientational Clock in Birds. Cold Spring Harbor Symp. Quant. Biol., 25:379–387.
———— (1965) Overt Circadian Frequencies and Circadian Rule. In ASCHOFF, J. (ed.), Circadian Clocks. Amsterdam (North-Holland), 87–94.
———— (1968) Synchronisation der circadianen Aktivitätsperiodik von Eidechsen durch Temperaturcyclen verschiedener Amplitude. Z. Vergleich. Physiol., 58:225–228.
HOFSTÄTTER, P. R. (1959) Psychologie. 3rd ed. Frankfurt (Fischer-Bücherei).
HOHORST, W., and GRAEFE, G. (1961) Ameisen—obligatorische Zwischenwirte des Lanzettegels (Dicrocoelium dendriticum). Naturwiss., 48:229–230.
HOKANSON, J. E., and BURGESS, M. (1962) The Effects of Three Types of Aggression on Vascular Processes. J. Abnorm. Soc. Psychol., 64:446–449.
HOKANSON, J. E., and SHETLER, S. (1961) The Effect of Overt Aggression on Physiological Tension Level. J. Abnorm. Soc. Psychol., 63:446–448.
HÖLLDOBLER, B. (1967) Zur Physiologie der Gast-Wirt-Beziehungen (Myrmecophilie) bei Ameisen. Z. Vergleich. Physiol., 56:1–21.
HOLST, D. v. (1969) Sozialer Stress bei Tupajas (Tupaia belangeri) Z. vergleich. Physiol., 63:1–58.
HOLST, E. v. (1932) Untersuchungen über die Funktionen des Zentralnervensystems beim Regenwurm (Lumbricus terrestris L.). Zool. Jb. (Physiol.), 51, 4:547–588.

——— (1933) Weitere Versuche zum nervösen Mechanismus der Bewegungen beim Regenwurm (*Lumbricus terrestris* L.). Zool. Jb. (Physiol.), 53, 1:67–100.

——— (1935) Über den Prozeß der zentralen Koordination. Arch. Ges. Physiol., 236:149–158.

——— (1936) Versuche zur Theorie der relativen Koordination. Arch. Ges. Physiol., 237:93–121.

——— (1937) Baustein zu einer vergleichenden Physiologie der lokomotorischen Reflexe bei Fischen II. Z. Vergleich. Physiol., 24:532–562.

——— (1938) Neuere Versuche zur Deutung der relativen Koordination bei Fischen. Arch. Gcs. Physiol., 240:1–43.

——— (1939) Die relative Koordination als Phänomen und als Methode zentralnervöser Funktionsanalyse. Ergeb. Physiol., 42:228–306.

——— (1943) Über die relative Koordination bei Arthropoden. Pflüg. Arch., 246:847–865.

——— (1950a) Quantitative Messung von Stimmungen im Verhalten der Fische. In Physiological Mechanisms in Animal Behavior. Symp. Soc. Exp. Biol., 4. Cambridge (Univ. Press), 143–172.

——— (1950b) Die Tätigkeit des Statolithenapparates der Wirbeltiere. Naturwiss., 12:265–272.

——— (1955) Regelvorgänge in der optischen Wahrnehmung. 5th Conf. Soc. Biol. Rhythm, Stockholm, 26–34.

——— (1956) Optische Wahrnehmungen, die wir selbst erzeugen, und ihre Bedeutung für unser Dasein. Jahrb. Max-Planck-Ges., 121–149.

——— (1957) Die Auslösung von Stimmungen bei Wirbeltieren durch "punktförmige" elektrische Erregung des Stammhirns. Die Naturwiss., 44:549–551.

——— (1969) Zur Verhaltensphysiologie bei Tieren und Menschen. I. und II. Munich (Piper).

HOLST, E. v. KAISER, M., RÖBIG, G., and GÖLDNER, G. (1950) Die Arbeitsweise des Statolithen-Apparates bei Tieren. Z. Vergleich. Physiol., 32:60–120.

HOLST, E. v., and MITTELSTAEDT, H. (1950) Das Reafferenz-Prinzip. Naturwiss., 37:464–476.

HOLST, E. v., and SAINT-PAUL, U. v. (1960) Vom Wirkungsgefüge der Triebe. Naturwiss., 18:409–422.

HOLZAPFEL, M. (1938) Über Bewegungsstereotypien bei gehaltenen Säugern. I. und II. Z. Tierpsychol., 2:46–72.

——— (1939) Über Bewegungsstereotypien bei gehaltenen Säugern. III. Analyse der Bewegungsstereotypie eines Gürteltieres. Zool. Garten, 10:184–193.

——— (1940) Triebbedingte Ruhezustände als Ziel von Appetenzhandlungen. Naturwiss., 28:273–280.

——— (1949) Die Beziehungen zwischen den Trieben junger und erwachsener Tiere. Schweiz. Z. Psychol., 8:32–60.

HOOGLAND, R., MORRIS, D., and Tinbergen, N. (1957) The Spines of the Sticklebacks *(Gasterosteus* and *Pygosteus)* as Means of Defence against Predators *(Perca* and *Esox).* Behaviour, 10:205–236.

HOPPENHEIT, M. (1964) Beobachtungen zum Beutefangverhalten der Larve von *Aeschna cyanea* Müll. *(Odonata).* Zool. Anz., 172:216–232.

HÖRMANN, L. v. (1912) Genuß- und Reizmittel in den Ostalpen, eine volkskundliche Skizze. Z. Deut. Oesterr. Alpenver., 43:78–100.

HÖRMANN-HECK, S. v. (1957) Untersuchungen über den Erbgang einiger Verhaltensweisen bei Grillenbastarden *(Gryllus campestris* x *Gryllus bimaculatus).* Z. Tierpsychol., 14:137–183.

HOWARD, H. E. (1920) Territory in Bird Life. New York (Dutton).

HSIAO, H. H. (1929) An Experimental Study of the Rat's "Insight" within a

Spatial Complex. Univ. Calif. Publ. Psychol., 4:57–70.

HUBEL, D. H., and WIESEL, T. N. (1959) Receptive Fields of Single Neurons in the Cats Striate Cortex. J. Physiol. (London), 148:574–591.

—— (1962) Receptive Fields, Binocular Interactions and Functional Architecture in the Cats Visual Cortex. J. Physiol. (London), 160:106–154.

—— (1963) Receptive Fields of Cells in Striate Cortex of Very Young, Visually Inexperienced Kittens. J. Neurophysiol., 24:994–1002.

HUBER, F. (1955) Sitz und Bedeutung nervöser Zentren für Instinkthandlungen beim Männchen von Gryllus campestris. Z. Tierpsychol., 12:12–48.

HÜCKSTEDT, B. (1965) Experimentelle Untersuchungen zum "Kindchenschema." Z. Exptl. Angew. Psychol., 12:421–450.

HUDGENS, G. A., DENENBERG, V. H., and ZARROW, M. X. (1968) Mice Reared with Rats: Effects of Preweaning and Postweaning Social Interaction upon Adult Behaviour. Behaviour, 30:259–274.

HUET, M. (1952) Dix années de pisciculture aux Congo Belge et aux Ruanda. Traité de pisciculture. Brussels.

HUIZINGA, J. (1956) Homo ludens. Rowohlts Deut. Encycl., 21.

HULL, C. L. (1943) Principles of Behaviour. New York.

HUNSAKER, D. (1962) Ethological Isolating Mechanisms in the Sceloporus torquatus Group of Lizards. Evolution, 16:62–74.

HUNTER, W. S. (1913) The Delayed Reaction in Animals and Children. Behavior Monogr., 2:21–30.

HUXLEY, J. S. (1923) Courtship Activities in the Red-Throated Diver (Colymbus stellatus Pontopp): together with a Discussion of the Evolution of Courtship in Birds. J. Linnean Soc. London Zool., 53:253–292.

HYDÉN, H. (1961) Satellite Cells in the Nervous System. Sci. Am., 205(6): 62–70.

HYDÉN, H., and EGYHAZI, E. (1962) Nuclear RNA Changes of Nerve Cells during a Learning Experiment in Rats. Proc. Natl. Acad. Sci. U.S., 48:1366–1373.

IERSEL, J. V. (1953) An Analysis of the Parental Behavior of the Three-Spined Stickleback (Gasterosteus aculeatus). Behaviour, Suppl., 3.

—— (1958) Some Aspects of Territorial Behavior of the Male Three-Spined Stickleback. Arch. Neerl. Zool., 13:383–400.

IMMELMANN, K. (1959) Experimentelle Untersuchungen über die biologische Bedeutung artspezifischer Merkmale beim Zebrafinken (Taeniopygia castanotis Gould). Zool. Jahrb. Abt. System, 86:438–592.

—— (1961) Beitrag zur Biologie und Ethologie australischer Honigfresser (Meliphagidae). J. Ornithol., 102:164–207.

—— (1962a) Vergleichende Beobachtungen über das Verhalten domestizierter Zebrafinken in Europa und ihrer wilden Stammform in Australien. Z. Tierzücht., 77:198.

—— (1962b) Beiträge zu einer vergleichenden Biologie australischer Prachtfinken (Spermestidae). Zool. Jahrb. System, 90:1–196.

—— (1965) Prägungserscheinungen in der Gesangsentwicklung junger Zebrafinken. Naturwiss., 52:169–170.

—— (1966) Zur Irreversibilität der Prägung. Naturwiss., 53:209.

—— (1967) Zur ontogenetischen Gesangsentwicklung bei Prachtfinken. Zool. Anz., Suppl., 30:320–332.

INHELDER, E. (1955) Zur Psychologie einiger Verhaltensweisen, besonders des Spiels von Zootieren. Z. Tierpsychol., 12:88–144.

ITANI, J. (1958) On the Acquisition and Propagation of a New Food Habit in the Troop of Japanese Monkeys at Takasakiyama. Primates, 1:84–98 (in Japanese with English summary).

Jacobs, W. (1953a) Vergleichende Verhaltensstudien an Feldheuschrecken *(Orthoptera, Acrididae)* und einiger anderer Insekten. Zool. Anz., Suppl., 19:115–138.

———— (1953b) Verhaltensbiologische Studien an Feldheuschrecken. Z. Tierpsychol., Suppl., 1.

———— (1966) Die Gesänge der Heuschrecken. In BURCKHARDT, D., SCHLEIDT, W. and ALTNER, H. (eds.): Düfte, Farben und Signale. Munich. (Moos).

JACOBS-JESSEN, U. F. (1959) Zur Orientierung der Hummeln und einiger anderer Hymenopteren. Z. Vergleich. Physiol., 41:597–641.

JACOBSON, A. L., BABICH, F. R., BUBASH, S., and JACOBSON, A. (1965) Differential-Approach Tendencies Produced by Injection of RNA from Trained Rats. Science, 150:636–637.

JAMES, W. (1890) Principles of Psychology. New York (Holt, Rinehart and Winston Inc.).

JANDER, R. (1966a) Die Phylogenie von Orientierungsmechanismen der Arthropoden. Zool. Anz., Suppl., 29:266–306.

———— (1966b) Die Hauptentwicklungen der Lichtorientierung bei den tierischen Organismen. Verh. Verb. Deut. Biol., 3:28–34.

JENKINS, D., WATSON, A., and MILLER, G. R. (1967) Population Fluctuations in the Red Grouse *Lagopus lagopus scoticus.* J. Animal Ecol., 36:97–122.

JENNINGS, H. S. (1906) The Behavior of the Lower Organisms. New York.

JESPERSEN, O. (1925) Die Sprache. Heidelberg.

JOHNSON, D. L. (1967) Honeybees: Do They Use the Direction Information Contained in Their Dance Maneuver? Science, 155: 844–847.

JOLLY, A. (1966) Lemur Social Behavior and Primate Intelligence. Science, 153:501–506.

KAADA, B. (1967) Brain Mechanisms Related to Aggressive Behaviour. UCLA Forum in Medical Science, 7:95–133. In CLEMENTE, C. D., and LINDSLEY, D. B. (eds.), Aggression and Defense. Neural Mechanisms and Social Pattern. Berkeley (Univ. California Press).

KAESTNER, A. (1965) Lehrbuch der speziellen Zoologie. I. Wirbellose. 2nd ed. Jena (G. Fischer).

KAGAN, J., and MOSS, H. A. (1962) Birth to Maturity: A Study in Psychological Development. New York (Wiley).

KAHN, M. W. (1951) The Effect of Severe Defeat at Various Levels on the Aggressive Behavior of Mice. J. Gen. Psychol., 79:117–130.

———— (1954) Infantile Experience and Mature Aggressive Behavior of Mice: Some Maternal Influences. J. Gen. Psychol., 84:65–75.

KANT, I. (1960) Werke in 6 Bänden. Wiesbaden.

KAPUNE, T. (1966) Untersuchungen zur Bildung eines "Wertbegriffes" bei niederen Primaten. Z. Tierpsychol., 23:324–363.

KÄSTLE, W. (1963) Zur Ethologie des Grasanolis *(Norops auratus).* Z. Tierpsychol., 20:16–33.

———— (1965) Zur Ethologie des Anden-Anolis *Phenacosaurus richteri.* Z. Tierpsychol., 22:751–769.

KATZ, D. (1926) Sozialpsychologie der Vögel. Ergeb. Biol., 1:447–478.

KAUFMANN, S. A. (1967) Social Relations of Adult Males in a Free-Ranging Band of Rhesus Monkeys. In ALTMAN, S. A. (ed.) Social Communication among Primates. Chicago (Univ. Press), 73–98.

KAWAI, M. (1958) On the Rank System in a Natural Group of Japanese Monkeys. Primates, 1:84–98 (in Japanese with English summary).

———— (1965) Newly Acquired Pre-Cultural Behavior of the Natural Troop of Japanese Monkeys on Koshima Island. Primates, 6:1–30.

KAWAI, M. and MIZUHARA, H. (1959) An Ecological Study on the Wild Mountain Gorilla *(Gorilla gorilla beringei)*. Primates, 2:1–42.

KAWAMURA, S. (1963) The Process of Sub-Cultural Propagation among Japanese Macaques. In SOUTHWICK, C. H. (ed.), Primate Social Behavior. Princeton, N.J. (Van Nostrand), 82–90.

KEARTON, C. (1935) Die Insel der fünf Millionen Pinguine. Stuttgart.

KEENLEYSIDE, M. H. A. (1955) Some Aspects of Schooling of Fish. Behaviour, 8:183–248.

KELLOGG, W. N. (1961) Porpoises and Sonar. Chicago (Univ. Press).

—— (1968) Communication and Language in the Home-Raised Chimpanzee. Science, 165:423–427.

KEMPENDORFF, W. (1942) Über das Fluchtphänomen und die Chemorezeption von *Heliosoma nigricans*. Arch. Molluskenk., 74.

KENNEDY, J. S. (1951) The Migration of the Desert Locust *(Schistocerca gregaria Forsk)*. Phil. Trans. Roy. Soc. London, B235:163–290.

KING, J. A. (1955) Social Behavior, Social Organisation and Population Dynamics in a Black-Tailed Prairiedog Town in the Black Hills of South Dakota. Contrib. Lab. Vert. Biol. Univ. Mich., 67.

—— (1957) Relationships between Early Social Experience and Adult Aggressive Behavior in Inbred Mice. J. Gen. Psychol., 90:151–166.

KING, J. A., and GURNEY, N. L. (1954) Effect of Early Social Experience on Adult Aggressive Behavior in C 57 BL/10 Mice. J. Comp. Physiol. Psychol., 47:326–336.

KIRCHSHOFER, R. (1960) Über das "Harnspritzen" des Großen Mara. Z. Säugetierkde., 25:112–127.

KISLAK, J. W. and BEACH, F. A. (1955) Inhibition of Aggression by Ovarian Hormones. Endocrinology, 56:684–692.

KITZLER, G. (1942) Die Paarungsbiologie einiger Eidechsen. Z. Tierpsychol., 4:353–402.

KLAUSEWITZ, W. (1961) Einige systematisch und ökologisch bemerkenswerte Meergrundeln *(Pisces, Gobiidae)*. Senckenbergiana Biol., 41:149–162.

—— (1965) Osteichthyes, Knochenfische. Handb. d. Biol., Frankfurt, 6 (2), 542–628.

KLAUSEWITZ, W., and EIBL-EIBESFELDT, I. (1959) Neue Röhrenaale von den Malediven und Nikobaren *(Pisces, Apodes, Heterocongridae)*. Senckenbergiana Biol., 40:135–153.

KLINGEL, H. (1967) Soziale Organisation und Verhalten freilebender Steppenzebras. Z. Tierpsychol., 24:580–624.

KLINGELHÖFFER, W. (1956) Terrarienkunde. 2.: Lurche. SCHERPNER, C. (ed.), 2nd ed. Stuttgart (Kernen).

KLINGHAMMER, E. (1967) Factors Influencing Choice of Mate in Altricial Birds. In STEVENSON, H. W. (ed.), Early Behavior. New York (Wiley), 5–42.

KLINGHAMMER, E., and HESS, E. H. (1964) Parental Feeding in Ring Doves *(Streptopelia roseogrisea)*, Innate or Learned? Z. Tierpsychol., 21:338–347.

KLOOT, W. G. van der, and WILLIAMS, C. M. (1953) Cocoon Construction by the Cecropia Silkworm. II. The Role of the Internal Environment. Behaviour, 5:157–174.

KLOPFER, P. H. (1957) An Experiment on Emphatic Learning in Ducks. Am. Naturalist, 91:61–63.

—— (1962) Aspects of Ecology. Englewood Cliffs, N.J. (Prentice-Hall).

—— (1963) Behavioral Aspects of Habitat Selection. Wilson Bull., 75:15–22.

KLOPFER, P. H., and GAMBLE, J. (1966) Maternal "Imprinting" in Goats. The Role of Chemical Senses. Z. Tierpsychol., 23:588–592.

KLOPFER, P. H., and HAILMAN, J. P. (1965) Habitat Selection in Birds. Advan. Study Behavior, 1:279–303.

KLUYVER, H. N. (1947) Over het gedrag van een jonge Grauwe Vliegenvanger en van een troep Pestvogels in de winter. Ardea, 35:131–135.

KNEUTGEN, J. (1964) Beobachtungen über die Anpassung von Verhaltensweisen an gleichförmige akustische Reize. Z. Tierpsychol., 21:763–779.

——— (in preparation) Eine Musikform und ihre biologische Funktion.

KOEHLER, O. (1943) "Zähl"—Versuche an einem Kolkraben und Vergleichsversuche an Menschen. Z. Tierpsychol., 5:575–712.

——— (1949) "Zählende" Vögel und vorsprachliches Denken. Zool. Anz., Suppl., 13:129–238.

——— (1950) Die Analyse der Taxisanteile instinktartigen Verhaltens. Symp. Soc. Exptl. Biol. Cambridge, 4:269–302.

——— (1952) Vom unbenannten Denken. Zool. Anz., Suppl., 16:202–211.

——— (1953) Orientierungsvermögen von Mäusen: Versuche im Hochlabyrinth. Wiss. Film B635. Göttingen (Inst. wiss. Film).

——— (1954a) Das Lächeln als angeborene Ausdrucksbewegung. Z. menschl. Vererb. u. Konst.lehre, 32:330–334.

——— (1954b) Vorbedingungen und Vorstufen unserer Sprache bei Tieren. Zool. Anz., Suppl., 18:327–341.

——— (1955) Zählende Vögel und vergleichende Verhaltensforschung. Acta 11, Congr. Intern. Ornith., Basel, 588–598.

——— (1966a) Vom Spiel bei Tieren. Freiburger Dies Universitatis, 13:1–32.

——— (1966b) HOCKETT, C. F. (1960) Logical Considerations in the Study of Animal Communication. Am. Inst. Biol. Sci. Publ., 7:392–430. Also Z. Tierpsychol., 23:250–254.

KOEHLER, O., and ZAGARUS, A. (1937) Beiträge zum Brutverhalten des Halsbandregenpfeifers (Charadrius hiaticula L.). Beitr. Fortpfl. Biol. Vögel, 13:1–9.

KOENIG, L. (1951) Beiträge zu einem Aktionssystem des Bienenfressers (Merops apiaster L.). Z. Tierpsychol., 8:169–210.

——— (1953) Beobachtungen am afrikanischen Blauwangenspint (Merops superciliosus chrysocercus) in freier Wildbahn und Gefangenschaft, mit Vergleichen zum Bienenfresser (Merops apiaster L.). Z. Tierpsychol., 10:180–204.

KOENIG, O. (1951a) Das Aktionssystem der Bartmeise (Panurus biarmicus L.). Oesterr. Zool. Z., 1:1–82.

——— (1951b) Das Aktionssystem der Bartmeise (Panurus biarmicus L.). Oesterr. Zool. Z., 3:247–325.

——— (1968) Biologie der Uniform. Naturwiss. Medizin., 5(22):3–19 and 5(23):40–50.

KOFORD, C. B. (1963a) Rank of Mothers and Sons in Bands of Rhesus Monkeys. Science, 141:356–357.

——— (1963b) Group Relations in an Island Colony of Rhesus Monkeys. In SOUTHWICK, C. (ed.), Primate Social Behavior. Princeton, N.J. (Van Nostrand), 136–152.

——— (1966) Population Changes in Rhesus Monkeys: Cayo Santiago 1960–1964. Tulane Stud. Zool., 13:1–7.

KÖHLER, W. (1921) Intelligenzprüfungen an Menschenaffen. Berlin.

KOHTS, N. (1935) Infant Ape and Human Child (Instincts, Emotions, Play, Habits). Sci. Mem. Mus. Darwinianum, 3 (with Russian and English summaries).

KOLLER, G. (1955) Hormonale und psychische Steuerung beim Nestbau weisser Mäuse. Zool. Anz., Suppl., 19:123–132.

KOMISARUK, B. R., and OLDS, J. (1968) Neuronal Correlates of Behaviour in Freely Moving Rats. Science, 161:810–812.

KONISHI, M. (1963) The Role of Auditory Feedback in the Vocal Behavior in the Domestic Fowl. Z. Tierpsychol., 20:349–367.

—— (1964) Effects of Deafening on Song Development in Two Species of Juncos. Condor, 66:85–102.

—— (1965a) Effects of Deafening on Song Development of American Robins and Black-Headed Grosbeaks. Z. Tierpsychol., 22:584–599.

—— (1965b) The Role of Auditory Feedback in the Control of Vocalisation in the White-Crowned Sparrow. Z. Tierpsychol., 22:770–783.

—— (1966) The Attributes of Instinct. Behaviour, 27:316–328.

KORRINGA, P. (1957) Lunar Periodicity. In HEDGEPETH, J. W. (ed.), Treatise on Marine Ecology and Palaeoecology: 1 Ecology. Memoire 67, Geol. Soc. Am. Baltimore (Waverly Press), 917–934.

KORTLANDT, A. (1940) Eine Übersicht über die angeborenen Verhaltensweisen des mitteleuropäischen Kormorans. Arch. Neerl. Zool., 4:401–442.

—— (1955) Aspects and Prospects of the Concept of Instinct. Arch. Neerl. Zool., 11:155–284.

—— (1962) Chimpanzees in the Wild. Sci. Am., 206:128–138.

—— (1965) How Do Chimpanzees Use Weapons when Fighting Leopards? Yearbook Am. Phil. Soc., 327–332. Philadelphia.

—— (1967a) Experimentation with Chimpanzees in the Wild. In STARCK, D., SCHNEIDER, R. und KUHN, H. J. (eds.): Neue Ergebnisse der Primatologie. Stuttgart (Fischer) 208–224.

—— (1967b) Handgebrauch bei freilebenden Schimpansen. In RENSCH, B. (ed.) Handgebrauch und Verständigung bei Affen und Frühmenschen, Bern (Huber), 59–102.

KORTLANDT, A., and KOOIJ, M. (1963) Prohominid Behavior in Primates. Symp. Zool. Soc. London, 10:61–88.

KORTMULDER, K. (1968) An Ethological Theory of the Incest Taboo and Exogamy. Current Anthropology 9:437–449.

KRAFFT-EBING, R. v. (1924) Psychopathia sexualis. 17th ed. Stuttgart.

KRAMER, G. (1933) Untersuchungen über die Sinnesleistungen und das Orientierungsverhalten von Xenopus laevis. Zool. Jahrb. Abt. Phys., 52:629–676.

—— (1949) Macht die Natur Konstruktionsfehler? Wilhelmshavener Vorträge, Schriftenreihe d. Nordwestdtsch. Universitätsges., 1:1–19.

—— (1950) Über individuell und anonym gebundene Gemeinschaften der Tiere und Menschen. Studium Gen., 3:564–572.

—— (1952) Die Sonnenorientierung der Vögel. Zool. Anz., Suppl., 16:72–84.

—— (1957) Experiments on Birds Orientation and Their Interpretation. Ibis., 96:173–185.

—— (1959) Recent Experiments on Bird Orientation. Ibis., 101:399–416.

—— (1961) Long-Distance Orientation. In MARSHALL, A. J. (ed.), Biology and Comparative Physiology of Birds, 2. New York (Academic Press), 341–371.

KRECHEVSKI, I. (1932) "Hypotheses" in Rats. Psychol. Rev., 39:516–532.

KROTT, P., and KROTT, K. (1963) Zum Verhalten der Braunbären (Ursus arctos) in den Alpen. Z. Tierpsychol., 20:160–206.

KRUEGER, F. (1948) Lehre vom Ganzen. Bern (Huber).

KRUIJT, J. P. (1958) Speckling of the Herring Gull Egg in Relation to Brooding Behavior. Compt. Rend. Soc. Neerl. Zool., 12:565–567.

—— (1964) Ontogeny of Social Behavior in Burmese Red Jungle Fowl (Gallus gallus spadiceus). Behaviour, Suppl., 12.

KRUMBIEGEL, I. (1940) Die Persistenz physiologischer Eigenschaften in der Stammesgeschichte. Z. Tierpsychol., 4:249–258.

KRUSHINSKII, L. V. (1962) Animal Behavior, Its Normal and Abnormal Development. In WORTIS, J. (ed.), The International Behavioral Sciences Series. New York (Consultants Bureau).

KUENZER, E., and KUENZER, P. (1962) Untersuchungen zur Brutpflege der Zwergcichliden. Z. Tierpsychol., 19:56–83.

KUENZER, P. (1968) Die Auslösung der Nachfolgereaktion bei erfahrungslosen Jungfischen von Nannacara anomala (Cichlidae). Z. Tierpsychol., 25:257–314.

KÜHLMANN, D. H. H. (1966) Putzerfische säubern Krokodile. Z. Tierpsychol., 23:853–854.

KÜHME, W. (1965) Freilandstudien zur Soziologie des Hyänenhundes. Z. Tierpsychol., 22:495–541.

——— (1966) Beobachtungen zur Soziologie des Löwen in der Serengeti-Steppe. Z. Säugetierkde, 31:205–213.

KÜHN, A. (1919) Die Orientierung der Tiere im Raum. Jena (G. Fischer).

——— (1955) Vorlesungen über Entwicklungsphysiologie. Heidelberg (Springer).

KÜHNELT, W. (1965) Grundriß der Ökologie unter besonderer Berücksichtigung der Tierwelt. Jena (G. Fischer).

KULLENBERG, B. (1956) Field Experiments with Chemical Sexual Attractants on Aculeate Hymenoptera Males. Zool. Bidr. Uppsala, 31:253–354.

KULZER, E. (1954) Untersuchungen über die Schreckreaktion bei Erdkrötenquappen. Z. Vergleich. Physiol., 36:443–463.

KUMMER, H. (1957) Soziales Verhalten einer Mantelpavian-Gruppe. Schweiz. Z. Psychol., Suppl., 33. Schweiz. Z. Psychol. Bern (Huber).

——— (1968) Social Organization in Hamadryas Baboons—a Field Study. Bibliotheca Primat., Basel (Karger).

KUMMER, H., and KURT, F. (1965) A Comparison of Social Behavior in Captive and Wild Hamadryas Baboons. In VOGTBERG, H. (ed.), The Baboon in Medical Research. Austin (Univ. Texas Press), 1–16.

KUNKEL, P. (1959) Zum Verhalten einiger Prachtfinken. Z. Tierpsychol., 16:302–350.

KUO, Z. Y. (1930) The Genesis of the Cats Responses to the Rat. J. Comp. Psychol., 11:1–35.

——— (1932) Ontogeny of Embryonic Behavior in Aves. J. Exptl. Biol., 61:395–430, 453–489.

——— (1960–1961) Studies on the Basic Factors in Animal Fighting J. Genet. Psychol., 96:201–239 and 97:181–209.

——— (1967) The Dynamics of Behavior Development. An Epigenetic View. New York (Random House).

LACK, D. (1943) The Life of the Robin. London (Cambridge Univ. Press).

——— (1947) Darwin's Finches. London (Cambridge Univ. Press).

LAGERSPETZ, K. (1964) Studies on the Aggressive Behaviour of Mice Suomalaisen Tiedeakatemian Toimituksia. Ann. Acad. Sci. Fennice, B131 Helsinki.

LAGERSPETZ, K., and TALO, S. (1967) Maturation of Aggressive Behaviour in Young Mice. Rept. Inst. Psychol. Univ. of Turku, 28:1–9.

LAGERSPETZ, K., and WORINEN, K. (1965) A Cross-fostering Experiment with Mice Selectively Bred for Aggressiveness and Non aggressiveness. Rept. Psychol. Inst. Univ. Turku, No 17, 1–6.

LANCASTER, J. B. (1966) Primate Communication Systems and the Emergency of Human Language. In JAY, P. (ed.), Primate Social Behavior. New York (Holt, Rinehart and Winston).

LANG, E. M. (1961) Goma, das Gorillakind. Zürich (A. Müller).
———— (1964) Jambo—First Gorilla Raised by Its Mother in Captivity. Natl. Geogr. Mag., 125(3):446–453.
LARSSON, K. (1959) Experience and Maturation in the Development of Sexual Behavior in Male Puberty Rats. Behaviour, 14:101–107.
LASHLEY, K. S. (1915) Notes on the Nesting Activity of the Noddy and Sooty Terns. Pap. Dept. Marine Biol. Carnegie Inst. Wash., 7:61–84.
———— (1929) Brain Mechanisms and Intelligence: A Quantitative Study of Injuries to the Brain. Chicago.
———— (1931) Mass Action in Cerebral Function. Science, 73:245–254.
———— (1935) The Behavior of Rats in Latch Box Situations. Comp. Psychol. Monogr., 11(2):1–42.
———— (1938) Experimental Analysis of Instinctive Behavior. Psychol. Rev., 45:445–471.
LAUDIEN, H. (1966) Untersuchungen über das Kampfverhalten der Männchen von Betta splendens. Z. wiss. Zool., 172:134–178.
LAWICK-GOODALL, J. van (1965) New Discoveries among Africa's Chimpanzees. Natl. Geogr. Mag., 128(6):802–831.
———— (1968) The Behavior of Free-living Chimpanzees in the Gombe Stream Reserve. Animal Behaviour Monographs 1(3):161–311.
LAWICK-GOODALL, J. VAN and LAWICK, H. VAN (1966) Use of Tools by the Egyptian Vulture Neophron percnopterus. Nature, 212:1468–1469.
LEAKEY, L. S. B. (1963) Adventures in the Search of Man. Natl. Geogr. Mag., 123:132–152.
LECOMTE, J. (1961) Le comportement agressif des ouvrières d'Apis mellifica L. Ann. Abeille, 4:165–275.
LEHMANN, F. E. (1958) Gestaltung sozialen Lebens bei Tier und Mensch. Bern (Francke).
LEHR, E. (1967) Experimentelle Untersuchungen an Affen und Halbaffen über Generalisation von Insekten- und Blütenabbildungen. Z. Tierpsychol., 24:208–244.
LEHRMAN, D. S. (1953) A Critique of Konrad Lorenz's Theory of Instinctive Behavior. Quart. Rev. Biol., 28:337–363.
———— (1955) The Physiological Basis of Parental Feeding Behavior in the Ring Dove (Streptopelia risoria). Behaviour, 7:241–286.
———— (1961) The Presence of the Mate and of Nesting Material as Stimuli for the Development of Incubation Behavior and for Gonadotropic Secretion in the Ring Dove. Endocrinology, 68:507–516.
LEMAGNEN, J. (1952) Les phénomenes olfacto-sexuels chez l'homme. Arch. Sci. Physiol., 6:125–160.
LENNEBERG, E. H. (1964) A Biological Perspective of Language. In LENNEBERG, E. (ed.), New Directions in the Study of Language. Cambridge, Mass. (M. I. T. Press), 65–88.
LENNEBERG, E. H., REBELSKY, F. G., and NICHOLS, J. A. (1965) The Vocalisation of Infants to Deaf and to Hearing Parents. Human Develop., 8:23–27.
LERSCH, P. (1951) Gesicht und Seele. 2nd ed. Munich.
LEUTHOLD, W. (1966) Variations in Territorial Behavior of Uganda Kob, Adenota kob thomasi. Behaviour, 27:215–258.
———— (1967) Beobachtungen zum Jugendverhalten von Kob-Antilopen. Z. Säugetierkde, 32:59–63.
LEVY, D. M. (1934) Experiments on the Suckling Reflex and Social Behavior of Dogs. Am. J. Orthopsychol., 4:203.
LEYHAUSEN, P. (1954a) Die Entdeckung der relativen Koordination. Studium Gen., 7:45–60.

———— (1954b) Vergleichendes über die Territorialität bei Tieren und den Raumanspruch des Menschen. Homo, 5:68–76.

———— (1956a) Das Verhalten der Katzen (Felidae). In KUKENTHAL, Handb. d. Zool., 8 (10), Berlin (de Gruyter).

———— (1956b) Verhaltensstudien an Katzen. Z. Tierpsychol., Suppl., 2.

———— (1965a) Über die Funktion der relativen Stimmungshierarchie (dargestellt am Beispiel der phylogenetischen und ontogenetischen Entwicklung des Beutefangs von Raubtieren). Z. Tierpsychol., 22:412–494.

———— (1965b) The Communal Organisation of Solitary Mammals. Symp. Zool. Soc. London, 14:249–263.

LIMBAUGH, C. (1961) Cleaning Symbiosis. Sci. Am., 205(8):42–49.

LIMBAUGH, C., PEDERSON, H., and CHACE, F. A. (1961) Shrimps that Clean Fishes. Bull. Marine Sci. Gulf Caribbean, 11:237–257.

LINDAUER, M. (1952) Ein Beitrag zur Frage der Arbeitsteilung im Bienenstaat. Z. Vergleich. Physiol., 34:299–345.

———— (1957) Sonnenorientierung der Bienen unter der Äquatorsonne und zur Nachtzeit. Naturwiss., 44:1–6.

———— (1961) Communication among Social Bees. Cambridge, Mass. (Harvard Univ. Press).

———— (1963) Allgemeine Sinnesphysiologie, Orientierung im Raum. Fortschr. Zool., 16:58–140.

LISSMANN, H. W. (1946) The Neurological Basis of the Locomotory Rhythm in the Spinal Dogfish. J. Exptl. Biol., 23:143–176.

LISSMANN, H. W., and MACHIN, K. G. (1958) The Mechanism of Object Location in Gymnarchus niloticus and Similar Fish. J. Exptl. Biol., 35:451.

LIVINGSTONE, F. B. (1969) Genetics, Ecology and the Origins of Incest and Exogamy. Current Anthropology 10:45–61.

LLOYD, J. E. (1965) Aggressive Mimikry in Photuris: Fireflies femme fatales. Science, 149:653–654.

LOEB, J. (1913) Die Tropismen. Handb. Vergleich. Physiol., 4.

LORENZ, K. (1931) Beiträge zur Ethologie sozialer Corviden. J. Ornithol., 79:67–127.

———— (1935) Der Kumpan in der Umwelt des Vogels. J. Ornithol., 83:137–413.

———— (1937) Über die Bildung des Instinktbegriffes. Naturwiss., 25:289–300, 307–318, 325–331.

———— (1939) Vergleichende Verhaltensforschung. Zool. Anz., Suppl., 12:69–102.

———— (1940) Durch Domestikation verursachte Störungen arteigenen Verhaltens. Z. Angew. Psychol. Charakt.kde, 59:2–81.

———— (1941) Vergleichende Bewegungsstudien bei Anatiden. J. Ornithol., 89:194–294.

———— (1943) Die angeborenen Formen möglicher Erfahrung. Z. Tierpsychol., 5:235–409.

———— (1949) Er redete mit dem Vieh, den Vögeln und den Fischen. Vienna (Borotha-Schoeler).

———— (1950a) Ganzheit und Teil in der tierischen und menschlichen Gemeinschaft. Studium Gen., 3:455–499.

———— (1950b) The Comparative Method in Studying Innate Behaviour Patterns. Symp. Soc. Exptl. Biol., 4:221–268.

———— (1950c) So kam der Mensch auf den Hund. Vienna (Borotha-Schoeler).

———— (1951) Ausdrucksbewegungen höherer Tiere. Naturwiss., 38:113–116.

———— (1953) Die Entwicklung der vergleichenden Verhaltensforschung in den letzten 12 Jahren. Zool. Anz., Suppl., 16:36–58.

———— (1954a) Das angeborene Erkennen. Natur u. Volk, 84:285–295.

—— (1954b) Morphology and Behavior Patterns in Allied Species. 1st Conf. on Group Proc. Josiah Macy Jr. Found., New York, 168–220.

—— (1956) The Objectivistic Theory of Instinct. In GRASSE, P. P. (ed.), L'instinct dans le comportement des animaux et da l'homme. Paris (Fondation Singer-Polignac), 51–76.

—— (1957) Methoden der Verhaltensforschung. In KÜKENTHAL, Handb. d. Zool., 8, 10(1), 1–22.

—— (1958) The Evolution of Behaviour. Sci. Am., 199(6):67–78.

—— (1959) Psychologie und Stammesgeschichte. In HEBERER, G. (ed.), Evolution der Organismen. Stuttgart (G. Fischer).

—— (1961) Phylogenetische Anpassung und adaptive Modifikation des Verhaltens. Z. Tierpsychol., 18:139–187.

—— (1962) Naturschönheit und Daseinskampf. Kosmos, 58:340–348.

—— (1963a) Das sogenannte Böse. Vienna (Borotha-Schoeler).

—— (1963b) Die "Erfindung" von Flugmaschinen in der Evolution der Wirbeltiere. Therap. Monats, 13:138–148. Mannheim (Boehringer)

—— (1965a) Über tierisches und menschliches Verhalten. Aus dem Werdegang der Verhaltenslehre (Ges. Abhandl.). I u. II. Munich (Piper).

—— (1965b) Evolution and Modification of Behavior. Chicago (Univ. Press).

—— (1966) Stammes- und kulturgeschichtliche Ritenbildung. Mitt. Max-Planck-Ges., 1:3–30 and Naturwiss. Rundschau, 19:361–370.

LORENZ, K., and SAINT-PAUL, U. v. (1968) Die Entwicklung des Spießens und Würgens bei den drei Würgerarten *Lanius collurio, L. senator* u. *L. excubitor.* J. Ornithol., 109:137–156.

LORENZ, K., and TINBERGEN, N. (1938) Taxis und Instinkthandlung in der Eirollbewegung der Graugans. Z. Tierpsychol., 2:1–29.

LUDWIG, J. (1965) Beobachtungen über das Spiel von Boxern. Z. Tierpsychol., 22:813–838.

LÜLING, V. H. (1958) Morphologisch-anatomische und histologische Untersuchungen am Auge des Schützenfisches *Toxotes jaculatrix,* nebst Bemerkungen zum Spuckgehaben. Z. Morphol. Ökol. Tiere, 47:529–610.

LUTHER, W. (1958) Symbiose von Fischen *(Gobiidae)* mit einem Krebs *(Alpheus)* im Roten Meer. Z. Tierpsychol., 15:175–177.

MCCONNELL, J. V. (1962) Memory Transfer through Cannibalism. J. Neuropsychiat., 3:542–548.

MCCONNELL, J. V., JACOBSON, A. L., and KIMBLE, D. P. (1959) The Effects of Regeneration upon Retention of a Conditioned Response in the Planarian. J. Comp. Physiol. Psychol., 52:1–5.

MCDERMOTT, F. A. (1917) Observations on the Light-Emission of American Lampyridae. Can. Entomol., 49:53–61.

MCDOUGALL, W. (1936) An Outline of Psychology. 7th ed. London.

MACFARLANE, D. A. (1930) The Role of Kinaestesis in Maze Learning. Univ. Calif. Publ. Psychol., 4:277–305.

MCGILL, T. (1965) Readings in Animal Behavior. New York (Holt, Rinehart and Winston Inc.).

MCKINNEY, F. (1965) The Comfort Movements of *Anatidae.* Behaviour, 25:120–220.

MCNEIL, E. (1959) Psychology and Aggression. J. Conflict Resolution, 3:195–293.

MACHEMER, H. (1966) Versuche zur Frage nach der Dressierbarkeit hypotricher Ciliaten unter Einsatz hoher Individuenzahlen. Z. Tierpsychol., 23:641–654.

MACKENSEN, G. (1965) Zur Verhaltensweise blinder Kinder. Studium Gen., 18:9–14.

MACKINTOSH, J. H., and GRANT, E. C. (1966) The Effect of Olfactory Stimuli on the Agonistic Behaviour of Laboratory Mice. Z. Tierpsychol., 23:584–587.

MAGNUS, D. (1954) Zum Problem der "überoptimalen" Schlüsselreize. Zool. Anz., Suppl., 18:317–325.

—— (1958) Experimentelle Untersuchungen zur Bionomie und Ethologie des Kaisermantels *Argynnis paphia* L. I. Über optische Auslöser von Auffliegereaktionen und ihre Bedeutung für das Sichfinden der Geschlechter. Z. Tierpsychol., 15:397–426.

—— (1964) Zum Problem der Partnerschaften mit Diademseeigeln. Zool. Anz., Suppl., 27:404–417.

MAIER, N. R. F., and SCHNEIRLA, T. C. (1935) Principles of Animal Psychology. New York.

MAKKINK, G. F. (1936) An Attempt at an Ethogram of the European Avocet (*Recurvirostra avosetta* L.) with Ethological and Psychological Remarks. Ardea, 25:1–60.

MANLEY, G. (1960) The Agonistic Behavior of the Black-Headed Gull. (Dissertation, Oxford, quoted by R. A. STAMM, 1964).

MANNING, A. (1956) Some Aspects of the Foraging Behaviour of Bumblebees. Behaviour, 9:164–201.

—— (1961) The Effects of Artificial Selection for Mating Speed in *Drosophila melanogaster*. Animal Behavior, 9:82–92.

—— (1965) *Drosophila* and the Evolution of Behaviour. Viewpoints Biol., 4:125–169.

MARLER, P. R. (1955a) Studies of Fighting in Chaffinches (1). Behaviour in Relation to the Social Hierarchy. Brit. J. Animal Behavior, 3:111–117.

—— (1955b) Studies on Fighting in Chaffinches (2). The Effect on Dominance Relations of Disguising Females as Males. Brit. J. Animal Behavior, 3:137–146.

—— (1956a) Studies of Fighting in Chaffinches (3). Proximity as a Cause of Aggression. Brit. J. Animal Behavior, 4:23–30.

—— (1956b) Über einige Eigenschaften tierlicher Rufe. J. Ornithol., 97: 220–227.

—— (1957a) Specific Distinctness in the Communication Signals of Birds. Behaviour, 11:13–39.

—— (1957b) Studies of Fighting in Chaffinches (4). Appetitive and Consummatory Behaviour. Brit. J. Animal Behavior, 5:29–37.

—— (1959) Developments in the Study of Animal Communication. In BELL, P. R. (ed.), Darwin's Biological Work. London (Cambridge Univ. Press), 150–206.

MARLER, P. R., and BOATSMAN, D. J. (1951) Observations on the Birds of Pico, Azores. Ibis, 93:90–99.

MARLER, P. R., and HAMILTON, W. J. (1966) Mechanisms of Animal Behavior. New York (Wiley).

MARLER, P. R., and TAMURA, M. (1964) Culturally Transmitted Patterns of Vocal Behavior in Sparrows. Science, 146:1483–1486.

MARQUENIE, J. (1950) De balts van de kleine Watersalamander. Levende Natuur, 53:147–155.

MARTIN, R. D. (1966a) Tree Shrews: Unique Reproductive Mechanism of Systematic Importance. Science, 152:1402–1404.

—— (1966b) Sind Spitzhörnchen wirklich Vorfahren der Affen? Umschau, 13:437–438.

—— (1968) Reproduction and Ontogeny in Tree-shrews *(Tupaia belangeri)*, with Reference to Their General Behaviour and Taxonomic Relationships. Z. Tierpsychol., 25:409–495, 505–532.

MARTINS, T. and VALLE, J. R. (1948) Hormonal Regulation of Micturition Behavior. J. Comp. Physiol. Psychol., 41:301–311.

MASNE, G. le (1950) Classification et caractéristiques des principaux types de groupements sociaux réalisés chez les invertébrés. Colloq. Intern. Centre Natl. Rech. Sci. (Paris), 34:19–70.

MASON, W. A. (1965) The Social Development of Monkeys and Apes. In DE VORE, I. (ed.), Primate Behavior. New York (Holt, Rinehart and Winston), 514–543.

MATTHEWS, G. V. T. (1955) Bird Navigation. New York (Cambridge Univ. Press).

MATURANA, H. R., LETTVIN, J. Y., McCULLOCH, W. S., and PITTS, W. H. (1960) Anatomy and Physiology of Vision in the Frog (Rana pipiens). J. Gen. Physiol., Suppl., 6:129–175.

MAXWELL, S. S. (1923) Labyrinth and Equilibrium. Philadelphia.

MAY, E., and BIRUKOW, G. (1966) Lunar-periodische Schwankungen des Lichtpreferendums bei der Bachplanarie. Naturwiss., 53:182.

MAYER, G. (1952) Untersuchungen über Herstellung und Struktur des Radnetzes von Aranea diadema und Zilla x- notata mit besonderer Berücksichtigung des Unterschiedes von Jugend- und Altersnetzen. Z. Tierpsychol., 9:337–362.

MAYER, J., and THOMAS, D. W. (1967) Regulation of Food Intake and Obesity. Science, 156:328–337.

MAYR, E. (1950) Ecological Factors in Speciation. Evolution, 1:263–288.

———— (1958) Behavior and Systematics. In ROE, A. and SIMPSON, G. (eds.), Behavior and Evolution. New Haven (Yale Univ. Press), 341–366.

MEAD, M. (1935) Sex and Temperament in Three Primitive Societies. New York (Morrow).

———— (1956) Birth of a Baby in New Guinea. In SODDY, K. (ed.), Mental Health and Infant Development. I. New York (Basic Books).

———— (1965) Leben in der Südsee. Munich (Szczesny).

MEISENHEIMER, J. (1921) Geschlecht und Geschlechter im Tierreich. Jena (G. Fischer).

MELCHERS, M. (1960) Cupiennius salei (Ctenidae): Kokonbau und Eiablage. Encycl. cinem., E363, Göttingen (Inst. wiss. Film).

———— (1963) Zur Biologie und zum Verhalten von Cupiennius salei (Keyserling), einer amerikanischen Ctenidae. Zool. Jahrb. System, 91:1–90.

———— (1964) Cupiennius salei (Ctenidae): Spinnhemmung beim Kokonbau. Encycl. cinem., E364. Publ. zu wiss. Filmen, 1A, 21–24. Göttingen (Inst. wiss. Film).

MERKEL, F. W., and WILTSCHKO, W. (1965) Magnetismus und Richtungsfinden zugunruhiger Rotkehlchen (Erithacus rubecula). Vogelwarte, 23:71–77.

MERTENS, R. (1959) La Vie des Amphibiens et Reptiles. Horizons de France. Paris.

MERZ, F. (1965) Aggression und Aggressionsantrieb. In THOMAE, H. (ed.), Handb. d. Psychol., 2:569–601.

MEYER-HOLZAPFEL, M. (1956) Das Spiel bei Säugetieren. In KUKENTHAL, Handb. d. Zool., 8(10):1–36.

———— (1958) Soziale Beziehungen bei Säugetieren. In LEHMANN, F. E. (ed.), Gestaltungen sozialen Lebens bei Tier und Mensch. Bern (Francke), 86–109.

MILGRAM, S. (1963) Behavioral Study of Obedience. J. Abnorm. Social Psychol., 67:372–378.

—— (1965a) Liberating Effects of Group Pressure. J. Personality. Social Psychol., 1:127–134.

—— (1965b) Some Conditioning Obedience and Disobedience. Human Relations, 18:57–76.

—— (1966) Einige Bedingungen von Autoritätsgehorsam und seiner Verweigerung. Z. Exptl. Angew. Psychol., 13:433–463.

MILL, J. S. (1843) A System of Logic, II. London (Parker).

MILNE, L. J., and MILNE, M. (1963) Die Sinneswelt der Tiere und Menschen. Hamburg (Parey).

MISTSCHENKO, M. N. (1933) Über die mimische Gesichtsmotorik der Blinden. Folia Neuropathol. Estonia, 13:24–43.

MITSCHERLICH, A. (1957) Aggression und Anpassung. I. Psyche, 10:177–193.

—— (1959) Aggression und Anpassung. II. Psyche, 12:523–537.

MITTELSTAEDT, H. (1953) Über den Beutefangmechanismus der Mantiden. Zool. Anz., Suppl., 16:102–106.

—— (1954) Regelung und Steuerung bei der Orientierung der Lebewesen. Regelungstechnik, 10:226–232.

MIYADI, D. (1965) Social Life of Japanese Monkeys. Science in Japan. American Association for the Advancement of Science.

—— (1967) Differences in Social Behavior among Japanese Macaque Troops. Neue Ergebnisse der Primatologie, (First Congress Int. Primatol. Soc.) 228–231. Stuttgart (G. Fischer).

MÖHRES, F. P. (1953) Über die Ultraschallorientierung der Hufeisennasen. Z. Vergleich. Physiol., 34:547–588.

—— (1961) Die elektrischen Fische. Natur u. Volk, 91:1–12.

MONNIER, M., and WILLI, H. (1953) Die integrative Tätigkeit des Nervensystems beim meso-rhombo-spinalen Anencephalus (Mittelhirnwesen). Monatschr. Psychiat. Neurol., 126:239–258.

MONTAGU, M. F. A. (1962) Culture and the Evolution of Man. New York (Oxford Univ. Press).

—— (1968) Man and Aggression. New York (Oxford Univ. Press).

MONTESSORI, M. (1952) Kinder sind anders. 5th ed. Stuttgart (Klett).

MORGAN, C. Lloyd (1894) Introduction to Comparative Psychology. London.

—— (1900) Animal Behavior. London.

MÖRRIS, D. (1954) The Reproductive Behavior of the River-Bullhead (*Cottus gobio* L.), with Special Reference to the Fanning Activity. Behaviour, 7:1–32.

—— (1956) The Feather Postures of Birds and the Problem of the Origin of Social Signals. Behaviour, 9:75–113.

—— (1957) "Typical Intensity" and Its Relation to the Problem of Ritualization. Behaviour, 11:1–12.

—— (1958) The Reproductive Behaviour of the Ten-Spined Stickleback (*Pygosteus pungitius* L.). Behaviour, Suppl., 6.

—— (1963) Biologie der Kunst. Düsseldorf (Rauch).

—— (1968) Der nackte Affe. Munich (Droemer.)

MORSE, D. H. (1968) The Use of Tools by Brown-headed Nuthatches. Wilson Bull., 80:220–224.

MOSS, A. M. (1920) Sphingidae of Para, Brazil. Nov. Zool., 27:333–424.

MOYNIHAN, M. (1955) Some Aspects of Reproductive Behavior in the Black-headed Gull (*Larus ridibundus ridibundus* L.) and Related Species. Behaviour, Suppl., 4.

—— (1964) Some Behavior Patterns of Platyrrhine Monkeys. I. The Night Monkey (*Aotes trivirgatus*). Smithsonian Misc. Collections, 146(5):1–84.

MÜLLER, D. (1961) Quantitative Luftfeind-Attrappenversuche bei Auer- und

Birkhühnern (*Tetrao urogallus* L. und *Lyrurus tetrix* L.). Naturforsch., 16b: 551–553.

MÜLLER-USING, D. (1952) Über einige bisher unbeachtete Übersprunghandlungen bei höheren Säugern. Z. Tierpsychol., 9:479–481.

MUNN, N. L. (1950) Handbook of Psychological Research on the Rat. New York (Houghton-Mifflin).

MURIE, A. (1944) The Wolves of Mount McKinley, Fauna of the National Parks of the U.S.A. Fauna Ser., 5.

MYERS, R. E. (1956) Functions of Corpus callosum in Interocular Transfer. Brain, 79:358–363.

MYKYTOWYCZ, R. (1955) Further Observations on the Territorial Function and Histology of the Submandibular Cutaneous (Chin) Glands in the Rabbit (*Oryctolagus cuniculus*). Animal Behavior, 13:400–412.

———— (1960) Social Behaviour of an Experimental Colony of Wild Rabbits. CSIRO Wildlife Res. (Canberra), 5(1):1–20.

———— (1966) Observations on Odoriferous and Other Glands in the Australian Wild Rabbit, *Oryctolagus cuniculus* L., and the Hare, *Lepus europaeus* P. I. The Anal Gland. II. The Inguinal Gland. III. Harders Lacrimal and Submandibular Glands. CSIRO Wildlife Res. (Canberra), 11:11–29, 49–90.

MYKYTOWYCZ, R., and DUDZINSKI, M. L. (1966) A Study of the Weight of Odoriferous and other Glands in Relation to Social Status and Degree of Sexual Activity in the Wild Rabbit, *Oryctolagus cuniculus.* CSIRO Wildlife Res. (Canberra), 11:31–47.

MYRBERG, A. A. (1964) An Analysis of Preferential Care of Eggs and Young by Adult Cichlid Fishes. Z. Tierpsychol., 21:53–98.

———— (1965) Sound Production by Cichlid Fishes. Science, 149:555–558.

NAPIER, J. (1962) The Evolution of the Hand. Sci. Am., 207(6):56–63.

NAYLOR, E. (1958) Tidal and Diurnal Rhythms of Locomotor Activity in *Carcinus maenas.* J. Exptl. Biol., 35.602–610.

NELSON, K. (1964) Behavior and Morphology in the Glandulocaudine Fishes (*Osteriophysi, Characidae*). Univ. Calif. Publ. Zool., 75:59–152.

NICE, M. M. (1941) The Role of Territory in Bird Life. Am. Midland Naturalist, 26:441–487.

———— (1962) Development of Behavior in Precocial Birds. Trans. Linnean Soc. New York, 8.

NICOLAI, J. (1956) Zur Biologie und Ethologie des Gimpels. Z. Tierpsychol., 13:93–132.

———— (1959a) Familientradition in der Gesangstradition des Gimpels (*Pyrrhula pyrrhula* L.). J. Ornithol., 100:39–46.

———— (1959b) Verhaltensstudien an einigen afrikanischen und paläarktischen Girlitzen. Zool. Jahrb. System, 87:317–362.

———— (1964) Der Brutparasitismus der *Viduinae* als ethologisches Problem. Prägungsphänomene als Faktoren der Rassen- und Artbildung. Z. Tierpsychol., 21:129–204.

———— (1965a) Vogelhaltung und Vogelpflege. Das Vivarium. Stuttgart (Franckh).

———— (1965b) *Columba livia:* Flug von Haustaubenrassen. I. Der Flug des Birmingham Rollers. II. Der Flug des Steller-Kröpfers, Filme, Göttingen (Inst. wiss. Film).

———— (1965c) Der Brutparasitismus der Witwenvögel, Naturwiss. Med. 2(7): 3–15.

NIEBOER, H. J. (1960) Ethological Observations on the Ant Lion (*Euroleon nostras* Fourcroy). Arch. Neerl. Zool., 13:609–611.

Nitschke, A. (1953) Über Eigenart und Ausdrucksgehalt frühkindlicher Motorik. Deut. Med. Wochschr., 78:1787–1792.

Noble, G. K. (1927) The Value of Life History Data in the Study of the Evolution of Amphibia. Ann. N.Y. Acad. Sci., 30.

—— (1931) Biology of Amphibia. New York.

—— (1934) Experimenting with the Courtship of Lizards. Nat. Hist., 34: 1–15.

Noble, G. K., and Bradley, H. T. (1933) The Mating Behavior of Lizards. Nat. Hist., 34:1–15.

Noble, G. K., and Curtis, B. (1939) The Social Behavior of the Jewel Fish, Hemichromis bimaculatus Gill. Bull. Am. Mus. Nat. Hist., 76:1–46.

Oehlert, B. (1958) Kampf und Paarbildung einiger Cichliden. Z. Tierpsychol., 15:141–174.

Ohm, D. (1964) Die Entwicklung des Kommentkampfverhaltens bei Jungcichliden. Z. Tierpsychol., 21:308–325.

Ohm, T. (1948) Die Gebetsgebärden der Völker und das Christentum.

Olds, J. (1958) Self-Stimulation of the Brain. Science, 127:315–324.

Oppenheim, R. (1966) Amniotic Contractions and Embryonic Motility in the Chick Embryo. Science, 152:528–529.

Osche, G. (1952) Die Bedeutung der Osmoregulation und des Winkverhaltens für freilebende Nematoden. Z. Morphol. Ökol. Tiere, 41:54–77.

—— (1962) Ökologie des Parasitismus und der Symbiose. Fortschr. Zool., 15:125–164.

—— (1966) Die Welt der Parasiten. Verständl. Wiss., 89, Heidelberg (Springer).

Papi, F. (1959) Sull'orientamento astronomico in specie del gen. Arctosa. Z. Vergleich. Physiol., 41:481–489.

Papi, F., and Pardi, L. (1959) Nuovi reperti sull'orientamento lunare die Talitrus saltator. Z. Vergleich. Physiol., 41:583–596.

Pardi, L. (1960) Innate Components in the Solar Orientation of Littoral Amphipods. Cold Spring Harbor Symp. Quant. Biol., 25:395–401.

Pawlow, I. P. (1927) Conditioned Reflexes (Oxford).

Peckham, G., and Peckham, E. (1904) Instinkt und Gewohnheiten der solitären Wespen. Berlin.

Peiper, A. (1951) Instinkt und angeborenes Schema beim Säugling. Z. Tierpsychol., 8:449–456.

—— (1953) Schreit- und Steigbewegungen beim Neugeborenen. Arch. Kinderheilkde., 147:135.

—— (1961) Die Eigenart der kindlichen Hirntätigkeit. 3rd ed. Leipzig.

Peiponen, V. A. (1960) Verhaltensstudien am Blaukehlchen. Ornis Fennica, 37:69–83.

Penfield, W. (1952) Memory Mechanisms. Arch. Neurol. Psychiat. (Chicago), 67:178–191.

Pengelley, E. T., and Fisher, K. C. (1963) The Effect of Temperature and Photoperiod on the Yearly Hibernating Behavior of Captive Golden-Mantled Ground Squirrels (Citellus lateralis tescorum). Can. J. Zool., 41:1103–1120.

Pennycuick, C. J. (1960) The Physical Basis of Astronavigation in Birds: Theoretical Considerations. J. Exptl. Biol., 37:573–593.

Perdeck, A. C. (1958a) The Isolating Patterns in two Sibling Species of Grasshoppers (Chorthippus brunneus and Chorthippus biguttulus). Behaviour, 12: 11–75.

—— (1958b) Two Types of Orientation in Migrating Starlings, Sturnus vul-

garis and Chaffinches, *Fringilla coelebs,* as Revealed by Displacement Experiments. Ardea, 46:1–37.

PERNAU, F. A. v. (1716) Unterricht, was mit dem lieblichen Geschöpff denen Vögeln, auch außer dem Fang, nur durch die Ergründung Deren Eigenschafften, und Zahmmachung, oder anderer Abrichtung, Man sich vor Lust und Zeit-Vertreib machen können. Nürnberg.

PETERS, II. M. (1937a) Experimentelle Untersuchungen über die Brutpflege von *Haplochromis multicolor,* einem maulbrütenden Knochenfisch. Z. Tierpsychol., 1:201–218.

——— (1937b) Studien am Netz der Kreuzspinne. Z. Morphol. Ökol. Tiere, 32:613–649 and 33:128–150.

——— (1939) Die Probleme des Kreuzspinnennetzes. Z. Morphol. Ökol. Tiere, 36:179–266.

——— (1953) Weitere Untersuchungen über den strukturellen Aufbau des Radnetzes der Spinnen. Z. Naturforsch., 8b:355–370.

——— (1956) Gesellungsformen der Tiere. In Handb. d. Soziologie, II. Stuttgart.

PETERSEN, B., LUNDGREN, L., and WILSON, L. (1957) The Development of Flight Capacity in a Butterfly. Behaviour, 10:324–339.

PETERSEN, E. (1965) Biologische Beobachtungen über Verhaltensweisen einiger einheimischer Nager beim Öffnen von Nüssen und Kernen. Z. Säugetierkde, 30:156–162.

PFEIFFER, W. (1960) Über die Schreckreaktion bei Fischen und die Herkunft des Schreckstoffes. Z. Vergleich. Physiol., 43:578–614.

——— (1963) Vergleichende Untersuchungen über die Schreckreaktion und den Schreckstoff der Ostariophysen. Z. Vergleich. Physiol., 47:111–147.

PILLERI, G. (1960a) Über das Auftreten von "Kletterbewegungen" im Endstadium eines Falles von Morbus Alzheimer. Arch. Psychiat. Nervenkde, 200:455–461.

——— (1960b) Kopfpendeln ("Leerlaufendes Brustsuchen") bei einem Fall von Pickscher Krankheit. Arch. Psychiat. Nervenkde, 200:603–611.

——— (1961) Orale Einstellung nach Art des Klüver-Bucy-Syndroms bei hirnatrophischen Prozessen. Schweiz. Arch. Neurol. Neurochirg. Psychiat., 87: 286–298.

PLACK, A. (1968) Die Gesellschaft und das Böse. 2nd ed. Munich (List).

PLOOG, D. W. (1964a) Verhaltensforschung und Psychiatrie. In GRUHLE, H. W., JUNG, R., MAYER-GROSS, W., and MÜLLER, M. (eds.), Psychiatrie der Gegenwart. 1. Berlin (Springer), 291–443.

——— (1964b) Verhaltensforschung als Grundlagenwissenschaft für die Psychiatrie. 6. Psychiatertagg. d. Landschaftsverb. Rheinland—123. wiss. Vers. d. Rhein. Ver. f. Psychiatrie, 1–23.

——— (1964c) Über experimentelle Grundlagen der Gedächtnisforschung. Nervenarzt, 35:377–386.

PLOOG, D. W., BLITZ, J., and PLOOG, F. (1963) Studies on Social and Sexual Behavior of the Squirrel Monkey (*Saimiri sciureus*). Folia Primat., 29–66.

PÖPPEL, E. (1968) Desynchronisation circadianer Rhythmen innerhalb einer isolierten Gruppe. Arch. Ges. Physiol., 299:364–370.

PORTMANN, A. (1953) Das Tier als soziales Wesen. 2nd ed. Zürich (Rhein).

POULSEN, H. (1953) A Study of Incubation Responses and Some Other Behavior Patterns in Birds. Vidensk. Medd. Danks. Naturh. Foren.

PRECHTL, H. F. R. (1951) Zur Paarungsbiologie einiger Molcharten. Z. Tierpsychol., 8:337–348.

——— (1953a) Zur Physiologie des Angeborenen Auslösemechanismus. Behaviour, 5:32–50.

────── (1953b) Die Kletterbewegungen beim Säugling. Monatsschr. Kinder-heilkde, 12:519–521.

────── (1953c) Stammesgeschichtliche Reste im Verhalten des Säuglings. Um-schau, 21:656–658.

────── (1955) Die Entwicklung der frühkindlichen Motorik. I-III. Wiss. Filme, C. 651, C 652, C 653. Göttingen (Inst. wiss. Film).

────── (1956) Neurophysiologische Mechanismen des formstarren Verhal-tens. Behaviour, 9:243–319.

────── (1958) The Directed Head-Turning Response and Allied Movements of the Human Baby. Behaviour, 13:212–242.

PRECHTL, H. F. R., and KNOL, A. R. (1958) Fußsohlenreflexe beim neuge-borenen Kind. Arch. Psychiat. Z. Ges. Neurol., 196:542–553.

PRECHTL, H. F. R., and SCHLEIDT, W. M. (1950) Auslösende und steuernde Mechanismen des Saugaktes. I. Z. Vergleich. Physiol., 32:252–262.

────── (1951) Auslösende und steuernde Mechanismen des Saugaktes. II. Z. Vergleich. Physiol., 33:53–62.

PREVOST, J. (1961) Écologie du Manchot empereur: *Aptenodytes forsteri* Gray. Expeditions polaires Françaises. Miss. Paul-Emile Victor. Publ. 222. Actu-alités Sci. Ind. No. 1291. Paris (Hermann).

PREYER, W. (1885) Spezielle Physiologie des Embryo. Leipzig (Grieben).

RABER, H. (1948) Analyse des Balzverhaltens eines männlichen Truthahns (*Meleagris*). Behaviour, 1:237–266.

RANDALL, J. E. (1958) A Review of the Labrid Fish genus Labroides with Des-criptions of Two New Species and Notes on Ecology. Pacific Sci., 12: 327–347.

RANDALL, J. E., and RANDALL, H. E. (1960) Examples of Mimikry and Protec-tive Resemblance in Tropical Marine Fishes. Bull. Marine Sci. Gulf Carib-bean, 10:444–480.

RAO, K. S. (1954) Tidal Rhythmicity of Rate of Water Propulsion in *Mytilus* and Its Modificability by Transplantation. Biol. Bull., 106:353–359.

RASA, O. A. E. (1969) The Effect of Pair Isolation on Reproductive Success in *Etroplus maculatus* (Cichlidae). Z. Tierpsychol., 26:846–852.

RASMUSSEN, K. (1908) People of the Polar North. London.

RAY, C. (1966) Stalking Seals under Antarctic Ice. Natl. Geogr., 129:54–65.

REAUMUR, R. A. F. (1734–1742) Mémoires pour servir à l'histoire des insectes. Paris (Impr. Royale), 1–6.

REESE, E. S. (1962a) Submissive Posture as an Adaptation to Aggressive Be-havior in Hermit Crabs. Z. Tierpsychol., 19:645–651.

────── (1962b) Shell Selection Behavior of Hermit Crabs. Animal Behavior, 10:347–360.

────── (1963a) The Behavioral Mechanisms Underlying Shell Selection by Hermit Crabs. Behaviour, 21, 78–126.

────── (1963b) A Mechanism Underlying Selection or Choice Behavior Which Is Not Based on Previous Experience. Am. Zool., 3:508.

────── (1968) Shell Use: An Adaptation for Emigration from the Sea by the Coconut Crab. Science, 161:385–386.

REGEN, J. (1924) Über die Orientierung des Grillenweibchens nach dem Stri-dulationsschall des Männchens. Sitz. Ber. Akad. Wiss. Wien, Math. Nat. Kl., 132.

REIMARUS, H. S. (1762) Allgemeine Betrachtungen über die Triebe der Thiere, hauptsächlich über ihre Kunsttriebe. Hamburg.

REMANE, A. (1952) Die Grundlagen des natürlichen Systems der vergleichen-den Anatomie und der Phylogenetik. Leipzig.

—— (1960) Das soziale Leben der Tiere. Rowohlts Deut. Enzykl., 97. Hamburg (Rowohlt).

REMMERT, H. (1965) Biologische Periodik. Handb. d. Biol., 5. Frankfurt (Athenaion), 335–411.

RENSCH, B. (1957) Ästhetische Faktoren bei Farb- und Formbevorzugungen von Affen. Z. Tierpsychol., 14:71–99.

—— (1958) Die Wirksamkeit ästhetischer Faktoren bei Wirbeltieren. Z. Tierpsychol., 15:447–461.

—— (1961) Malversuche mit Affen. Z. Tierpsychol., 18:347–364.

—— (1962) Gedächtnis, Abstraktion und Generalisation bei Tieren. Arbeitsgemeinschaft Forsch. Landes Nordrhein-Westfalen, Köln (Westdeutscher Verlag).

—— (1963) Versuche über menschliche "Auslöser-Merkmale" beider Geschlechter. Z. Morphol. Anthropol., 53:139–164.

—— (1965) Die höchsten Lernleistungen der Tiere. Naturwiss. Rundschau, 18:91–101.

RENSCH, B. and DÖHL, J. (1968) Wahlen zwischen zwei überschaubaren Labyrinthwegen durch einen Schimpansen. Z. f. Tierpsychol. 25:216–231.

—— and DÜCKER, G. (1959) Versuche über visuelle Generalisation bei einer Schleichkatze. Z. Tierpsychol., 16:671–692.

REUTER, O. M. (1913) Lebensgewohnheiten und Instinkte der Insekten. Berlin.

RICHARD, P. B. (1955) Bièvres constructeurs de barrages. Mammalia. 19:293–301.

—— (1964) Les matériaux de construction du Castor (*Castor fiber*), leur signification pour ce rongeur. Z. Tierpsychol., 21:592–601.

RICHTER, R. (1927) Die fossilen Fährten und Bauten der Würmer. Palaeont. Z., 9:193–240.

RIESEN, A. H. (1960) Effects of Stimulus Deprivation on the Development and Atrophy of the Visual Sensory System. J. Orthopsychiat., 30:23–26.

RIESS, B. F. (1954) The Effect of Altered Environment and of Age on the Mother-Young Relationships among Animals. Ann. N.Y. Acad. Sci., 57:606–610.

RISLER, H. (1953) Das Gehörorgan des Männchens von *Anopheles stephensi* Liston (Culicidae). Zool. Jahrb. (Anat.), 73:165–186.

—— (1955) Das Gehörorgan der Stechmücken. Mikrokosmos, 44:217–220.

RITTINGHAUS, H. (1963) *Sterna hirundo* (Laridae): Balz und Kopulation. Encycl. cinem., E 659, Göttingen (Inst. wiss. Film).

ROBERTS, W. W. and BERGQUIST, E. H. (1968) Attack elicited by Hypothalamic Stimulatin in Cats Raised in Social Isolation. J. Comp. Physiol. Psychol. 66:590–595.

ROBERTS, W. W., and CAREY, R. J. (1965) Rewarding Effect of Performance of Gnawing Aroused by Hypothalamic Stimulation in the Rat. J. Comp. Physiol. Psychol., 59:317–324.

ROBERTS, W. W., and KIESS, H. O. (1964) Motivational Properties of Hypothalamic Aggression in Cats. J. Comp. Physiol. Psychol., 58:187–193.

ROBERTS, W. W., STEINBERG, M. L., and MEANS, L. W. (1967) Hypothalamic Mechanisms for Sexual, Aggressive, and other Motivational Behaviors in the Opossum, *Didelphis virginiana*. J. Comp. Physiol. Psychol. 64:1–15.

ROBINSON, J. (1962) *Pilobolus* sp. and the Translocation of Infective Larvae of *Dictyocaulus viviparus* from Faeces to Pastures. Nature, 193:353–354.

ROBSON, K. S. (1967) The Role of Eye-to-eye Contact in Maternal-Infant Attachment. J. Child Psychol. 8:13–25.

ROEDER, K. D. (1935) An Experimental Analysis of the Sexual Behavior of the Preying Mantis. Biol. Bull. 69:203–220.

——— (1937) The Control of Tonus and Locomotory Activity in the Preying Mantis (*Mantis religiosa* L.). J. Exptl. Zool., 76:353–374.

——— (1955) Spontaneous Activity and Behavior. Sci. Monthly. Wash. 80: 362–370.

——— (1963a) Nerve Cells and Insect Behavior. Cambridge, Mass. (Harvard Univ. Press).

——— (1963b) Ethology and Neurophysiology. Z. Tierpsychol., 20:434–440.

ROEDER, K. D., and TREAT, E. A. (1961) The Reception of Bat Cries by the Tympanic Organ of Noctuid Moths. In ROSENBLITH, (ed.), Sensory Communication. Cambridge, Mass. (M. I. T. Press) and New York (Wiley).

RÖSCH, A. G. (1925) Untersuchungen über die Arbeitsteilung im Bienenstaat. Z. Vergleich. Physiol., 6:571–631.

RÖSEL V. ROSENHOF, A. J. (1746–1761) Insekten-Belustigungen, I–IV. Nuremberg (Fleischmann).

——— (1758) Die natürliche Historie der Frösche hiesigen Landes etc. Nuremberg (Fleischmann).

ROSENKÖTTER, L. (1966) Auf Excursion in die Menschenkunde. (Frankfurter Hefte) 21.

ROSS, D. M. (1960) The Association between the Hermit Crab *Eupagurus bernhardus* L. and the Sea Anemone *Calliactis parasitica*. Proc. Zool. Soc. London, 134:43–47.

ROSS, S. (1951) Sucking Behavior in Neonate Dogs. J. Abnorm. Social. Psychol., 46:142–149.

ROTH, L. M. (1948) An Experimental Laboratory Study of the Sexual Behaviour of *Aedes aegypti*. Am. Midland Naturalist., 40:265–352.

ROTHBALLER, A. B. (1967) Aggression, Defense and Neurohumors. U.C.L.A. Forum Medical Science, 7:135–170. In CLEMENTE, C. D. and LINDSLEY, D. B. (eds.), Aggression and Defense. Berkeley (Univ. Calif. Press).

ROTHENBUHLER, W. C. (1964) Behavior Genetics of Nest Cleaning in Honeybees. IV. Responses of F_1 and Backcross Generations to Disease Killed Brood. Am. Zoologist, 4:111–123.

ROTHMANN, M., and TEUBER, E. (1915) Einzelausgabe der Anthropoidenstation auf Teneriffa. I. Ziele und Aufgaben der Station sowie erste Beobachtungen an den auf ihr gehaltenen Schimpansen. Abhandl. Preuss. Akad. Wiss. Berlin, 1–20.

ROUBAUD, E. (1916) Recherches biologiques sur les guepes solitaires et sociales d'Afrique. La genese de la vie sociale et l'evolution de l'instinct maternal chez les vespids. Ann. Sci. Nit., 1:1–160.

RUITER, L. de (1952) Some Experiments on the Camouflage of Stick-Caterpillars. Behaviour, 4:222–232.

——— (1955) Countershading in Caterpillars. Arch. Neerl. Zool., 11:1–57.

——— (1963) The Physiology of Vertebrate Feeding Behaviour: Towards a Synthesis of the Ethological and Physiological Approaches to Problems of Behaviour. Z. Tierpsychol., 20:498–516.

RUSSELL, Cl. and RUSSEL, W. W. S. (1968) Violence, Monkeys and Man. London (Macmillan).

RUSSELL, E. S. (1938) The Behaviour of Animals: An Introduction to Its Study. London.

SACKETT, G. P. (1966) Monkeys Reared in Isolation with Pictures as Visual Input. Evidence for an Innate Releasing Mechanism. Science, 154:1468–1473.

SADE, D. S. (1967) Determinants of Dominance in a Group of Freeranging Rhesus Monkeys. In ALTMAN, S. A. (ed.), Social Communication among Primates. Chicago (Univ. Chicago Press), 99–114.

SAINT PAUL, U. v. (1967) *Lanius collurio*—Ontogense des Beutespießens. Encycl. cinem., E 1241. Göttingen (Inst. wiss. Film).

SALLER, K. (1963) Die Aufrichtung des Menschen und ihre Folgen. Z. Morphol. Anthropol., 54:82–111.

SAUER, F. (1954) Die Entwicklung der Lautäußerungen vom Ei ab schalldicht gehaltener Dorngrasmücken (*Sylvia c. communis* Latham). Z. Tierpsychol., 11:1–93.

——— (1956) Zugorientierung einer Mönchsgrasmücke (*Sylvia atricapilla*) unter künstlichem Sternenhimmel. Naturwiss., 43:231–232.

——— (1957) Die Sternenorientierung nächtlich ziehender Grasmücken (*Sylvia atricapilla, borin* und *curruca*). Z. Tierpsychol., 14:29–70.

——— (1961) Further Studies on the Stellar Orientation of Nocturnally Migrating Birds. Psychol. Forsch., 26:224–244.

SAUER, F., and SAUER, E. (1955) Zur Frage der nächtlichen Zugorientierung von Grasmücken. Rev. Suisse Zool., 62:250–259.

——— (1960) Star Navigation of Nocturnal Migrating Birds. Cold Spring Harbor Symp. Quant. Biol., 25:463–473.

SCHAFER, W. (1965) Aktualpaläontologische Beobachtungen. Natur Mus., 95.83–90.

SCHALLER, F. (1962) Die Unterwelt des Tierreiches. Verständl. Wiss., 78. Berlin (Springer).

SCHALLER, F., and SCHWALB, H. (1961) Attrappenversuche mit Larven und Imagines einheimischer Leuchtkäfer. Zool. Anz., Suppl., 24:154–166.

SCHALLER, G. B. (1963) The Mountain Gorilla. Chicago (Univ. Chicago Press).

SCHEIN, W. M. (1963) On the Irreversibility of Imprinting. Z. Tierpsychol., 20:462–467.

SCHENKEL, R. (1947) Ausdrucksstudien an Wölfen. Behaviour, 1:81–129.

———(1956) Zur Deutung der Phasianidenbalz. Ornithol. Beobacht., 53:182.

——— (1958) Zur Deutung der Balzleistungen einiger Phasianiden und Tetraoniden. Ornithol. Beobacht., 55:65–95.

——— (1964) Zur Ontogenese des Verhaltens bei Gorilla und Mensch. Z. Morphol. Anthropol., 54:233–259.

——— (1966) Zum Problem der Territorialität und des Markierens bei Säugern — am Beispiel des Schwarzen Nashorns und des Löwen. Z. Tierpsychol., 23:593–626.

——— (1967) Submission: Its Features and Function in the Wolf and Dog. Am. Zoologist, 7:319–329.

SCHEVILL, W. E. (1955) Evidence for Echolocation by Cetaceans. Deep Sea Res., 3:153.

SCHIFTER, H. (1965) Beobachtungen am Großmaulwels, *Chaca chaca*. Natur Mus., 95:465–469.

SCHILDKRAUT, J. J. (1965) The Catecholamine Hypothesis of Affective Disorders: A Review of Supporting Evidence. Am. J. Psychiat., 122:509–522.

SCHILDKRAUT, J. J., and KETY, S. S. (1967) Biogenic Amines and Emotion. Science, 156:21–30.

SCHJELDERUP-EBBE, T. (1922a) Soziale Verhältnisse bei Vögeln. Z. Psychol., 90:106–107.

——— (1922b) Beiträge zur Sozialpsychologie des Haushuhns. Z. Psychol., 88:225–252.

——— (1935) Social Behavior of Birds. In MURCHISON, A. (ed.), A Handbook of Social Psychology, 947–972.

SCHLEIDT, M. (1954) Untersuchungen über die Auslösung des Kollerns beim Truthahn. Z. Tierpsychol., 11:417–435.

SCHLEIDT, W. M. (1961a) Über die Auslösung der Flucht vor Raubvögeln bei Truthühnern. Naturwiss., 48:141–142.

——— (1961b) Reaktionen von Truthühnern auf fliegende Raubvögel und Versuche zur Analyse ihrer AAMs. Z. Tierpsychol., 18:534–560.

——— (1962) Die historische Entwicklung der Begriffe "Angeborenes auslösendes Schema" und "Angeborener Auslösemechanismus." Z. Tierpsychol., 19:697–722.

——— (1964a) Über die Spontaneität von Erbkoordinationen. Z. Tierpsychol., 21:235–256.

——— (1964b) Wirkungen äußerer Faktoren auf das Verhalten. Fortschr. Zool., 16:469–499.

——— (1966) Aus dem Signal-Inventar der Truthühner. In BURKHARDT, D., SCHLEIDT, W. M., and ALTNER, H. (eds.), Signale in der Tierwelt. Munich (H. MOOS), 130–134.

SCHLEIDT, W. M. and SCHLEIDT, M. (1962) Meleagris gallapavo silvestris (Meleagrididae)-Kampfverhalten der Hähne. Wiss. Film E 487, Göttingen (Inst. wiss. Film).

SCHLEIDT, W. M., SCHLEIDT, M., and MAGG, M. (1960) Störungen der Mutter-Kind-Beziehung bei Truthühnern durch Gehörverlust. Behaviour, 16:254–260.

SCHLICHTER, D. (1968) Das Zusammenleben von Riffanemonen und Anemonenfischen. Z. Tierpsychol., 25:933–954.

SCHLOSSER, K. (1952a) Körperliche Anomalien als Ursache sozialer Ausstoßung bei Naturvölkern. Z. Morphol. Anthropol., 44.

——— (1952b) Der Signalismus in der Kunst der Naturvölker. Biologisch-psychologische Gesetzlichkeiten in den Abweichungen von der Norm des Vorbildes. Arbeiten a. d. Mus. f. Völkerkde Univ. Kiel, I. Kiel (Mühlau).

——— (1952c) Der Rangkampf biologisch und ethnologisch gesehen. Act. IV Congr. Intern. Sci. Anthropol. Ethnol. Vienna, 2:43–50.

SCHMIDT, H. D. (1960) Bigotry in School Children. Commentary, 29:253–257.

SCHMIDT, L. (1952) Gestaltheiligkeit im bäuerlichen Arbeitsmythos. Veröffentl. Öst. Mus. Volkskde, 1:1–240.

SCHMIDT, R. S. (1957) The Evolution of Nestbuilding Behavior in Apicotermes. Evolution, 9:157–181.

——— (1958) The Nests of Apicotermes trägardhi: New Evidence on the Evolution of Nest-Building. Behaviour, 12:76–94.

SCHNEIDER, D. (1962) Electrophysiological Investigation on the Olfactorical Specifity of Sexual Attracting Substances in Different Species of Moths. J. Insect. Physiol., 8:15–30.

——— (1966) Vergleichende Neurophysiologie. Nervenarzt, 37:454–457.

——— (1967) Wie arbeitet der Geruchsinn bei Mensch und Tier? Mitt. Max-Planck-Ges., 294–314.

SCHNEIDER, F. (1961) Beeinflussung der Aktivität des Maikäfers durch Veränderung der gegenseitigen Lage magnetischer und elektrischer Reize. Mitt. Schweiz. Entomol. Ges., 33:223–237.

SCHNEIDER, H. (1963) Bioakustische Untersuchungen an Anemonenfischen der Gattung Amphiprion. Z. Morphol. Ökol. Tiere, 53:453–474.

SCHNEIRLA, T. C. (1946) Problems of Biopsychology and Social Organization. J. Abnorm. Social Psychol., 41:385–402.

——— (1956) Interrelationships of the "Innate" and the "Acquired" in Instinctive Behavior. In GRASSÉ, P. P. (ed.), L'Instinct dans le Comportement des Animaux. Paris. 387–452.

—— (1959) An Evolutionary and Developmental Theory of Biphasic Processes Underlying Approach and Withdrawal. Nebraska Symp. on Motivation. Lincoln (Univ. Nebraska Press), 1–41.

—— (1965) Aspects of Stimulation and Organisation in Approach Withdrawal Processes Underlying Vertebrate Behavioral Development. Advan. Animal Behavior, 1:1–74.

—— (1966) Behavioral Development and Comparative Psychology. Quart. Rev. Biol., 41:283–302.

SCHÖNE, H. (1951) Die Lichtorientierung der Larven von *Acilius sulcatus* und *Dytiscus marginalis.* Z. Vergleich. Physiol., 33:63–98.

—— (1959) Die Lageorientierung mit Statolithenorganen und Augen. Ergeb. Biol., 21:161–209.

—— (1962) Optisch gesteuerte Lageänderungen (Versuche an Dytiscidenlarven zur Vertikalorientierung). Z. Vergleich. Physiol., 45:590–604.

—— (1965–1966) Vorlesung über Orientierungsvorgänge, gehalten im Wintersem. 1965–1966, Univ. Munich.

SCHÖNE, H., and EIBL-EIBESFELDT, I. (1965) *Grapsus grapsus (Brachyura):* Drohen. Encycl. cinem., E599. Publ. z. wiss. Filmen, 1 A, 391–396. Göttingen (Inst. wiss. Film).

SCHÖNE, H, and SCHONE, H. (1963) Balz und andere Verhaltensweisen der Mangrovekrabbe, *Goniopsis cruentata,* und das Winkverhalten der eulitoralen Brachyuren. Z. Tierpsychol., 20:641–656.

SCHREMMER, F. (1960) Beobachtungen über die Bestäubung der Blüten von *Ophrys fuciflora* durch Männchen der Bienenart *Eucera nigrilabris.* Österr. Botan. Z., 107:6–17.

SCHULTZ, J. H. (1966) Organstörungen und Perversionen im Liebesleben. Munich (E. Reinhardt).

SCHULTZE-WESTRUM, T. (1965) Innerartliche Verständigung durch Düfte beim Gleitbeutler *Petaurus breviceps papuanus* Thomas (Marsupialia, Phalangeridae). Z. Vergleich. Physiol., 50:151–220.

—— (1967) Biologische Grundlagen zur Populationsphysiologie der Wirbeltiere. Naturwiss., 54:576–579.

—— (1968) Ergebnisse einer zoologisch-völkerkundlichen Expedition zu den Papuas. Umschau, 68, 295–300.

SCHULTZE-WESTRUM, T., and BRAUN, B. (1967) Soziale Abhängigkeiten der innerartlichen Aggression bei Flugbeutlern. Z. Vergleich. Physiol.

SCHUTZ, F. (1956) Vergleichende Untersuchungen über die Schreckreaktion bei Fischen und deren Verbreitung. Z. Vergleich. Physiol., 38:84–135.

—— (1964) Über geschlechtlich unterschiedliche Objektfixierung sexueller Reaktionen bei Enten im Zusammenhang mit dem Prachtkleid des Männchens. Zool. Anz., Suppl., 27:282–287.

—— (1965a) Sexuelle Prägung bei Anatiden. Z. Tierpsychol., 22:50–103.

—— (1965b) Homosexualität und Prägung bei Enten. Psychol. Forsch., 28:439–463.

—— (1968) Sexuelle Prägungserscheinungen bei Tieren. In GIESE, H. (ed.), Die Sexualität des Menschen. Handb. d. Med. Sexualforschung. Stuttgart (Enke), 284–317.

SCHÜZ, E. (1952) Vom Vogelzug. Frankfurt (Schöps-Verlag).

SCHWARTZKOPFF, J. (1960) Vergleichende Physiologie des Gehörs. Fortschr. Zool., 12:206–264.

—— (1962) Vergleichende Physiologie des Gehörs und der Lautäußerungen. Fortschr. Zool., 15:214–336.

SCHWIDETZKY, I. (1950) Grundzüge der Völkerbiologie. Stuttgart (Enke).

SCHWINCK, I. (1955) Weitere Untersuchungen zur Frage der Geruchsorientierung der Nachtschmetterlinge: partielle Fühleramputation bei Spinner-männchen, insbesondere des Seidenspinners. Z. Vergleich. Physiol., 37: 439–458.

SCOTT, J. P. (1960) Aggression. Chicago (Univ. Chicago Press).

―――― (1962) Critical Periods in Behavioral Development. Science, 138:949–958.

―――― (1963) The Process of Primary Socialization in Canine and Human Infants. Monogr. Soc. Res. Child Develop., 28.

―――― (1964) The Effects of Early Experience on Social Behavior and Organization. In ETKIN, W. (ed.), Social Behavior and Organization among Vertebrates. Chicago (Univ. Chicago Press).

SCOTT, J. P., and FREDERICSON, E. (1951) The Causes of Fighting in Mice. Physiol. Zool., 24:273–309.

SCOTT, J. P., and FULLER, J. L. (1965) Genetics and Social Behavior of the Dog. Chicago (Univ. Chicago Press).

SEILACHER, A. (1967) Fossil Behavior. Sci. Am., 217(2):72–80.

SEISS, R. (1965) Beobachtungen zur Frage der Übersprungbewegungen im menschlichen Verhalten. Psychol. Beitr., 8:1–97.

SEITZ, A. (1940) Die Paarbildung bei einigen Zichliden I. Z. Tierpsychol., 4:40–84.

―――― (1941) Die Paarbildung bei einigen Zichliden. II. Z. Tierpsychol., 5:74–101.

SEVENSTER, P. (1961) A Causal Analysis of a Displacement Activity: Fanning in *Gasterosteus aculeatus*. Behaviour, Suppl., 9.

―――― (1968) Motivation and Learning in Sticklebacks (*Gasterosteus aculeatus* L.). In INGLE, D. (ed.), The Central Nervous System and Fish Behavior. Chicago (Univ. Chicago Press), pp. 233–245.

SEVENSTER-BOL, A. C. A. (1962) On the Causation of Drive Reduction after a Consummatory Act. Arch. Neerl. Zool., 15:175–236.

SEXTON, O. J. (1960) Experimental Studies on Artificial Batesian Mimics. Behaviour, 15:244–252.

SHARPE, R. S. and JOHNSGARD, P. A. (1966) Inheritance of Behavioral Characters in F_2 Mallard × Pintail (*Anas platyrhynchos* L. × *Anas acuta* L.) Hybrids. Behaviour, 37:259–272.

SHAW, C. E. (1948) The Male Combat "Dance" of Some Crotalid Snakes. Herpetologica, 4:137–145.

SHEFFIELD, F. D., and ROBY, T. B. (1950) Reward Value of a Non-Nutritive Sweet Taste. J. Comp. Physiol. Psychol., 43:471–481.

SHERRINGTON, C. S. (1931) Quantitative Management of Contraction in Lowest Level Coordinations. Brain, 54:1–28.

SHINKMAN, P. G. (1963) Visual depth discrimination in day old chick. J. Comp. Physiol. Psychol., 56:410–414.

SIEBENALER, J. B., and CALDWELL, D. K. (1956) Cooperation among Adult Dolphins. J. Mammal., 37:126–128.

SIELMANN, H. (1955) Brutbiologie des Schwarzspechtes. Wiss. Film, C697. Göttingen (Inst. wiss. Film).

―――― (1958) Das Jahr mit den Spechten. Berlin (Ullstein).

―――― (1967) *Ptilonorhynchus violaceus* — Bauen an der Laube, E1075. *Ptilonorhynchus violaceus* — Balz und Kopulation, E1076. *Chlamydera nuchalis* — Verteilen von Sammlungsstücken an der Laube, E1077. *Chlamydera nuchalis* — Balz und Kopulation, E1078. *Chlamydera lauterbachi* — Bauen an der Laube, E1080. *Chlamydera lauterbachi* — Balz, E1081. *Prionodura newtonia* — Behängen der Reisigtürme, E1082. *Amblyornis macgregoriae* — Behängen des "Maibaumes" und Balz, E1083. *Amblyornis subalaris* — Schmücken der Laube, E1084. Encycl. cinem. Göttingen (Inst. wiss. Film).

SKINNER, B. F. (1938) The Behavior of Organisms. New York.
——— (1953) Science and Human Behavior. New York (Macmillan).
——— (1966) Phylogeny and Ontogeny of Behavior Contiguencies of Reinforcement throw Light on Contiguencies of Survival in the Evolution of Behavior. Science, 153:1203–1213.
SLUCKIN, W. (1965) Imprinting and Early Learning. Chicago (Aldine).
SMALL, W. S. (1900) An Experimental Study of the Mental Processes of the Rat. Am. J. Psychol., 11:133–165.
SMITH, R. I. (1958) On Reproductive Patterns as a Specific Characteristic among Nereid Polychaetes. Systematic Zool., 7:60–73.
SOMMER, R. (1966) Man's proximate Environment. J. Social Issues, 22:59–70.
SORENSON, E. R., and GAJDUSEK, D. C. (1966) The Study of Child Behavior and Development in Primitive Cultures. Pediatrics, Suppl., 37:149–243.
SOUTHWICK, C. H. (1963) Challenging Aspects of the Behavioral Ecology of Howling Monkeys. In SOUTHWICK, C. H. (ed.), Primate Social Behavior. Princeton, N.J. (Van Nostrand), 185–191.
SPALDING, D. A. (1873) Instinct with Original Observation on Young Animals. MacMillans Mag., 27:282–283 (reprinted in Brit. J. Animal Behavior, 2:1–11, 1954).
SPANNAUS, G. (1961) Der wissenschaftliche Film als Forschungsmittel der Völkerkunde. Der Film im Dienste der Wissenschaft. Festschr. z. Einweihung d. Neubaus f. d. Inst. wiss. Film, Göttingen, 67–83.
SPERRY, R. W. (1958) Physiological Plasticity and Brain Circuit Theory. In HARLOW, H. F. and WOOLSEY, C. C. N. (eds.), Biological and Biochemical Bases of Behavior. Madison (Univ. Wisconsin Press), 401–424.
——— (1964) The Great Cerebral Commissure. Sci. Am., 210(1):42–52.
SPINDLER, M., und BLUHM, E. (1934) Kleine Beiträge zur Psychologie des Seelöwen (Eumetopias calif.). Z. Vergleich. Physiol., 21:616–631.
SPINDLER, P. (1958) Studien zur Vererbung von Verhaltensweisen. 1. Verhalten auf einen starken akustischen Reiz. Anthropol. Anz., 22(2):137–155.
——— (1961) Studien zur Vererbung von Verhaltensweisen. 3. Verhalten gegenüber jungen Katzen. Anthropol. Anz., 25(1):60–80.
SPITZ, R. A. (1945) Hospitalism. The Psychoanalytical Study of the Child, 1. New York (International Univ. Press), 53–74.
——— (1946) Anaclitic Depression. An Inquiry into the Genesis of Psychiatric Conditions in Early Childhood. The Psychoanalytical Study of the Child. 2. New York (International Univ. Press), 313–342.
——— (1951) Psychogenic Diseases in Infancy. The Psychoanalytical Study of the Child. 6. New York (International Univ. Press), 255–275.
——— (1957) Die Entstehung der ersten Objektbeziehungen. Stuttgart (Klett).
——— (1965) The First Year of Life. New York (International Univ. Press).
SPITZ, R. A., and WOLF, K. M. (1946) The Smiling Response: A Contribution to the Ontogenesis of Social Relations. Gen. Psychol. Monogr., 34:57–125.
STAEHELIN, B. (1953) Gesetzmäßigkeiten im Gemeinschaftsleben schwer Geisteskranker. Schweiz. Arch. Neurol. Psychiat., 72:277–298.
——— (1954) Gesetzmäßigkeiten im Gemeinschaftsleben Geisteskranker, verglichen mit tierpsychologischen Ergebnissen. Homo, 5:113–116.
STAMM, J. S. (1954) Control of Hoarding Activity in Rats by the Median Cerebral Cortex. J. Comp. Physiol. Psychol., 47:21–27.
——— (1955) The Function of the Median Cerebral Cortex in Maternal Behaviour of Rats. J. Comp. Physiol. Psychol., 48:347–356.
STAMM, R. A. (1960) Paarintimität und Streitigkeiten bei Agapornis personata. Verhandl. Naturforsch. Ges. Basel, 71:1–14.
——— (1962) Aspekte des Paarverhaltens von Agapornis personata Reichenow. Behaviour, 29:1–56.

―――― (1964) Perspektiven zu einer vergleichenden Ausdrucksforschung. In KIRCHHOFF, R. (ed.), Handb. d. Psychol., 5:255–288.

STARCK, D. (1959) Neuere Ergebnisse der vergleichenden Anatomie und ihre Bedeutung für die Taxonomie. J. Ornithol., 100:47–59.

STEINEN, K. von den (1894) Unter den Naturvölkern Zentralbrasiliens. Reiseschilderungen und Ergebnisse der zweiten Schingu-Expedition 1887–1888. Berlin (Geogr. Verlagshandlg., D. Reimer). Reprinted in BÖLSCHE, W., Neue Welten. Berlin (Deutsche Bibliothek), 1917.

STEINIGER, F. (1950) Zur Soziologie und sonstigen Biologie der Wanderratte. Z. Tierpsychol., 7:356–379.

―――― (1951) Revier- und Aktionsraum bei der Wanderratte. Z. Hyg. Zool., 39:33–51.

STOKES, B. (1955) Behaviour as a Means of Identifying Two Closely-Allied Species of Gall Midges. Brit. J. Animal Behaviour, 9:154–157.

STRASSEN, O. zur (1952) Zweckdienliches Sprechen beim Graupapagei. Verh. Dt. Zool. Ges., Freiburg, 84–89.

STRESEMANN, E. (1934) Aves. In KÜKENTHAL, Handb. d. Zool., 7 (2). Berlin (de Gruyter).

STRUHSAKER, T. T. (1967) Auditory Communication among Vervet Monkeys, *Cercopithecus aethiops.* In ALTMAN, ST. (ed.), Social Communication in Primates. Chicago (Univ. Chicago Press), 281–334.

TAUB, E., and BERMAN, A. J. (1964) The Effect of Massive Somatic Deafferentiation on Behavior and Wakefulness in Monkeys. Papers presented at Psychonomic Science Meeting, Niagara, Ont., October.

TAUB, E., ELLMAN, S. J., and BERMAN, A. J. (1965) Deafferentiation in Monkeys. Effects on Conditioned Grasp Response. Science, 151:593–594.

TAUERN, O. D. (1918) Patasiwa und Patalima, vom Molukkeneiland Seran und seinen Bewohnern. Leipzig (Voigtländer).

TEITELBAUM, P. (1961) Disturbances in Feeding and Drinking Behavior after Hypothalamic Lesions. Nebraska Symp. Motivation, 9:39–65.

TELLE, H. J. (1966) Beitrag zur Kenntnis der Verhaltensweise bei Ratten, vergleichend dargestellt bei *Rattus norvegicus* und *Rattus rattus.* Angew. Zool., 9:129–196.

TEMBROCK, G. (1954) Rotfuchs und Wolf. Z. Säugetierkde, 19:152–159.

―――― (1960) Spielverhalten und vergleichende Ethologie. Z. Säugetierkde, 25:1–14.

―――― (1964) Verhaltensforschung. 2nd ed. Jena (G. Fischer).

TETS, G. P. v. (1965) A Comparative Study of Some Social Communication Patterns in the *Pelecaniformes.* Am. Ornithol. Union, Ornithol. Monogr., 2.

THIBAUT, J. W., and COWLES, J. T. (1952) The Role of Communication in the Reduction of Interpersonal Hostility. J. Abnorm. Social Psychol., 47:770–777.

THIELCKE, G. (1961) Stammesgeschichte und geographische Variation des Gesanges unserer Baumläufer (*Certhidea familiaris* L. u. *C. brachydactyla* Brehm). Z. Tierpsychol., 18:188–204.

―――― (1965) Gesangsgeographische Variation des Gartenbaumläufers *(Certhidea brachydactyla)* in Hinblick auf das Artbildungsproblem. Z. Tierpsychol., 22:542–566.

THOMAS, E. (1961) Fortpflanzungskämpfe bei Sandottern *(Vipera ammodytes).* Zool. Anz., Suppl., 24:502–505.

THOMPSON, J. (1941) Development of Facial Expression of Emotion in Blind and Seeing Children. Arch. Psychol. N.Y., 264:1–47.

THOMPSON, T. I. (1963) Visual Reinforcement in Siamese Fighting Fish. Science, 141:55–57.

—— (1964) Visual Reinforcement in Fighting Cocks. J. Exptl. Analysis Behavior, 7:45–49.

THORNDIKE, E. L. (1911) Animal Intelligence. New York (Macmillan).

THORPE, W. H. (1938) Further Experiments on Pre-Imaginal Conditioning in Insects. Proc. Roy. Soc. London, B126:370–397.

—— (1939) Further Studies on Pre-Imaginal Olfactory Conditioning in Insects. Proc. Roy. Soc. London, B127:424–433.

—— (1951) The Definition of Some Terms Used in Animal Behavior Studies. Animal Behavior, 9:34–49.

—— (1954) The Process of Song Learning in the Chaffinch as Studied by Means of the Sound Spectograph. Nature, 173:465.

—— (1958a) The Learning of Song Patterns by Birds, with Special Reference to the Song of the Chaffinch. Ibis, 100:535–570.

—— (1958b) Further Studies on the Process of Song Learning in the Chaffinch (Fringilla coelebs). Nature, 182:554–557.

—— (1961a) Sensitive Periods in the Learning of Animals and Men: A Study of Imprinting with Special Reference to the Introduction of Cyclic Behavior. In THORPE, W. II., and ZANGWILL, O. L. (eds.), Current Problems in Animal Behavior. Cambridge (Univ. Press), 194–224.

—— (1961b) Bird Song. The Biology of Vocal Communication and Expression in Birds. Cambridge Monogr. Exptl. Biol., 12.

—— (1963) Learning and Instinct in Animals. London (Methuen).

—— (1966) Ritualisation in the Individual Development of Bird Song. Phil. Trans. Roy. Soc. London, 551(772):351–358.

THORPE, W. H., and JONES, F. H. W. (1937) Olfactory Conditioning in a Parasitic Insect and Its Relation to the Problem of Host Selection. Proc. Roy. Soc. London, B124:56–81.

THORPE, W. H., and NORTH, M. E. W. (1965) Origin and Significance of the Power of Vocal Imitation: with Special Reference to the Antiphonal Singing of Birds. Nature, 208:219–223.

TIGER, L., and FOX, R. (1966) The Zoological Perspective in Social Science. Man, 1:75–81.

TINBERGEN, N. (1935) Über die Orientierung des Bienenwolfes II: Die Bienenjagd. Z. Vergleich. Physiol., 21:699–716.

—— (1940) Die Übersprungbewegung. Z. Tierpsychol., 4:1–40.

—— (1948) Social Releasers and the Experimental Method Required for Their Study. Wilson Bull., 60:6–52.

—— (1951) The Study of Instinct. London (Oxford Univ. Press).

—— (1952) "Derived" Activities, Their Causation, Biological Significance and Emancipation during Evolution. Quart. Rev. Biol., 27:1–32.

—— (1953) Social Behaviour in Animals, with Special Reference to Vertebrates. London (Methuen).

—— (1955) Tiere untereinander. Berlin (Parey).

—— (1957) The Functions of Territory. Bird Study, 5:14–27.

—— (1959) Einige Gedanken über "Beschwichtigungsgebärden." Z. Tierpsychol., 16:651–665.

—— (1963) The Herring Gull's World. London (Collins).

—— (1965) Behavior and Natural Selection. In MOORE, J. A. (ed.), Ideas in Modern Biology. Proc. 16th Inst. Zool. Congr. Washington, 1963, 6:521–542.

TINBERGEN, N., BROEKHUYSEN, G. J., FEEKES, F., HOUGHTON, J. C. W., KRUUK, H., and SZULC, E. (1962) Egg-Shell Removal by the Black-Headed Gull Larus ridibundus L. A Behaviour Component of Camouflage. Behaviour, 19:74–118.

TINBERGEN, N., and KRUYT, W. (1938) Über die Orientierung des Bienenwolfes. III: Die Bevorzugung bestimmter Wegmarken. Z. Vergleich. Physiol., 25:292–334.

TINBERGEN, N., and KUENEN, D. J. (1939) Über die auslösenden und richtung-
gebenden Reizsituationen der Sperrbewegung von jungen Drosseln *(Turdus
m. merula* L. und *T. e. ericetorum* Turton). Z. Tierpsychol., 3:37–60.

TINBERGEN, N., MEEUSE, B. J. D., BOEREMA, L. K., and VAROSSIEAU, W. W.
(1943) Die Balz des Samtfalters *(Eumenis semele* L.). Z. Tierpsychol., 5:182–
226.

TINBERGEN, N., and PERDECK, A. C. (1950) On the Stimulus Situation Releasing the Begging Response in the Newly-Hatched Herring Gull Chick *(Larus
argentatus)*. Behaviour, 3:1–38.

TOLMAN, E. C. (1932) Purposive Behaviour in Animals and Men. New York
(Appleton).

TOLMAN, E. C., and HONZIK, C. H. (1930a) Insight in Rats. Univ. Calif. Publ.
Psychol., 4:215–232.

———— (1930b) Introduction and Removal of Reward, and Maze Performance
in Rats. Univ. Calif. Publ. Psychol., 4:257–275.

TOMPA, F. S. (1962) Territorial Behavior. The Main Factor Controlling a Local
Song Sparrow Population. Auk, 79:687–697.

TOWBIN, E. J. (1949) Gastric Distension as a Factor in the Satiation of Thirst
in Esophagostomised Dogs. Am. J. Physiol., 159:533–541.

TRACY, H. C. (1926) The Development of Motility and Behavior in the Toad-
Fish *(Opsanus tau)*. J. Comp. Neurol., 40:253–369.

TRUMLER, E. (1959) Das "Rossigkeitsgesicht" und ähnliches Ausdrucksverhal-
ten bei Einhufern. Z. Tierpsychol., 16:478–488.

TRYON, R. C. (1940) Genetic Differences in Maze Learning in Rats. In 39th
Yearbook National Society for the Study of Education, Bloomington, Ill.
Publ. School, 1:111–119.

TSCHANZ, B. (1962) Über die Beziehungen zwischen Muttertier und Jungen
beim Mufflon. Experientia, 18:187.

———— (1965) Beobachtungen und Experimente zur Entstehung der "persönlichen" Beziehung zwischen Jungvogel und Elterntier bei Trottellummen.
Verhandl. Schweiz. Naturforsch. Ges. Zürich, 211–216.

———— (1968) Trottellummen: Die Entstehung der persönlichen Beziehungen
zwischen Jungvogel und Eltern. Beih. Z. Tierpsychol., 103 pp.

TURHAN, B. M. (1960) Über die Bedeutung des Gesichtsausdrucks. Psychol.
Beitr., 5:440–454.

TWEEDIE, M. (1966) Butterfly Mimics. Animals, 8:318–321.

UEXKÜLL, J. v. (1921) Umwelt und Innenwelt der Tiere. 2nd ed. Berlin.

———— (1937) Umweltsforschung. Z. Tierpsychol., 1:33–34.

UEXKÜLL, J. v., and KRISZAT, G. (1934) Streifzüge durch die Umwelten von
Tieren und Menschen. Ein Bilderbuch unsichtbarer Welten. Verständl.
Wiss.; new ed., 1963.

UHRICH, J. (1938) The Social Hierarchy in Albino Mice. J. Comp. Physiol.
Psychol., 25:373–413.

VALENSTEIN, E. S., RISS, W., and YOUNG, W. C. (1955) Experimental and
Genetic Factors in the Organisation of Sexual Behavior in Male Guinea
Pigs. J. Comp. Physiol. Psychol., 48:397–403.

VERWEY, J. (1930) Die Paarungsbiologie des Fischreihers. Zool. Jahrb. Allgem. Zool. Physiol., 48:1–120.

WAGNER, H. O. (1938) Beobachtungen über die Balz des Paradiesvogels *Paradisea guilielmi*. J. Ornithol., 86:550–553.

———— (1954) Massenansammlungen von Weberknechten. Z. Tierpsychol.,
11:348–352.

WAHLERT, G. v. (1957) Weitere Untersuchungen zur Phylogenie der Schwanz-
lurche. Zool. Anz., Suppl., 20:347–352.
———— (1962) Beobachtungen und Bemerkungen zum Putzverhalten von Mit-
telmeerfischen. Veröffentl. Inst. Meeresforsch., 7:71–78.
———— (1963) Die ökologische und evolutorische Bedeutung der Fisch-
schwärme. Veroeffentl. Inst. Meeresforsch. Meeresbiol. Symp., 3:197–213.
WALK, R. D. (1966) The Development of depth Perception in animals and
Human Infants. Monographs of the Society for Research in Child Development, 31:82–108.
WALKER, B. W. (1952) A Guide to the Grunion. Calif. Fish Game, 38:409–420.
WALLRAFF, H. G. (1959) Örtlich und zeitlich bedingte Variabilität des Heimkehrverhaltens von Brieftauben. Z. Tierpsychol., 16:513–544.
———— (1960a) Können Grasmücken mit Hilfe des Sternenhimmels navigie-
ren? Z . Tierpsychol., 17:165–177.
———— (1960b) Über Zusammenhänge des Heimkehrverhaltens von Brieftauben mit meteorologischen und geophysikalischen Faktoren. Z. Tierpsychol., 17:82–114.
———— (1967) The Present Status of our Knowledge about Pigeon Homing.
Proc. 14th Intern. Ornithol. Congr. Oxford. (Blackwell), 331–358.
WALTHER, F. R. (1958) Zum Kampf- und Paarungsverhalten einiger Anti-
lopen. Z. Tierpsychol., 15:340–380.
———— (1961) Entwicklungszüge im Kampf- und Paarungsverhalten der Horn-
tiere. Jahrb. G. v. Opel — Freigehege Tierforsch., 3:90–115.
———— (1964) Verhaltensbeobachtungen an Thomsongazellen. Z. Tierpsychol.,
21:871–890.
———— (1965) Verhaltensstudien an der Grant Gazelle (Gazella granti Brooke)
im Ngorongoro Krater. Z. Tierpsychol., 22:166–208.
———— (1966) Mit Horn und Huf. Berlin (Parey).
WASHBURN, S. L., and DE VORE, I. (1961) The Social Life of Baboons. Sci.
Am., 204:62–71.
WASMANN, E. (1920) Die Gastpflege der Ameisen. Abhandl. Theoret. Biol.,
2:4. Berlin.
———— (1925) Ameisenmimikry. Abhandl. Theoret. Biol., 19:1–164.
WATSON, A. (1966) Social Status and Population Regulation in the Red
Grouse (Lagopus lagopus scoticus). The Royal Soc. Pop. Study Group,
Proc. 2. Royal Society, London, 22–30.
WATSON, J. B. (1919) Psychology from the Standpoint of a Behaviorist. Phila-
delphia (Lippincott).
———— (1930) Der Behaviorismus. Stuttgart.
WEHMER, F. (1965) Effects of Prior Experience with Objects on Maternal
Behaviors in the Rat. J. Comp. Physiol. Psychol., 60:294–296.
WEIDKUHN, P. (1968/69) Aggressivität und Normativität. Über die Vermittler-
rolle der Religion zwischen Herrschaft und Freiheit. Ansätze zu einer
kulturanthropologischen Theorie der sozialen Norm. Anthropos, 63/64.
WEIDMANN, U. (1951) Über den systematischen Wert von Balzhandlungen bei
Drosophila. Rev. Suisse Zool., 54:502–511.
———— (1955) Some Reproductive Activities of the Common Gull Larus canus
L. Ardea, 43:85–132.
———— (1956) Verhaltensstudien an der Stockente. Z. Tierpsychol., 13:209–
271.
———— (1959) The Begging Response of the Black-Headed Gull Chick. Bericht
vorgetr. 6. Intern. Ethologenkonf., Cambridge.
———— (1965) Colour and Behaviour. In Colour and Life. Inst. Biol. London,
79–100.
WEIH, A. S. (1951) Untersuchungen über das Wechselsingen (Anaphonie) und

über das angeborene Lautschema einiger Feldheuschrecken. Z. Tierpsychol., 8:1–41.

WEISS, P. (1939) Principles of Development. New York (Holt, Rinehart and Winston).

——— (1941a) Self-Differentiation of the Basic Patterns of Coordination. Comp. Psychol. Monogr., 17:1–96.

——— (1941b) Autonomous versus Reflexogenous Activity of the Central Nervous System. Proc. Am. Phil. Soc., 84:53–64.

WELTY, J. C. (1934) Experiments on Group Behaviour of Fishes. Physiol. Zool., 7:85–127.

WENDLER, G. (1964) Laufen und Stehen der Stabheuschrecke *Carausius morosus:* Sinusborstenfelder in den Beinen als Glieder von Regelkreisen. Z. Vergleich. Physiol., 48:198–250.

——— (1965) The Coordination of Walking Movements in Arthropods. Nerv. and Horm. Mech. of Integration, 20th Symp. Soc. Exptl. Biol., 229–249.

——— (1966) Der Regelkreis gezielter Bewegungen. In BURKHARDT, D., SCHLEIDT, W. M., and ALTNER, H. (eds.), Düfte, Farben und Signale. Munich (Moos).

——— (1968) Ein Analogmodell der Beinbewegungen eines laufenden Insekts. Kybernetik, Beih. zu "Elektron. Rechenanlagen." Munich (Oldenbourg), 18, 67–74.

WENDT, H. (1964) Erfolgreiche Zucht des Baumwollköpfchens oder Pincheäffchens *Leontocebus (Oedipomidas) oedipus.* Säugetierkdl. Mitt., 12:49–52.

WENNER, A. M. (1967) Honey Bees: Do They Use the Distance Information Contained in Their Dance Maneuver? Science, 155:847–849.

WESENBERG-LUND, C. (1939) Biologie der Süßwassertiere. Wien (Springer).

——— (1943) Biologie der Süßwasser-Insekten. Berlin (Springer).

WEVER, R. (1968) Einfluß schwacher elektromagnetischer Felder auf die circadiane Periodik der Menschen. Naturwiss., 55:29–32.

WHEELER, W. M. (1928) Social Life among the Insects. New York.

WHITMAN, C. O. (1899) Animal Behavior. Biol. Lect. Marine Biol. Lab., Woods Hole, 285–338.

——— (1919) The Behavior of Pigeons. Publ. Carnegie Inst., 257:1–161.

WICKLER, W. (1957) Vergleichende Verhaltensstudien an Grundfischen. I. Beiträge zur Biologie, besonders zur Ethologie von *Blennius fluviatilis* Asso im Vergleich zu einigen anderen Bodenfischen. Z. Tierpsychol., 14:393–428.

——— (1958) Vergleichende Verhaltensstudien an Grundfischen. II. Die Spezialisierung des *Steatocranus.* Z. Tierpsychol., 15:427–446.

——— (1959) Vergleichende Verhaltensstudien an Grundfischen. III. Die Umspezialisierung von *Noemacheilus kuiperi.* Z. Tierpsychol., 16:410–423.

——— (1960a) Belegexemplare zu Ethogrammen. Z. Tierpsychol., 17:141–142.

——— (1960b) Die Stammesgeschichte typischer Bewegungsformen der Fisch-Brustflosse. Z. Tierpsychol., 17:31–66.

——— (1961a) Ökologie und Stammesgeschichte von Verhaltensweisen. Fortschr. Zool., 13:303–365.

——— (1961b) Über das Verhalten der Blenniiden *Runula* und *Aspidontus.* Z. Tierpsychol., 18:421–440.

——— (1961c) Über die Stammesgeschichte und den ökologischen Wert einiger Verhaltensweisen der Vögel. Z. Tierpsychol., 18:320–342.

——— (1962a) Ei-Attrappen und Maulbrüten bei afrikanischen Cichliden. Z. Tierpsychol., 18:129–164.

—— (1962b) Das Züchten von Aquarienfischen. Das Vivarium. Stuttgart (Franckh).

—— (1963) Zum Problem der Signalbildung, am Beispiel der Verhaltens-mimikry zwischen *Aspidontus* und *Labroides*. Z. Tierpsychol., 20:657–679.

—— (1964a) Das Problem der stammesgeschichtlichen Sackgassen. Natur-wiss. Med., 1(2):6–29.

—— (1964b) Phylogenetisch-vergleichende Verhaltensforschung mit Hilfe von Enzyklopädie-Einheiten. Res. Film, 5:109–118.

—— (1964c) Signalfälschung, natürliche Attrappen und Mimikry. Um-schau, 64:581–585.

—— (1964d) *Antennarius nummifer (Antennariidae):* Beutefang. Encycl. cinem., E141, Publ. zu wiss. Filmen, 1A, 41–47, Göttingen (Inst. wiss. Film).

—— (1964e) *Emblemaria pandionis:* Kampfverhalten. Encycl. cinem., E517, Publ. zu wiss. Filmen, 1A, 176–180. Göttingen (Inst. wiss. Film).

—— (1965a) Über den taxonomischen Wert homologer Verhaltensmerk-male. Naturwiss., 52:441–444.

—— (1965b) Die äußeren Genitalien als soziale Signale bei einigen Prima-ten. Naturwiss., 52:269–270.

—— (1965c) *Gastromyzon borneensis (Gastromyzonidae):* Kriechen und Schwimmen. Encycl. cinem., E611, Publ. zu wiss. Filmen, 1A, 421–426. Göttingen (Inst. wiss. Film).

—— (1965d) Neue Varianten des Fortpflanzungsverhaltens afrikanischer Cichliden *(Pisces, Perciformes)*. Naturwiss., 52:219.

—— (1965e) Die Evolution von Mustern der Zeichnung und des Verhaltens. Naturwiss., 52:335–341.

—— (1965f) Signal Value of the Genital Tassel in the Male *Tilapia macro-chir* Blgr. *(Pisces, Cichlidae)*. Nature, 208:595 596.

—— (1965g) Mimicry and the Evolution of Animal Communication. Nature, 208:519–521.

—— (1966a) Specialization of Organs Having a Signal Function in Some Marine Fish. Intern. Conf. Tropical Oceanogr., Miami Lab. 1965.

—— (1966b) Über die biologische Bedeutung des Genitalanhanges der männlichen *Tilapia macrochir*. Senckenbergiana Biol., 47:419–427.

—— (1966c) Ursprung und biologische Deutung des Genitalpräsentierens männlicher Primaten. Z. Tierpsychol., 23:422–437.

—— (1966d) Orchideen und Mimikry. In BURKHARDT, D., et al. (eds.), Signale in der Tierwelt. Munich (H. Moos).

—— (1966e) Natürliche "Übersexualisierung" des Soziallebens beim Bra-bantbuntbarsch. Umschau, 17:571–572.

—— (1967a) Socio-Sexual Signals and Their Intraspecific Imitation among Primates. In MORRIS, D. (ed.), Primate Ethology. London (Weidenfeld and Nicolson), 69–147.

—— (1967b) Vergleichende Verhaltensforschung und Phylogenetik. In HEBERER, G. (ed.), Die Evolution der Organismen. I. 3rd ed. Jena (G. Fischer), 420–508.

—— (1968a) Mimikry-Signalfälschung in der Natur. Munich (Kindler).

—— (1968b) Mutter-Kind-Signale- Ursprung und Bedeutungswandel. Um-schau, 23:718–719.

—— (1968c) Das Mißverständnis der Natur des ehelichen Aktes in der Moraltheologie. Stimmen d. Zeit, 182:289–303.

WIEPKEMA, P. R. (1961) An Ethological Analysis of the Reproductive Be-haviour of the Bitterling (*Rhodeus amarus* Bloch). Arch. Neerl. Zool., 14: 103–199.

WIESEL, T. N., and HUBEL, D. H. (1963a) Effects of Visual Deprivation on Morphology and Physiology of Cells in the Cats Lateral Geniculate Body. J. Neurophysiol., 24:978–993.

——— (1963b) Single-Cell Responses in Striate Cortex of Kittens Deprived of Vision in One Eye. J. Neurophysiol., 24:1003–1017.

——— (1965) Comparison of the Effects of Unilateral and Bilateral Eye Closure on Cortical Unit Responses in Kittens. J. Neurophysiol., 28:1029–1040.

WIESER, S. (1955) Die motorischen Schablonen des Oralsinnes. Fortschr. Neurol. Psychiat., 23:94–184.

WIESER, S., and ITIL, T. (1954) Die Aufbaustufen der primitiven Motorik. Arch. Psychiat. Nervenkr., 191:450–462.

WIGGLESWORTH, V. B. (1964) The Life of Insects. London (Weidenfeld and Nicolson).

WILLIAMS, L. (1967) Man and Monkey. London (Deutsch).

WILSON, A. P. (1968) Social Behavior of Free-ranging Rhesus Monkeys with an Emphasis on Aggression. Dissertation. Berkeley (Univ. of California).

WILSON, D. M. (1961) The Central Nervous Control of Flight in Locust. J. Exptl. Biol., 38:471–490.

——— (1964) Relative Refractoriness and Patterned Discharge of Locust Flight Motor Neurones. J. Exptl. Biol., 41:191–205.

——— (1965) Motor Output Patterns during Random and Rhythmic Stimulation of Locust Thoracic Ganglia. Biophys. J., 5:121.

——— (1966) Insect Walking. Ann. Rev. Entomol., 11:103–123.

——— (1968) The Flight Control System of the Locust. Sci. Am., 218(5):83–90.

WILSON, E. O. (1963) Pheromones. Sci. Am., 208(5):100–114.

——— (1965) Chemical Communication in Social Insects. Science, 149:1064–1071.

WILSONCROFT, W. E., and SHIPE, D. U. (1965) Tail, Paw and Pup Retrieving in the Rat. Psychon. Sci., 3:494.

WILTSCHKO, W. (1968) Über den Einfluß statischer Magnetfelder auf die Zugorientierung der Rotkehlchen (Erithacus rubecula). Z. Tierpsychol., 25:537–558.

WINDLE, W. F. (1940) Physiology of the Fetus. Philadelphia (Saunders).

——— (1944) Genesis of Somatic Motor Function in Mammalian Embryo: a Synthesizing Article. Physiol. Zool., 17:247–260.

WINKELSTRÄTER, K. H. (1960) Das Betteln der Zoo-Tiere. Beih. Schweiz. Z. Psychol. Suppl., 39.

WINN, H. F., SALMON, M., and ROBERTS, N. (1964) Sun-Compass Orientation by Parrot Fishes. Z. Tierpsychol., 21:798–812.

WINTER, P., PLOOG, D. W., and LATTA, J. (1966) Vocal Repertoire of the Squirrel Monkey (Saimiri sciureus), Its Analysis and Significance. Exptl. Brain Res., 1:359–384.

WOLF, A. V. (1958) Thirst, Physiology of the Urge to Drink. Springfield, Ill. (Charles C Thomas, Publisher).

WOLF, G. (1957a) Der wissenschaftliche Film. Naturwiss., 44:477–482.

——— (1957b) Encyclopaedia Cinematographica. Forschungsfilm, 2:304–310.

WOLFE, J. B. (1936) Effectiveness of Token-Rewards in Chimpanzees. Comp. Psychol. Monogr., 12:1–72.

WUNSCHMANN, A. (1963) Quantitative Untersuchungen zum Neugierverhalten von Wirbeltieren. Z. Tierpsychol., 20:80–109.

WYNNE-EDWARDS, V. C. (1962) Animal Dispersion in Relation to Social Behaviour. London (Oliver & Boyd).

YERKES, R. M. (1948) Chimpanzees, a Laboratory Colony. 4th ed. New Haven (Yale Univ. Press).

YERKES, R. M., and ELDER, J. H. (1936) Oestrus, Receptivity, and Mating in Chimpanzees. Comp. Psychol. Monogr., 13(5):1–39.

YOUNG J. Z. (1961) Learning and Discrimination in the Octopus. Biol. Rev., 36:32–96.

YOUNG, W. C. (1965) The Organisation of Sexual Behavior by Hormonal Action during Prenatal and Larval Periods in Vertebrates. In BEACH, F. A. (ed.), Sex and Behavior. New York (Wiley).

ZANGWILL, O. L. (1961) Lashley's Concept of Cerebral Mass Action. In THORPE, W. H., and ZANGWILL, O. L. (eds.), Current Problems in Animal Behaviour. London (Cambridge Univ. Press), 59–87.

ZEEB, K. (1964) Zirkusdressur und Tierpsychologie. Mitt. Naturforsch. Ges. Bern., 21.

ZIEGLER, H. E. (1920) Der Begriff des Instinkts einst und jetzt. Jena (G. Fischer).

ZIPPELIUS, H. M. (1949) Untersuchungen über das Balzverhalten heimischer Molche. Zool. Anz. Suppl., 12:127–130.

ZUCKERMANN, S. (1932) The Social Life of Monkeys and Apes. London.

ZUMPE, D. (1964) *Chelmon rostratus:* Kampfverhalten. Encycl. cinem., E 207, Publ. wiss. Film 1 A, 335–339, Göttingen (Inst. Wiss. Film).

——— (1965) Laboratory Observations on the Aggressive Behaviour of Some Butterfly Fishes. Z. Tierpsychol., 22:226–236.

AUTHOR INDEX

SUBJECT INDEX

Absolute coordination, 40
Absolute thresholds, 65
Abstraction, 261–268
Action systems, 10
Adaptations, 185–186, 218; early ontogenetic, 202–206; to nonbiological factors, 270–273
Afferent throttling, 81
Aggregations, 350
Aggression, 134; the dynamic instinct, concept of, 326–334; intraspecific, 306–334, fighting behavior, 314–325, territoriality, 306–314
Alarm calls, 206
Alarm substances, 69–71
Alternation, 175, 176, 177
American Anthropologist, 445
Amplitude of movement, 101
Analogies, 190
Anonymous groups, 350–351; closed, 351; open, 350
Appeasement, 123, 125–134
Appetitive behavior, 45
Artificial selection, 185
Audio-Visual Services, 11
Auslöser (social releasers), 71
Automatisms, 40–41, 42
Auxiliary criterion, 189
Averaging, 175, 177

Bees, language of, 136–143
Begging behavior, 101, 112–113, 127
Behavior, 1–2, 8; appeasement, 123, 125–134; appetitive, 45; begging, 101, 112–113, 127; biology of, 2; blind or deaf-blind children, 403–408; care, 106; cause of, 2; comparative study of, 8; complex, 3; conflict, 110, 175–184; contact, 334–362; courtship, 107–108, 117–125; curiosity, 238–248; deceptive, 282–283; eating, 179; ecology and, 269–305, behavior toward species members, 305–362, environment other than the species, relationships to, 270–304; embryology of, 201–202; escape, 283–285; expression of, 1; expressive, innate, 112; fighting, intraspecific, 314–325; function of, 1–2; greeting, 105–106, 125–134, 426–428; grooming, 94–95, 125–134; hierarchical organization of, 162–174; human, comparative method in study of, results of, 408–431; inciting, 107–108; insight, 261–268; intraspecific, releasers for, 116–148; communication about external environment, 135–147, group cohesion, signals that promote, 116–135, threat postures warding off nonspe-

cies members, 147–148; motivating factors, 44–59; play, 238–248; presenting, 100; protean, 283; response to stimulus, 60–90, innate releasing mechanism as basis for innate recognition, 60–64, key stimuli, releasers and, 64–90; sexual, 45, 53, 106; social, human, primary and secondary, means of, 442–457; spontaneity of, 4; temporal factors in, 389–397; territorial, 306–314; threat, 95–97, 130, 133, 311; toward species members, 305–362, aggression, intraspecific, 306–334, living in groups, 334–362
Behavioral inventory, 10–14
Behavior fossils, 199–200
Behaviorism, 9
Behavior patterns, 1; description of, 10–11; development of, phylogenetic, 185–200, behavior fossils, 199–200, general remarks, 185–187, historical rudiments, 192–195, homology, concept of, 187–192, process of domestication, 195–199, research on domesticated animals, 195–199; displacement activities, 177; expressive movements, origin of, 91–115; fixed action, 15–43, 53, chain of, 43, constancy of, 16, deprivation experiment, 19–32, psychological characteristics of, 32–43, release in infants, 399–403, taxis component and, 15–19; genetics of, 181–184; innate, 5, 16; maturation of, 5; maturation of, 5, 206–217; mutations of, 185; ontogeny of, 201–248, adaptations, early ontogenetics, 202–206, embryology of behavior, 201–202, innate disposition to learn, 217–248, maturation of, instinct-learning intercalation and, 206–217
Biology, 9
Blind or deaf-blind children, behavior of, 403–408
Body position and movement in space, control of, 366–371
British Journal for Animal Behavior, 8

Cancellation, 176, 177
Captive animals, behavior in, disturbances of, 11–13
Care behavior, 106
Chain reflexes, 32
Character displacement, 99
Chemical orientation, 365
Chemical stimuli, 69
Chemotaxis, 364
Classical conditioning, 250
Closed anonymous groups, 351